CAUSATION AND LAWS OF NATURE

AUSTRALASIAN STUDIES IN HISTORY AND PHILOSOPHY OF SCIENCE

VOLUME 14

The titles published in this series are listed at the end of this volume.

CAUSATION AND LAWS OF NATURE

Edited by

HOWARD SANKEY
Department of History and Philosophy of Science
University of Melbourne, Australia

KLUWER ACADEMIC PUBLISHERS
DORDRECHT / BOSTON / LONDON

Library of Congress Cataloging-in-Publication Data is available.

ISBN 0-7923-5914-3

Published by Kluwer Academic Publishers BV,
PO Box 17, 3300 AA Dordrecht, The Netherlands.

Sold and distributed in North, Central and South America
by Kluwer Academic Publishers, PO Box 358,
Accord Station, Hingham, MA 02018–0358, USA

In all other countries, sold and distributed
by Kluwer Academic Publishers, Distribution Centre
P.O. Box 322, 3300 AH Dordrecht, The Netherlands

Printed on acid-free paper

Printed and bound in Great Britain by Antony Rowe Limited.

TABLE OF CONTENTS

FOREWORD vii

ACKNOWLEDGEMENTS ix

HOWARD SANKEY / Introduction xi

I. LAWS AND CAUSES

ALAN CHALMERS / Making Sense of Laws of Physics 3

II. SCIENTIFIC ESSENTIALISM

BRIAN ELLIS / Causal Powers and Laws of Nature 19

D.M. ARMSTRONG / Comment on Ellis 35

BRIAN ELLIS / Response to David Armstrong 39

JOHN BIGELOW / Scientific Ellisianism 45

BRIAN ELLIS / Bigelow's Worries About Scientific Essentialism 61

MARTIN LECKEY / The Naturalness Theory of Laws 77

CAROLINE LIERSE / Nomic Necessity and Natural States: Comment on the Leckey-Bigelow Theory of Laws 83

III. LAWS, QUANTITIES AND DISPOSITIONS

ALAN BAKER / Are the Laws of Nature Deductively Closed? 91

JOHN FORGE / Laws of Nature as Relations Between Quantities? 111

CATHY LEGG / Real Law in Peirce's "Pragmaticism" (Or: How Scholastic Realism Met the Scientific Method) 125

DAVID LEWIS / Finkish Dispositions 143

BARRY TAYLOR / Comments on David Lewis: 'Finkish Dispositions' 157

J.J.C. SMART / Laws and Cosmology 161

D.M. ARMSTRONG / Comment on Smart 171

IV. CAUSATION AND THEORIES OF CAUSATION

D.M. ARMSTRONG / The Open Door: Counterfactual versus Singularist Theories of Causation 175

F. JOHN CLENDINNEN / Causal Dependence and Laws 187

JOHN D. COLLIER / Causation is the Transfer of Information 215

PHIL DOWE / Good Connections: Causation and Causal Processes 247

KEVIN B. KORB / Probabilistic Causal Structure 265

PETER MENZIES / Intrinsic versus Extrinsic Conceptions of Causation 313

HUW PRICE / The Role of History in Microphysics 331

KEITH HUTCHISON / No Interaction Without Prior Correlation: Comment on Huw Price 347

NOTES ON CONTRIBUTORS 349

INDEX 353

FOREWORD

Australia and New Zealand boast an active community of scholars working in the field of history, philosophy and social studies of science. 'Australasian Studies in History and Philosophy of Science' aims to provide a distinctive publication outlet for their work. Each volume comprises a group of essays on a connected theme, edited by an Australian or a New Zealander with special expertise in that particular area. In each volume, a majority of the contributors is from Australia or New Zealand. Contributions from elsewhere are by no means ruled out, however, and are indeed actively encouraged wherever appropriate to the balance of the volume in question. Earlier volumes in the series have been welcomed for significantly advancing the discussion of the topics they have dealt with. The present volume will I believe be greeted equally enthusiastically by readers in many parts of the world.

R.W. Home
General Editor
Australasian Studies in History
and Philosophy of Science

ACKNOWLEDGEMENTS

The papers collected in this volume are based on work presented at the symposium entitled *Causation and Laws of Nature*, which I organized at the annual conference of the Australasian Association of the History, Philosophy and Social Studies of Science, at the University of Melbourne, 27–30 June 1996. I wish to thank Rod Home for his encouragement and advice while I was organizing this symposium, as well as for his continued support as the idea for a symposium turned into the idea and later the reality of a book. To the participants in the symposium and subsequent contributors to this volume, I am extremely grateful for the overwhelming response with which my original call for papers was met, as well as for their enthusiastic participation in the symposium itself. Among these, I would single out Brian Ellis, in particular, for having introduced me to the problem of laws of nature, as well as for many helpful suggestions while both the symposium and the volume were in preparation. Conventions governing the anonymity of referees prevent me from identifying those philosophers, both here in Australia and overseas, who have assisted me in improving and insuring the quality of these papers by their critical comments: for their assistance I am deeply indebted.

I gratefully acknowledge the permission of Blackwell Publishers to include David Lewis's paper, 'Finkish Dispositions', which first appeared in *The Philosophical Quarterly* **17: 2** (1997), 143–58.

HOWARD SANKEY

HOWARD SANKEY

INTRODUCTION

Australian philosophy is distinctive for having produced a robust blend of realism, materialism and naturalism. Empiricism as an epistemological theory about the basis of our knowledge of the world is widely accepted among Australian philosophers. But the metaphysical version of empiricism which entails an anti-realist view of causation and laws of nature enjoys decidedly less currency. For the doctrine, customarily (but not uncontroversially) attributed to David Hume, that causal and nomological relationships reduce to phenomenal regularities, or constant conjunctions of objects, is one which is difficult to reconcile with a realist metaphysics of the kind espoused by Australian realists. Realists, who hold that there is a mind-independent reality the constituents of which are governed by underlying laws of nature, tend to see real causal processes and laws operative behind the empirical regularities and spatio-temporal contiguities which are evident at the merely observational level.

Just what is to be said, though, about the nature of laws of nature and causation remains a topic of considerable discussion, as the papers in this volume attest. Some think of laws of nature as grounded in the essential properties of members of natural kinds of things. Others balk at natural kinds, but see laws as necessitation relations between universals. Still others, despite being realists about the entities postulated by scientific theories, wish to treat laws of nature as large-scale, cosmic regularities. Debate about laws of nature proceeds to a certain extent on a separate front from debate about the issue of causation. Some treat causation as independent of laws of nature, as if an event might be caused without being subsumable under any law of nature. Others treat the two issues as inseparable, as though there may be no causal connection which is not the instantiation of a law of nature. As for causation itself, here also opinions diverge. Some seek to analyze causation in terms of the truth conditions of counterfactual assertions, while others see it as a matter of probability increase, conserved quantities or information transfer.

The papers collected in this volume reflect the intensity of interest and diversity of opinion which currently prevail among Australian philosophers with respect to the topics of causation and laws of nature. This intensity and diversity, moreover, is both harnessed by, and helps to fashion, a sharply focused and productive philosophical debate. As such, the papers in this volume are a powerful indication of the lively discussion, and remarkable creative energy, currently being invested in the metaphysics of causality and laws of nature by philosophers active in the Australasian region.

Rather than attempt an overview of the complex literature in this active area, in

H. Sankey (ed.), Causation and Laws of Nature, xi–xvi.

what follows I will give the briefest of outlines of the content of the individual contributions to the volume. The reader who particularly desires an introductory overview of some of the main topics under discussion may be immediately referred to Alan Chalmers' paper, 'Making Sense of Laws of Physics', which is the sole paper included in *Part One: Laws and Causes*. Chalmers' paper, which also served as the opening address of the *Causation and Laws of Nature* symposium, is an admirably clear discussion of the problem of laws and causes, which draws on both physics and the history of physics, while at the same time providing an overview of both historical and contemporary philosophical treatments of the problem.

The remaining three parts of the book consist of papers which have been grouped together according to common theme. Part Two is devoted to discussion of the scientific essentialist metaphysics currently defended by Brian Ellis. The papers in Part Three take up other approaches to, and aspects of, the issue of laws of nature. Part Four brings together a set of papers which deal with the topic of causation.

Part Two: Scientific Essentialism. In recent years, Brian Ellis has been developing a scientific essentialist metaphysics, akin to the powers ontology earlier espoused by Harré and Madden (1975). The central component of Ellis's essentialism is a theory of the nature of laws of nature which is strongly anti-Humean in character. Rather than empirical regularities, Ellis sees laws of nature as grounded in the intrinsic causal powers of members of natural kinds of things. The exercise of such powers is, in effect, the behavioural manifestation of the fundamental dispositional properties of objects which, Ellis maintains, cannot themselves be further reduced to a non-dispositional categorical base. These fundamental behavioural dispositions constitute the essential, identity-determining properties of natural kinds, since, according to Ellis, objects cannot belong to the kinds of things that they do unless they have the basic dispositions characteristic of those kinds. The necessity here is Kripkean: while members of a kind possess their essential properties necessarily, which properties these are is something that can only be discovered by *a posteriori* means.

Part Two opens with a position paper by Ellis, followed by exchanges with David Armstrong and John Bigelow, who raise a number of questions of detail about the essentialist position. Armstrong presses Ellis on the status of unmanifested dispositions, which seem to commit Ellis to relations to non-existent objects, reminiscent of Meinong. Further, in an attempt to show that Ellis must admit a categorical base for at least some powers, Armstrong raises the spectre of an infinite regress of inactive powers which arises if all properties are taken to reduce to powers. Bigelow, for his part, seeks clarification of several key aspects of Ellis's position. He asks, for example, "what *things* are there, which could be the bearers of the essential properties?", and suggests that only members of a kind – not the kind itself – may have essential properties. Bigelow also probes the relation between Ellis's view and its supposed Humean rival, as well as raising questions about the sense in which dispositions are intrinsic and the status of counterfactual conditionals with impossible antecedents.

For the essentialist, rather a lot of ordinary counterfactuals are rendered vacuous by the impossibility of their antecedents (e.g., "If we could walk on water, we would have less need for boats"). One way to reduce the extent of such vacuity is to allow exceptions to the laws of nature. Here a promising approach mentioned by Bigelow is offered by a variant of Ellis's account due to Martin Leckey, the "naturalness" theory of laws. According to this theory, as Leckey explains in his own contribution, the laws of nature are not grounded in the essential dispositional properties of things, but in a property of naturalness which things have a propensity to possess. When things possess the property of naturalness they are in a natural state, in which case they obey the laws of nature. However, the laws themselves may be violated, since things may lose or fail to possess the property of naturalness. In view of such violability, Caroline Lierse claims in her comments on Leckey, the naturalness theory seems at base to be a contingency theory of laws, since it denies an intrinsic connection between the properties of things and the behaviour which they manifest. In addition, Lierse objects that the property of naturalness remains unspecified, and responds to objections raised by Leckey against the essentialist theory of laws.

Part Three: Laws, Quantities and Dispositions. The papers in Part Three address a variety of issues pertaining to laws of nature, ranging from formal aspects of law statements, to the analysis of physical magnitudes and dispositions, and the nature and reality of laws themselves. In his contribution, Alan Baker considers the question of whether a statement entailed by a set of law statements is itself a law statement. Baker pays particular attention to the confirmability of disjunctive generalizations in arguing against both the maximal view that all such entailments are laws and the minimal view that none are. With John Forge's paper the focus shifts to the issue of what kinds of things enter into nomic relations. Forge claims that relations between physical quantities (e.g., pressure, temperature) make up an important class of laws, and argues that quantities are to be construed as relations, rather than as classes or properties. A historical tone enters with the paper by Cathy Legg, who discusses the work of the American pragmatist, C.S. Peirce, revealing interesting parallels between Peirce's "Scholastic realism" about universals and the nomological realism popular among Australian philosophers. David Lewis addresses yet another aspect of the issue of laws of nature: the analysis of dispositional statements. Lewis seeks to defend a counterfactual analysis of dispositions against an objection due to C.B. Martin involving "finkish dispositions" – dispositions which, when triggered, disappear unmanifested (Martin, 1994). In his comments on Lewis, Barry Taylor focuses on the causal bases of dispositions, and argues that Lewis must do more than he does to dispose of the possibility of dispositions which have no such base. In his paper, Jack Smart attempts to reconcile a regularity account of laws of nature with realism about theoretical entities, combining the regularity view with a minimal commitment to universals and a contextual notion of necessity weaker than that favoured by many realists. Commenting on Smart, David Armstrong allows that the explanation of regularities may be an "explanation too far", but holds nevertheless that the vision of regularities as explained by causal relations between universals is one that is "worth fighting for".

Part Four: Causation and Theories of Causation. The main focus of most of the papers in Part Four is on the nature of causation and theories of the nature of causation. While most of the authors argue in favour of one or another preferred theory of causation, they also have a good deal to say by way of comparison and critique of the various alternative theories of causation currently on offer. The chief exception is Huw Price, whose paper raises issues about causation within the context of a discussion of temporal asymmetry in physics.

David Armstrong's paper contrasts his own singularist theory of causation with the counterfactual theory of David Lewis. Armstrong rehearses objections to the Lewis account due to omissions, preemptions, preventions, and the like, before suggesting that causation is an intrinsic relation between events that may in some cases be directly observed, each instance of which instantiates a law as a matter of *a posteriori* necessity. In apparent opposition to singularist accounts of causation, John Clendinnen argues for an explication of the concept of causation which gives due weight to the role played by generalizations in identifying causal processes. The upshot, according to Clendinnen, is that causation fundamentally involves the nomic dependence of one event on another, so that causal connections are themselves nomic relations between events. In his paper, John Collier employs computational information theory to argue that causation is the transfer of items of information from one state of a system to another. His approach is based on a minimal metaphysics using only logic and concrete particulars, yet preserves key elements of other theories of causation. Phil Dowe seeks a non-circular analysis of causation in terms of causal processes, of which he offers a 'non-causal' account. In defence of his account, he attempts to forestall objections stemming from a causal theory of identity, and addresses the problems of preventions and omissions. Kevin Korb's article provides an overview of recent literature on probabilistic causality, and defends an objective homogeneity view of probabilistic causal relations against contextual unanimity approaches. Korb also surveys some recent attempts to model probabilistic causal reasoning within an artificial intelligence framework. Peter Menzies' paper proceeds somewhat at a meta-level, seeking to clarify what is at issue between singularist and non-singularist views of causation. The crux of the issue, according to Menzies, is the question of whether causation is an intrinsic or an extrinsic relation between events, rather than whether it reduces to Humean supervenience on spatiotemporal relations. Huw Price's paper probes the physical basis of the practice of explaining correlated events in terms of past rather than future occurrences. Price argues that the "No Teleology Principle" which grounds this practice cannot be sustained in the context of microphysics, despite support from the time-asymmetry of thermodynamics. The reason, put simply, is that microphysical laws are time-symmetrical in a manner which permits backwards causation. In his comment on Price, Keith Hutchison focuses on macro-level rather than micro-level correlations, in proposing counterexamples to Price's "No Teleology Principle".

Where, finally, does all this leave the issue of causation and laws of nature? Is there an emerging consensus amongst philosophers active in the Australasian region on these topics? At a general level, to be sure, Australian work on laws and causation is strongly anti-Humean, though significant Humean elements remain, for example in Smart's efforts to combine a regularity view of laws with entity realism. Apart from such generalities, however, it is difficult to discern an emerging consensus, for there is a considerable spread of opinion on various matters. Still, there are some significant convergences, as well as interesting points of divergence, which are worth noting as areas in which further developments may take place.

On the topic of scientific essentialism, for instance, opinion divides along a number of dimensions. Ellis and Lierse, and seemingly Bigelow, favour an essentialist theory of laws according to which laws hold necessarily, given the identity-determining character of the irreducible dispositions of things belonging to natural kinds. Chalmers should perhaps be included in this camp as well, since he also endorses a powers ontology.[1] By contrast, where the essentialist treats irreducible dispositions as crucial to laws of nature, others such as Armstrong and Lewis see dispositions as resting on a non-dispositional, categorical base. Moreover, where the essentialist employs an ontology of natural kinds, Smart and Armstrong express doubts about the need for natural kinds as an ontological category over and above concrete particulars and their properties and relations.

As for causation, here too one finds a spread of opinion. Clendinnen analyzes causation in terms of nomic dependence, Collier speaks of information transfer, Dowe of conserved quantities, while Korb favours probabilistic treatment of causality. Yet, despite appearances, there is perhaps some prospect of convergence among these alternative perspectives. Adopting a strategy employed elsewhere by Peter Menzies, Armstrong suggests that causation may be treated as a theoretical entity, to which reference is fixed by means of various "causal platitudes" (e.g., raises probability, supports counterfactuals) associated with one or another theory of causation (see Menzies, 1996). Similarly, Collier suggests that his information-theoretic approach preserves key aspects (e.g., regularity, conserved quantity) of other approaches to causation. The thought shared by this pair of suggestions is that the various theories of causation have all fastened onto genuine aspects of causation, which may eventually be brought together in a fully developed theory. However things ultimately fare with this latter thought, there is one definite tendency which runs through a number of the papers on causation: a singularist tendency. For singularists, causation is not to be approached *via* laws of nature, or circumstances extraneous to the causal situation; rather, it is an intrinsic relation obtaining in its own right between the events related as cause and effect. This line of thought is most evident in the papers by Armstrong, Dowe and Menzies. It also has some resonance in the papers by Collier and Korb.

As one might expect, a recurrent theme in the volume is the relationship between causation and laws of nature. While the question of their relationship receives little sustained discussion, a number of different views on the matter do emerge in the course of the volume. For example, Chalmers and Ellis speak of "causal laws", and they treat such laws as descriptions of the capacities of particular things to bring about events by their causal action. For Collier, laws of nature are the most abstract form of particular causal relations that include the necessary mathematical form for the preserved information to allow causation. Armstrong, while advocating singularism, holds that as a matter of empirically discoverable fact all causal relations do fall under laws of nature.[2] By contrast, Clendinnen sees a close conceptual connection between laws and causes, since, in his view, our concept of causation involves nomic connections between events. On the whole, therefore, no univocal message emerges about the relation between causation and laws of nature, though the space of possible approaches to the matter is displayed in the course of the discussion.

In a somewhat speculative vein, it is worth pondering the relation between the essentialist theory of laws and the singularist theory of causation. Is there much prospect of combining the two positions? For the essentialist, objects which belong to natural kinds possess certain dispositions or causal powers by virtue of their kind-membership. For the singularist, causation is an intrinsic relation obtaining between events. Let us suppose that any given instance of a particular cause producing a particular effect is the exercise of a causal power of an object which belongs to a natural kind. Then, given that the effect is produced by the action of a causal power of such an object, no extraneous circumstances are involved in bringing about the effect, precisely as required by singularism. Moreover, given that causal powers are what grounds laws of nature, any instance of singular causation, being the exercise of a causal power, just is an instance of a law of nature, as a matter of *a posteriori* necessity, which is consistent with scientific essentialism. It would seem, then, that two of the major trends evident in this volume, essentialism and singularism, are capable of being brought into closer proximity than is currently the case.

To end on a metaphilosophical note, the work presented in this volume reflects the extremely fruitful interaction currently underway between metaphysics and philosophy of science, two areas of philosophy which have for much of this century followed separate paths. Indeed, within the context of early twentieth century positivism fruitful interaction between these two areas of philosophy was scarcely conceivable. In my view, the project of developing a metaphysics of science, which this interaction serves, represents one of the most vital and promising areas of contemporary philosophical research. It is my hope that this collection will play a significant role in furthering this enterprise.

University of Melbourne

NOTES

1 Chalmers' endorsement is qualified, however, since he notes that there are non-causal laws to which the essentialist account does not apply; yet, since Ellis also notes the existence of such laws, this seems consistent with essentialism.

2 As Menzies points out in his paper, singularists divide amongst themselves over the proper stance to take on the relation between laws and causes. He contrasts anomic singularists, who deny a conceptual or a priori connection between laws and causes, with nomic singularists, for whom the connection is a priori. In Menzies' terms, Armstrong is an anomic singularist, but one who takes there to be an empirical connection between laws and causes.

REFERENCES

Harré, R. and Madden, E.H. (1975), *Causal Powers: A Theory of Natural Necessity*, Oxford: Blackwell
Martin, C.B. (1994), 'Dispositions and Conditionals', *The Philosophical Quarterly* **44**, 1–8
Menzies, P. (1996), 'Probabilistic Causation and the Pre-Emption Problem', *Mind* **105**, 85–117

I.

LAWS AND CAUSES

ALAN CHALMERS

MAKING SENSE OF LAWS OF PHYSICS

"I cannot conceive how a body devoid of understanding and sense, truly so called, can moderate and determine its own motions, especially so as to make them conformable to laws that it has no knowledge or apprehension of" (Boyle, [1686] 1979, pp. 181–2). With these words Robert Boyle questioned in a very fundamental way the appropriateness of applying the notion of a law to nature, since only intelligent beings are in a position to obey, or not to obey, laws.[1] In this paper I rise to the challenge embodied in Boyle's remark, and attempt to make sense of laws as they are employed in physics.

Within physics, the physical world is assumed to be governed by, or explicable in terms of, generalities such as Newton's laws of motion, the laws of thermodynamics, Maxwell's equations and a variety of conservation laws. My concern in this paper is to explore what kinds of claims are entailed in such generalisations. I take it for granted that physics is a highly successful and progressive body of knowledge and that those statements typically referred to as laws of nature play a crucial part in that success. With a close eye on the ways in which laws are actually used in physics I attempt to elucidate the kinds of claims about the world that they embody.

BOYLE'S ANALYSIS OF LAWS

Although the notion of a scientific law did not originate in the seventeenth century (see Ruby, 1986), it first gained wide currency at that time. Robert Boyle was both a productive experimental scientist and a leading mechanical philosopher of that era, and it is instructive to begin by examining his attempts to fashion a concept of law appropriate for science.

We have already seen that Boyle acknowledged the problematic character of the assumption that the material world acts in conformity with laws. For,

> to speak properly, a *law* being but a *notional rule of acting according to the declared will of a superior*, it is plain that nothing but an intellectual being can be properly capable of receiving and acting by a *law*. For if it does not understand, it cannot know what the will of the *legislator* is; nor can it have any intention to accomplish it, nor can it act with regard to it, or know when it does, in acting, either conform to or deviate from it. (Boyle, [1686] 1979, p. 181)

This leads to what I shall refer to as Boyle's question. "What is it that compels matter to behave in accordance to laws?"

The problem was an acute one for Boyle because, as a mechanical philosopher, he portrayed the material world as one which was particularly stark from an ontological

H. Sankey (ed.), Causation and Laws of Nature, 3–16.

point of view. For Boyle the material world is made up of fundamental particles or corpuscles which he referred to as natural minima, and which are all made of the one universal matter, characterised by its impenetrability, and only by that. Each natural minimum possesses a characteristic shape and size and is capable of motion. All phenomena in the material world are to be accounted for in terms of the motions and arrangements of such corpuscles.

Given this ontology, it is clear that the only source of activity and change in the world is the motion and collision of corpuscles. Even collisions are problematic, given the resources of Boyle's mechanical ontology. This can be highlighted by pushing the question, "by virtue of which of their properties do corpuscles rebound on colliding?". Their impenetrability ensures that colliding corpuscles cannot continue on their course, but that is all. It would be quite consistent with all of their properties as specified by Boyle for colliding particles to come to rest or to move off in any directions and with any speeds whatsoever. Boyle's answer to this conundrum is to explicitly invoke the active intervention of God. It is God who, "by his ordinary and general concourse", constantly maintains in corpuscles their power to transmit motion to one another in a law like way (Boyle, [1686] 1979, p. 181). Boyle refers to "the rules of action and passion among the parcels of the universal matter" as synonymous with "the fixed laws of nature" and observes that it would be difficult to give a satisfactory account of this "without acknowledging an intelligent Author or Disposer of things" (Boyle, [1663] 1979, p. 157).

For Boyle, then, laws do not result from or describe properties of matter. Rather, they are imposed on matter by God. In Boyle's own words, "the laws of motion . . . did not necessarily spring from the nature of matter, but depended upon the will of the divine author of things" (Boyle, [1690] 1772, p. 521). The laws of nature are contingent, then. They are imposed on matter by way of the will of the Creator, and could have been other than what they are had God chosen otherwise. Boyle even speculated about a possible world, alike ours in all respects other than that God imposed and sustained different laws of motion on it, so that

> the laws of propagation of motion among bodies may not be the same with those, that are established in our world; so that but one half, or some lesser part (as a third) of the motion, that is here communicated from a body of such a bulk and velocity, to another it finds at rest, or slowlier moved than itself, shall there pass from a movent to the body it impels, though all circumstances, except the laws of motion, be supposed to be the same. (Boyle, [1685] 1772, p. 140)

The main point I wish to emphasise from this look at Boyle's system is that, given the stark ontology from which he begins, with matter endowed with a few passive and no active properties, or, in modern parlance, with a few categorical and no dispositional properties, Boyle is faced with the problem of accommodating significant activity and law like behaviour in the world. God is invoked to do all of the work here. Without Him Boyle cannot even accommodate collisions, the main mechanism of change in his system. From Boyle's point of view this was a welcome, rather than a problematic aspect of his system, since the necessity to invoke God provided a bulwark against atheism (see Hutchison, 1983). I assume that, from a modern point of view, placing such a heavy, or indeed any, burden on the constant and wilful intervention of God is not acceptable. But eliminating God from the account leaves us with the problem. How can activity and law like behaviour be introduced into a world characterised in terms of passive, or categorical, properties only? How can Boyle's question be answered without invoking God?

THE HUMEAN DENIAL OF THE QUESTION

I take it for granted that there is much activity in the material world. Coconuts fall from trees and bounce when they hit the ground, puddles evaporate in the heat of the sun, running water erodes river banks, lightning strikes trees, pulsars radiate and so on. I also take it for granted that the explanation of such natural effects, and our capacity to purposefully and actively intervene in them, assumes that they are in some way governed by laws. Given this, I am led to ask how activity is to be accommodated and characterised and, with Boyle, what compels matter to behave in conformity to laws.

For Hume and his supporters, my questions presume too much. For them, to assume that nature is in some sense active, and that it is governed by laws that are other than observable regularities, is to go beyond what can be rationally justified. Their grounds for holding this are epistemological. We can observe the speeds and directions of two colliding billiard balls immediately before and immediately after they collide and we may discern a regularity in the way these speeds and directions are connected together. What we do not observe is something in addition to this which constitutes the ability or capacity of one billiard ball to move another. From this point of view, causation is nothing other than regular connection, and laws take the form, "events of type A are invariably accompanied or followed by events of type B". For instance, Galileo's law of fall is interpreted as making the claim that whenever a heavy object is released near the earth's surface it falls to the ground with a uniform acceleration. From this point of view, my assumption that nature is active, if it is taken to mean that bodies attract each other, that the heat of the sun has the power or capacity to evaporate water and running water the capacity to erode river banks, is not rationally justifiable. Consequently, the answer to Boyle's question is that nothing compels matter to move in accordance with laws, for those laws are nothing other than de facto regularities between observable events.

The Humean point of view involves a radical empiricism according to which the senses in some way provide human observers with facts about the world which constitute the building blocks from which our knowledge of the world is to be constructed. This approach ushers in a distinct ontology. The world, at least as far as it can be known, is comprised of logically distinct events, the existence of white sugar granules in the teaspoon, the sensation of sweetness that accompanies the transference of the sugar to the tongue, a configuration of billiard balls at some specific time and so on. Each individual act of perception delivers to us knowledge of an event of this kind. Since no set of events is seen as having any logical connection with events not in the set, causality and laws reduce to nothing other than regular associations between events, and any attempt to portray them as involving more than this is seen as epistemologically unwarranted and extravagant.

The Humean view can be challenged by questioning the epistemology it presupposes. Statements describing individual events are not given by way of the senses in as straightforward a way as the Humean supposes, and once the complications are appreciated, then they are not as epistemologically privileged compared with statements describing dispositions as the Humean supposes. No observation statement describing an observable event is securely established by way of the deliverances of the senses. As Popper famously observed, even a statement like "here is a glass of water" cannot be established with certainty on the basis of immediate experience since it involves universals, and has implications that go beyond what is immediately given, such as the implication that the water could spill or the glass break (Popper, 1977, pp. 94–5). Any doubts about the truth of the statement would typically be investigated by testing these and implications like it. This leads naturally to the position that observation statements are adequate to the extent that they are straightforwardly testable and withstand those tests. But once the account of the empirical base of our knowledge is construed in that way, then there is no qualitative difference between statements involving powers or dispositions and those that do not (or allegedly do not). Statements that attribute a degree of elasticity to billiard balls can be tested and supported by appropriate observations just as statements that attribute colour and speed to them can be tested and supported.

This is no place to enter into a detailed critique of Humean epistemology and its replacement by an alternative one. My main concern is the inadequacy of the Humean account of laws, which construes them as nothing other than empirically establishable regularities. That account fails to capture both the necessity and degree of generality that science presumes laws of nature to have. The occurrence of regularities is neither necessary nor sufficient for the instantiation of a law of nature.

The objection to the regularity view on the grounds that it fails to distinguish between accidental and law like regularities is well known.[2] It is illustrated by Popper's example of the regularity "no moa lives beyond fifty years". It may well be true that no moa, a species now extinct, ever lived beyond fifty years but some might well have done so had environmental conditions been more favourable, and for this reason we are inclined to disqualify the generalisation as a law of nature. But it qualifies as a law on the grounds that it is an exceptionless regularity. There is something more to law like behaviour than mere regularity. Another difficulty with the regularity view is that it fails to identify the direction of causal dependency. There is a regular connection between occurrences of smoking and lung cancer, but this is because smoking causes lung cancer, not the reverse. That is why we can hope to decrease the occurrence of cancer by eliminating smoking, but cannot hope to combat smoking by finding a cure for cancer.

Once we move away from the common-sense domain of ravens, moas, wristwatches and factory hooters, the domain well exploited by analytic philosophers to pose problems for the regularity view of laws, to take a closer look at laws as they function in science, then it becomes clear that, whether regularities are a sufficient condition for a law or not, they are not necessary and very rarely, if ever, obtain. It is not possible, in general, to discern amongst the happenings or events in the world exceptionless regularities that are candidates for laws of nature. If, for instance, we interpret the law of fall, mentioned above, as claiming that whenever an object is allowed to fall near the earth's surface it descends with a uniform acceleration, it is refuted by falling leaves, which do not do so. In a similar vein, from a strict regularity point of view, Archimedes' principle is refuted by floating needles, whilst Newton's first law, if true at all, becomes vacuously true, insofar as there never has been a body on which no forces act, as the law of gravitation ensures.

Whilst my appeal to Autumn leaves and the like poses problems for the regularity view, it need pose no problem for science, or even for common sense. For, after all, there is a ready explanation of why Autumn leaves do not fall with a uniform acceleration that appeals to variable air-resistance and winds. That is why it is necessary to test the law in specially contrived experimental conditions from which perturbing influences such as winds are excluded. The regularities of relevance to science, and which are indications of law like behaviour, are typically the hard-won results of detailed experimentation. Think, for example, of the lengths to which Henry Cavendish had to go to get attracting spheres to behave in a way that exemplified the inverse square law, and the path to J.J. Thomson's success in deflecting cathode rays by electric and magnetic fields in accordance with the Lorentz force law. It is of relevance to note that an experiment similar to that of Thomson's was performed by Heinrich Hertz over a decade earlier, and the cathode rays in his experimental set-up were not deflected at all by electric and magnetic fields. From a modern point of view we have a good idea of the countervailing influences that prevented them from being so. That is why we do not accept the failure of Hertz's experiment to exhibit the appropriate regularity as a refutation of the Lorentz force law.

There is an obvious response for the supporter of the regularity view of laws to these reflections on the importance of experiment in science, and of the scarcity of regularities of use to science amongst the happenings in the world at large. That response is to state laws in a conditional form. Laws then take the form "events of type A are followed, or accompanied by, events of type B provided . . . ", where the dots are to be filled in by appropriate conditions. The law of fall then becomes "whenever a heavy object is released close to the earth's surface it falls to the ground with a uniform acceleration provided it does not encounter a variable resistance or disturbances such as winds or magnetic attractions". An obvious difficulty for the regularity view here is how all the conditions needed for the law to hold are to be stated precisely, a difficulty I evaded in the previous sentence by my use of the phrase 'such as'. Things would be simpler if it was permissible to assert that the law will hold provided no non-gravitational forces or countervailing causes act, but such a move is only available within the regularity view if talk of forces and countervailing causes are themselves reduced to regularity talk. Those regularities will themselves need to be stated in conditional form and so the problem recurs. If the regularities that constitute laws for the Humean are to be stated in conditional form, then an adequate specification of the appropriate conditions becomes impracticable if not impossible.

However, a much stronger objection to this reformulated regularity view concerns the question of what governs the behaviour of the world outside of experimental conditions. If laws are identified with regularities, then they cannot be said to obtain outside of experimental situations where, in general, appropriate regularities do not occur. Stating laws in conditional form does not help here, because outside of experimental situations the conditions are not satisfied. But it might well be argued that the whole point of doing experiments is to find out, in the experimental context, laws that apply to the world outside of, as well as in, that context. Scientific knowledge is to be applied to the world in general. If we take the example of the falling leaf, then we have noted that it does not fall in a regular way that instantiates the law of fall, or the more general law of gravitational attraction to which the law of fall can now be reduced. Nevertheless, surely we must assume that the leaf is subject to gravity, otherwise we have no explanation of why Autumn leaves typically end up on the ground. Are we not inclined to say that the leaf is subject to the law of gravity but, at the same time, is subject to laws governing air resistance and hydrodynamics? It is the varying mix of these, and perhaps other, factors that jointly account for the irregular fall of the leaf. Such a natural move is not open to the defender of the view that laws are regularities subject to conditions, for, within such a view, laws do not apply when those conditions are not satisfied.

NANCY CARTWRIGHT'S SKEPTICISM CONCERNING LAWS

Nancy Cartwright is a contemporary philosopher who has been impressed by the lack of regularity amongst the data of relevance to science. This led her, in an influential book, *How the Laws of Physics Lie*, to argue that the fundamental laws of physics should not be taken in a realist sense, that is, they should not be interpreted as characterising some feature of the world. So long as the approach to laws is broadly Humean, Cartwright's case is a forceful one. However, there are serious problems with her position as we shall see.

Cartwright's focus is on the way in which science is made to work in the laboratory. She notes that the orderly behaviour achieved in a successful experiment is not usually of a kind that could be predicted from fundamental laws of nature plus appropriate initial conditions, and that any regularities that obtain do not exemplify those laws. She notes, for instance, that the theoretical treatment of an electronic amplifier does not yield a phenomenologically adequate description of the amplifier's performance. Such a description can only be obtained by modifying the formula in the light of empirical measurements on the amplifier. She notes that the reproducible scattering cross sections of fundamental importance in nuclear physics are not deduced from any theory, but are measured experimentally. She describes the various theoretical models devised to capture the various characteristics of lasers, but not one of them is capable of predicting or describing exactly how a laser behaves. Most theoretical predictions at best need to be corrected, or touched up, in the light of experimental measurements.

The gulf between what fundamental laws can predict and the regularities that are fabricated in the laboratory led Cartwright, in *How the Laws of Physics Lie*, to abandon a realist interpretation of fundamental laws. The strength of science, it would appear, is to be located, not in its establishment of fundamental laws of nature, but in the practical achievements of scientists in the experimental and technological context. The results of those endeavours do lead to regular and controllable behaviour and Cartwright coined the term 'phenomenological law' to refer to examples of that regular behaviour. She cites the performance characteristics of lasers, as specified by the manufacturers, as an example. "Spectra-physics Incorporated continuously runs a quarter of a million dollars' worth of lasers to death to test their performance characteristics. Nothing could be better confirmation than that" (Cartwright, 1983, p. 3). The ability of phenomenological descriptions of lasers to specify precisely and accurately what happens is contrasted by her with the failure of fundamental laws to do so.

Cartwright is right to stress that there is more to successful experimental activity than deriving behaviour from fundamental laws plus initial conditions. I spent three years as an experimental physicist attempting to scatter low energy electrons from molecules, hoping to produce regularities which would bear important information about the electron structure of those molecules. I failed, but not through ignorance of relevant fundamental laws and how to apply them. Hacking is right. Experiment has a life, or in my case a death, of its own (Hacking, 1983, p. 150). However, Cartwright's focus on phenomenological laws and her rejection of a realist interpretation of fundamental laws is inadequate for a range of reasons.

Cartwright's phenomenological laws cannot be taken to be significant laws of nature because, in general, they lie. To the extent that they are true they describe regularities, but the conditions that need to be satisfied for the descriptions to apply are extremely demanding and rarely achieved. After all, faulty lasers do occur, in spite of the stringent testing procedures carried out by manufacturers. Insofar as there are true generalisations about the sequences of events produced in some exemplary experiments, the reference of those generalisations, if they are to be true, is so specific that it seems inappropriate to refer to them as laws. If phenomenological laws are formulated with any reasonable degree of generality, then they become false descriptions of what happens, and Cartwright's position is open to the same kinds of objections as the regularity view in this respect.

Nancy Cartwright's skepticism with respect to the truth of fundamental laws does not stem from any qualms about making claims about the unobservable. She is a realist with respect to causes whether those causes are observable or not. She freely admits such things as electrons and electromagnetic fields amongst the causes that make things happen. "If there are no electrons in the cloud chamber, I do not know why the tracks are there" (Cartwright, 1983, p. 99). However, this realist position with respect to causes and the entities that are responsible for them sits uneasily with her denial of a realist interpretation of fundamental laws. It is as if we are allowed to admit electrons into our deliberations in physics but are not allowed to claim anything significant about them. When I attempted to produce a mono-energetic beam of electrons and direct them at target molecules using appropriate electron lenses I needed to assume that electrons are charged particles that are deflected by electromagnetic fields in accordance with the Lorentz force law, and the whole point of doing the experiment was based on some quantum mechanical reasons to suspect that doing this would give access to knowledge about energy levels in molecules not available through optical spectroscopy. The skeptic need derive no solace from my failures, because the technique has since been developed successfully by others, to the extent that low-energy electron spectroscopy is now a standard technique. The facility with which physicists strive to manipulate unobservable entities in the laboratory, fully acknowledged by Cartwright, is difficult to reconcile with a position that denies that the behaviour of those entities is governed by fundamental laws.

Once we follow Cartwright and deny a descriptive role to fundamental laws, then we run into the same problems as beset the regularity view that restricts the applicability of laws to those experimental situations where the appropriate regularities obtain. We are at a loss to say what governs the world outside of experimental situations. We are hard pressed, for example, to justify the use of radioactive dating in historical geology or palaeontology, or Newton's law of gravitation to comets. The orbit of

Halley's comet on its most recent return differed slightly from what had been predicted. The first sighting of the comet, by a Voyager spacecraft, enabled the predicted orbit to be *corrected*, thereby facilitating later terrestrial sightings. The regularity predicted on the basis of Newton's laws lied. Nevertheless, Halley's comet did return very nearly as scheduled. To echo Cartwright, if Halley's comet is not governed by Newton's laws, then I do not know why it returned as and when it did.

A detailed look at the nature of experiment and its epistemological implications has been long overdue, and those, like Cartwright, who have explored this area in recent years have greatly improved our knowledge of the nature of science. However, to focus exclusively on experiment is to confuse the domain of the testing and technological application of scientific theories with the domain of their applicability.

LAWS AS DESCRIPTIONS OF POWERS, TENDENCIES OR DISPOSITIONS

I suggest that the inadequacies of Cartwright's response to a situation she in many respects accurately and usefully describes, and the difficulties of the regularity view of laws in general, can be avoided in a straightforward way. It involves taking seriously what is implicit in much common sense as well as scientific talk, namely, that the material world is active. Things happen in the world of their own accord, and they happen because entities in the world possess the capacity or power or disposition to act or behave in the way they do. Balls bounce because they are elastic. Warnings on containers that declare the contents to be poisonous or inflammable or explosive tell us what the contents are capable of doing or how they are inclined to act. Specifying the mass and charge of the electron indicates how an electron will respond to forces and electromagnetic fields. An important element of what a thing is, is what it is capable of doing or becoming. We need to characterise things in terms of their potential as well as their actual being, as Aristotle correctly observed. Just as the ability to grow into an oak tree is an important part of what it is to be an acorn, so the capacity to attract unlike and repel like charges, and to radiate when accelerating, is part of what it is to be an electron.

Once such things as capacities, powers, tendencies, dispositions or propensities are included in our ontology, then laws can be taken as describing their mode of operation. If we interpret laws in this way, then we need no longer expect laws to describe sequences of happenings in the world, for those happenings will typically be the result of several capacities, powers, etc., acting in conjunction in complex ways. The fact that a tendency of a leaf to fall in accordance with the law of gravitation is swamped by the effect of a wind is no reason in itself to doubt that that tendency continued to act on the leaf in accordance with that law. From this point of view, we can readily understand why experiment is necessary to glean information relevant for the identification of laws. The tendencies corresponding to the law under investigation need to be separated from other tendencies, and this separation requires the practical intervention necessary to bring that about. Given the irregularities of ocean beds and the attraction of the sun and planets as well as the moon, we cannot hope to arrive at a precise account of the tides from Newton's theory plus initial conditions. Nevertheless, gravity is the major cause of the tides, and there are appropriate experiments that can help identify the precise law governing gravity.

From the point of view I advocate, causes and laws are intimately linked. Events are caused through the action of particulars that possess the power to act as causes. The gravitational attraction of the moon is the main cause of the tides, charged particles cause the ionisation that leads to the condensation that constitutes the track in a cloud chamber and oscillating charges cause the radio waves emitted from a transmitter. Descriptions of the mode of acting of the active powers involved in such cases constitute the laws of nature. The inverse square law of gravitation describes quantitatively the power to attract possessed by massive bodies, whilst the laws of classical electromagnetic theory describe, amongst other things, the capacity of charged particles to ionise and radiate. It is the active powers at work in the world that make laws true when they are true. We thus have a ready answer to Boyle's question. It is the powers and capacities possessed by particulars and operative when particulars interact that compel those particulars to behave in accordance with laws. Lawlike behaviour is brought about by efficient causation. Boyle faced the problem he did with laws, and needed to invoke God, just because he declined to ascribe dispositional properties to matter.

The majority of philosophers are reluctant to accept an ontology which includes dispositions as primitives. I do not understand their reluctance. Perhaps the reasons are in part historical. Powers were given a bad name by the near-mystical and obscure way they were employed in the magical tradition in the Renaissance, whilst they are alleged to have been exploited by the Aristotelians in a cavalier way under the guise of forms. Boyle's rejection of active properties in his mechanical philosophy can be seen as a reaction, and perhaps an over-reaction, to the excesses of those traditions as well as being motivated by theological concerns. However, there need be nothing mysterious or epistemologically suspect about invoking powers, tendencies and the like. Claims concerning them can be subject to stringent empirical tests to as great a degree as any other kind of claim. What is more, however much some philosophers may be averse to dispositional properties, scientists systematically invoke them and their work would be incapacitated without them. It is significant, in this respect, to note that Boyle, in his experimental science, as opposed to his mechanical philosophy, freely employed dispositional properties such as acidity and the spring of the air.[3] Elasticity in various forms was an embarrassment to the seventeenth century mechanical philosophers. William Petty described it as that "hard Rock in Philosophy . . . Which has long lain thwart us in the way of Our Enquiries"[4] and Hobbes complained that Boyle's attribution of elasticity to air was equivalent to the admission that air could move itself (see Shapin and Schaffer, 1985, p. 141). Boyle, and other seventeenth century scientists, continued to employ the concept of elasticity, and never succeeded in explaining it away by reference to non dispositional properties. Nor has anyone succeeded since. I do not understand what grounds philosophers have for questioning, or feeling the need to explain away, this common, indeed ubiquitous, usage by scientists of dispositional properties.

In the following I refer to my preferred view of laws as the causal view, and the laws that it describes as causal laws. This view, that laws describe the ways of acting of the dispositions, powers, capacities or propensities of particulars, is not peculiar to me. I was first convinced of it by R. Harré and E.H. Madden in their book *Causal Powers*, published in 1975, and learnt from Roy Bhaskar's *A Realist Theory of*

Science the importance of this view for understanding the role of experiment in science. Nancy Cartwright herself defends a similar view in *Nature's Capacities and Their Measurement*, published in 1989, although she does not quite acknowledge the way in which her revised position enables a realist view of fundamental laws to be rescued from the arguments of her previous book.[5] Closer to home, John Bigelow, Brian Ellis and Caroline Lierse have clearly articulated and extended the view of laws that I favour (Bigelow, Ellis and Lierse, 1992). The causal view has the merit that it acknowledges at the outset what is implicit in all scientific practice, that nature is active, it makes it clear what it is that compels particulars to obey laws, and it links causation and laws in a natural way. However, I cannot leave matters there with a good conscience because there are important laws of science that are difficult to fit into this scheme.

THERMODYNAMIC AND CONSERVATION LAWS

If ice is subjected to pressures higher than normal atmospheric pressures its melting point is lowered. This is why a wire from which weights are suspended will cut its way through a block of ice. The explanation of this at the molecular level is far from straightforward, and a precise, detailed account is probably not available. Since pressure tends to push molecules closer together, one might expect the forces of attraction between them to increase under such circumstances, leading to an increase in the thermal energy necessary to drag them apart and thus to an elevation in melting point. This is precisely what happens in a typical solid near melting point. But ice is not a typical solid. The water molecules in ice are rather loosely packed, more so than they are in the liquid state, which is why ice is less dense than water. (This is just as well, otherwise lakes and rivers would freeze from the bottom up, would freeze in their entirety during lengthy cold periods, thus eliminating fish and anything evolved from them as a viable life form.) If the molecules in ice are forced closer together than normal, the force between them decreases, so less thermal energy is needed to separate them, and the melting point falls. The precise way in which the forces depend on molecular positions is complicated, depending on fine quantum mechanical detail involving exchange as well as Coulomb forces, and is not known with precision.

Given the above complications it may come as a surprise that James Thomson was able to predict the depression of the freezing point of water with pressure in 1849 thereby anticipating the empirical discovery of the phenomenon. All he needed for his derivation was the laws of thermodynamics plus the empirically known fact that water is denser than ice. Thomson devised, in thought, a cyclic process that involved extracting heat from water at 0 C and converting it into ice at 0 C. It seemed as if this engine provided a means of extracting heat from water and converting all of it into the work done by the expansion involved when water is converted to ice, thus comprising a perpetual motion machine of the second kind, ruled out by the second law of thermodynamics. Thomson realised that this unacceptable conclusion could be blocked by assuming the depression of the melting point by an increase in pressure.

The feature of this case that I wish to highlight is that Thomson's prediction was made in ignorance of the details of the causal process at the molecular level. A characteristic feature, and a major strength, of thermodynamics is that it applies at the phenomenological level whatever the details of the causal process. It is precisely this feature of the laws of thermodynamics that makes it difficult to portray them as causal laws. What kind of generality can be claimed for a view of laws that cannot accommodate the law of conservation of energy, or any conservation laws for that matter!

The difficulties for the causal view of laws do not stop here. Newtonian mechanics can be formulated in its Lagrangian form. This formulation requires the potential and kinetic energy of a system to be known as a function of whatever set of coordinates of a system are necessary to fix them. Once these functions are known then the evolution of a system can be deduced from Lagrange's equations. This can be done without a detailed knowledge of the causal processes at work. James Clerk Maxwell, who attempted to cast his electromagnetic theory in Lagrangian form, illustrated this point in a characteristically vivid way (Maxwell, [1879] 1965, pp. 783–4). We imagine a belfry in which a complicated, interconnected piece of machinery is driven by bell ropes that drop to the bell ringers room below. We assume the number of ropes to be equal to the number of degrees of freedom of the system. The potential and kinetic energy of the system as a function of the position and velocity of the ropes can be determined by experiments done with the ropes. Once we have these functions we can write down Lagrange's equations for the system. It is then possible, given the positions and velocities of the ropes at any one instant, to derive their positions and velocities at any other instant. We can do this without needing to know the details of the causal story of what is happening in the belfry. Lagrange's equations do not state causal laws.

It might be objected that these observations about the Lagrangian formulation of mechanics do not constitute a fundamental objection to the causal view of laws. It might be pointed out, for example, that, whilst a Lagrangian treatment of the mechanism in the belfry can work as well as it does by ignoring the detailed causal story of the working of the mechanism, there is such a story to be had that can be formulated in Newtonian mechanical, and hence causal, terms once appropriate empirical access to the mechanism in the belfry is gained. After all, it might be observed, Lagrange's equations can be derived from Newton's.

This last claim is no longer true (if it ever was). In modern physics Lagrange's equations are interpreted in a more general way than the version of those equations that can be derived from Newton's laws. The energies involved are interpreted in a general way that includes all kinds of energy and not just that which results from the motion of massive bodies under the influence of forces. For instance, they can accommodate electromagnetic energy, which includes velocity dependent potential energies and necessitates such things as electromagnetic momentum of a field, which is a momentum different from that corresponding to mass times velocity. When pushed to the limit in modern physics, these Lagrangian (or related Hamiltonian) formulations are not such that they can be replaced by the causal accounts that underlie them. For instance, the various conservation principles, such as the conservation of charge and parity, intimately connected with symmetries in the Lagrangian function of the energies, are not explicable by reference to some underlying process. Consequently, if Boyle's question is pressed in this connection then no answer is forthcoming. I do not know what compels systems to act in accordance with conservation laws. They just do.

To explore this particular issue more thoroughly would require technicalities which would be out of place in this paper, so let me return to the law of conservation of energy, since I believe that the main difficulty I am raising for the view that construes laws of nature as causal laws can be well enough illustrated by this example. Since this law is assumed to apply, and with powerful effect, to systems in general, whatever the underlying causal details, it does not seem possible to construe it as a causal law. Bigelow et al (1992) have attempted to do so. They suggest that the world (by which they mean the whole universe) is one of a kind (even if it is the only one of its kind) just as an electron is one of a kind. Electrons obey Coulomb's law because part of what it is to be a thing of that kind, namely an electron, is to attract and repel other charges in accordance with the inverse square law. Particles that do not do so cannot be electrons. Extending this to the world as a whole, the suggestion is that it obeys the law of conservation of energy because part of what it is to be a world of the kind that ours is is to obey that law. The law of conservation of energy describes an essential feature of our world just as Coulomb's law describes an essential feature of electrons.

I do not consider that this attempt to construe the conservation of energy as a causal law, and thus assimilate it to the view of laws that I, along with Bigelow, Ellis and Lierse favour, succeeds. There are two reasons for this. Firstly, the conservation of energy, on those occasions when it is applied usefully and powerfully in physics, is applied to individual, local systems in the world, not to the world as a whole. It is true that, as far as local systems are concerned, the law of conservation applies exactly only to isolated systems, and none of the local systems on which physicists work are perfectly isolated. But this kind of reservation applies to the application of any law, as much of the foregoing discussion has stressed. The causal factors governed by any law never operate totally unhindered. The power of the law of conservation of energy and the evidence for its validity rests on its application to a variety of kinds of local systems, not to the universe as a whole. Consequently, it does not seem to me helpful or illuminating to construe the import of that law as making a claim about the essential properties of the world as a whole.

Even if it is conceded that the law of conservation of energy describes a property of the universe considered as one of a kind, there is still a gulf between that law so construed and the causal laws depicted in my preferred view. For that law, whilst it puts important constraints on any process in the universe, does not describe powers and capacities that are exercised in a law like way in bringing about the phenomena of the world as causal laws do. By talking of causal laws and the conservation of energy as both referring to essential properties of natural kinds, our authors convey the appearance of having offered a single unified account of laws of nature, but this appearance is deceptive. I do not see how Boyle's question can be answered in the case of the conservation of energy. It puts constraints on the kinds of mechanism that can exist in the world, but does not give a causal account of the operation of those mechanisms.

A NON-ESSENTIALIST VIEW OF LAWS

I have offered a view of laws that construes them as describing causal powers but have acknowledged that there are important and fundamental laws in physics that do not conform to this view. This fact seems to me to be interesting rather than worrying. I do not see why all the generalisations fundamental to science should be of the same kind. I do not see why we should hold an essentialist view of laws. Certainly scientists themselves use the term 'law' to refer to a variety of kinds of assertion. Maureen Christie has shown convincingly how a variety of claims referred to as laws by chemists, such as the law of multiple proportions, cannot be subsumed under any of the standard philosophical accounts of laws (see Christie, 1994). I remember being troubled at high school by the law of Dulong and Petit which, as I was taught it, reads "for any substance the atomic weight times its specific heat equals 6.4 approximately". (If my memory is correct, then my teacher should have been more careful and specified the units.) My worry about this 'law' was great enough because of the 'approximately' but was compounded when we were promptly informed that there are many common substances, such as carbon and silicon that do not obey it, not even approximately. The typical reaction of philosophers to examples such as these is that they are not so interested in low level 'laws' which perhaps amount to little more than useful rules of thumb for scientists in some limited set of circumstances, but with fundamental laws of nature. However, I have argued in my previous section that fundamental laws are themselves not of the same kind and so not subsumable under an essentialist account of laws of nature.

I finish with a challenge. Name one assertion that has been introduced into physics in the last fifty years that physicists refer to as a law. If you cannot meet this challenge then I think that that is of some significance for what I have argued in the closing sections of this paper.

University of Sydney

NOTES

[1] Boyle was certainly not the first to question the appropriateness of attributing laws to nature in a literal way. Similar doubts date back at least as far as Thomas Aquinas. See Jane Ruby (1986), especially p. 351.

[2] For a detailed critique of the regularity view of laws see Armstrong (1983, chapters 2–5).

[3] I have highlighted the disparity between Boyle's science and his mechanical philosophy in Chalmers (1993a).

[4] Cited by John Henry (1986, p. 350).

[5] I have argued this point in Chalmers (1993b).

REFERENCES

Armstrong, D.M. (1983), *What is a Law of Nature?*, Cambridge: Cambridge University Press

Bhaskar, Roy (1978), *A Realist Theory of Science*, Sussex: Harvester

Bigelow, John, Brian Ellis and Caroline Lierse (1992), 'The World as one of a kind: natural necessity and laws of nature', *British Journal for The Philosophy of Science* **43**, 371–88

Boyle, Robert ([1663] 1979), 'A requisite digression concerning those that would exclude the deity from intermeddling with matter', reprinted in M.A. Stewart (ed.), *Selected Philosophical Papers of Robert Boyle*, Manchester: Manchester University Press, 155–75

Boyle, Robert ([1685] 1772), 'High Veneration', in *The Works of the Honourable Robert Boyle, Volume V*, Thomas Birch (ed.), London: J & F Rivington, 130–57

Boyle, Robert ([1686] 1979), 'A Free Enquiry into the vulgarly received notion of nature', reprinted in M.A. Stewart (ed.), *Selected Philosophical Papers of Robert Boyle*, Manchester: Manchester University Press, 176–91

Boyle, Robert ([1690] 1772), 'Christian Virtuoso', in *The Works of the Honourable Robert Boyle, Volume V*, Thomas Birch (ed.), London: J & F Rivington, 508–40
Cartwright, Nancy (1983), *How The Laws of Physics Lie*, Oxford: Oxford University Press
Cartwright, Nancy (1989), *Nature's Capacities and Their Measurement*, Oxford: Oxford University Press
Chalmers, Alan (1993a), 'The lack of excellency of Boyle's mechanical philosophy', *Studies in History and Philosophy of Science* **24**, 1993, 541–64
Chalmers, Alan (1993b), 'So the laws of physics needn't lie', *Australasian Journal of Philosophy* 71, 1993, 196–205
Christie, Maureen (1994), 'Philosophers versus Chemists Concerning 'Laws of Nature'', *Studies in History and Philosophy of Science* **25**, 613–29
Hacking, Ian (1983), *Representing and Intervening*, Cambridge: Cambridge University Press
Harré, Rom, and E.H. Madden (1975), *Causal Powers*, Oxford: Blackwell
Henry, John (1986), 'Occult qualities and the experimental philosophy: active principles in pre-Newtonian matter theory', *History of Science* **24**, 1986, 335–81
Hutchison, Keith (1983), 'Supernaturalism and the mechanical philosophy', *History of Science* **21**, 1983, 297–333
Maxwell, James Clerk ([1879] 1965), 'Thomson and Tait's Natural Philosophy', *The Scientific Papers of James Clerk Maxwell, Volume 2*, W. D. Niven (ed.), New York: Dover, 776–85
Popper, Karl R. (1977), *The Logic of Scientific Discovery*, London: Hutchinson
Ruby, Jane E. (1986), 'The origins of scientific "law"', *Journal of the History of Ideas* **47**, 341–59
Shapin, S. and S. Schaffer (1985), *Leviathan and the Air Pump*, Princeton: Princeton University Press

II.

SCIENTIFIC ESSENTIALISM

BRIAN ELLIS

CAUSAL POWERS AND LAWS OF NATURE

In this paper, I motivate and defend an essentialist theory of causal laws. I argue that the theory of laws which philosophers mostly accept nowadays is a secularised version of the divine command theories of the Seventeenth and early Eighteenth Centuries. The metaphysical position which continues to inform this view is mechanism. According to mechanism, the world consists fundamentally of passive objects, having, in themselves, only the primary qualities. Today's neo-mechanists hold that the intrinsic properties on which the identities of things depend are all categorical, i.e. non-dispositional properties having a role and status similar to Locke's primary qualities. Thus, the ways in which things, constituted as they are, are disposed to act or interact are supposed to depend only on what the laws of nature happen to be. The latter, of course, are held to be contingent, and in another world, consisting of identical things, the ways in which they would behave would be quite different.

The viability of the mechanist and neo-mechanist viewpoints depends crucially on a distinction between how a thing is constituted, and what it characteristically does. At the level of ordinary-sized objects, this distinction seems easy enough to maintain. But, at a more basic level, it is not at all clear that it is defensible. For the most fundamental things in nature, whatever they may be, are presumably simple. That is, they have no ontologically distinct parts, and are not constituted by other things. The simplest things may have properties, and if mechanism or neo-mechanism were defensible, these would all be primary qualities, or other passive features. However, the most basic properties of things do not appear to be passive. On the contrary, they would all appear to be of the nature of powers, capacities and propensities,[1] rather than shape, size, impenetrability or the like. The question is then: What is the status of the basic properties in the actual world, i.e. of the real occurrent properties of the most basic kinds of things? Can the sort of distinction required for mechanism or neo-mechanism be defended at this level? Or do the identities of the most fundamental kinds of things depend, not on how they are constituted, but on how they are disposed to behave?

The most fundamental kinds of things in nature all seem to be both active and reactive. They have powers both to act and to interact with things of other kinds, and to be agents in a variety of causal processes. Moreover, their identities would seem to depend entirely on their roles in these processes. That being the case, what is evidently needed is a metaphysic which would allow us to identify kinds of things, not by their passive qualities, but by what they are disposed to do, or by how they are disposed to

H. Sankey (ed.), Causation and Laws of Nature, 19–34.

act or react. But any such metaphysic will be an essentialist one, because, if the identity of any kind of thing depends on its behavioural dispositions, then anything of that kind must be rigidly disposed to act or react in the required ways. Consequently, the laws of its behaviour must be necessarily true.

The proposed metaphysic is scientific essentialism.[2] This theory requires an ontology which includes primitive causal powers, capacities and propensities – and most of the paper is devoted to motivating the acceptance of such an ontology,[3] and exploring its consequences. However, there are some serious objections to scientific essentialism which still need to be answered.[4] The most obvious of these is that the laws of nature seem evidently to be contingent. For we have no difficulty in envisaging circumstances in which we should think that a causal law has been violated. So, however good the case for scientific essentialism may be, it has at least one obviously unacceptable consequence. This, and other objections to scientific essentialism, are dealt with in the final section of the paper.

I. SEVENTEENTH AND EARLY EIGHTEENTH CENTURY BACKGROUND

Belief in the essential passivity of matter derives from Seventeenth Century mechanism. According to this philosophy, two pieces of matter could, in themselves, differ only in respect of the primary qualities of their ultimate parts, i.e. their shapes, sizes, motions and configurations. These ultimate parts could collide with one another, and thus change their motions or configurations. But they had, in themselves, no causal powers, and their only capacities were the passive ones of impenetrability and (perhaps) resistance to acceleration.

In his essay *De Motu*, Berkeley (1720) expressed the doctrine of the essential passivity of matter:

> No known thing that we could call a *body* contains within itself anything that could be the origin or efficient cause of motion. For impenetrability, extension and shape do not include or mean any power of producing motion; on the contrary, not only they, but also all other bodily qualities whatsoever, gone through one by one, will be seen to be in fact passive, having in them nothing active, nothing that can in any way be understood as the fount and origin of motion. (p. 205)

According to Newton, whose authority on questions of natural philosophy dominated the Eighteenth Century, all causal power derives from God. It does not reside in matter, as many later Newtonians came to believe. So, if one object acts on, or is attracted by, another, then this is not because of the intrinsic natures of these objects. It is because this is what God commands.

In a letter to Mr. Bentley, in February 1692, Newton wrote:

> It is inconceivable that inanimate brute matter should, without the mediation of something else which is not material, operate on and affect other matter without mutual contact, as it must be if gravitation, in the sense of Epicurus, be essential and inherent in it. . . . That gravity should be innate, inherent, and essential to matter . . . is to me so great an absurdity that I believe that no man who has in philosophical matters a competent faculty of thinking can ever fall into it. Gravity must be caused by an agent acting constantly according to certain laws, but whether this agent be material or immaterial I have left to the consideration of my readers. (Thayer, 1953, p. 54)

This famous passage is vague about who or what the agent is whose constant action is responsible for gravitational attractions, but, from other sources, it is clear that Newton believed the ultimate source of power in the world to be God.[5]

Newton's theory of laws was one of several *Divine Command* theories, which were commonly accepted in the Seventeenth and Eighteenth Centuries. On all of these theories, the laws of nature were supposed to be God's commands, and consequently to have properties compatible with their divine origin. Specifically, the philosophers of the period supposed the laws to be constant, uniform and universal, and variable only by divine intervention. Most therefore took natural necessity for granted. Things *must* happen as God commands, they thought, and no exception to a law of nature could possibly occur, unless God specifically decreed that it should.

One philosopher who notoriously did not accept a Divine Command theory was Hume. If the usual reading of Hume is correct, then he rejected the notion of causal power altogether, including human and divine agency, and maintained that the laws of nature are just universal regularities.[6] They are not divine commands, he held; nor are they the ways that things have to behave because of their natures. They are just universal propositions which happen to be true, e.g. about how things move or change. If it seems to us that a cause *necessarily* produces its usual effect, then this *appearance* of necessity is just an illusion – a projection on to the events said to be causally related of our firm convictions or expectations concerning what sequences of events will occur. There are no necessary connections in nature, Hume thought, only the habits of mind established in those who observe it.

In this paper, I defend the essentialist view that the laws of nature depend on the natures of the things they relate. They are not independent of their natures, as the British Empiricists, and today's neo-mechanists, all maintain. The causal laws, i.e. those laws which describe the natural kinds of processes which can occur in the world, all hold, I argue, in virtue of the primitive causal powers, capacities and propensities of the things that take part in these processes. That is, they are grounded in their essential properties. They are not imposed by God on things that are essentially passive. Nor are they just regular patterns of behaviour which happen, miraculously and unaccountably, to be displayed by passive things, just *as if* they were imposed by God. A thing just does what a thing has got to do.

2. ELEMENTS OF AN ESSENTIALIST THEORY OF CAUSAL LAWS

On the essentialist theory of causal laws I wish to defend, the laws of nature depend on the natural kind of world we live in, and on the natural kinds of things and processes that occur in it. The species of laws of nature I focus upon are the so-called 'causal laws'. These are narrower in scope[7] than, say, the conservation laws, or the laws concerning the spacetime structure of the world, which I call 'global laws'. The causal laws are just the laws of action and interaction of the various kinds of things that exist in the world, and the most basic of these are concerned with natural kinds.

The causal laws, I hold, are grounded in the intrinsic properties and structures of the natural kinds. The intrinsic properties and structures of the kinds are not all passive, as the British Empiricists believed, and as some philosophers still believe, but many are active. They include the causal powers, capacities and propensities which determine how the various natural kinds of things are disposed to behave and interact with each other.

I postulate that amongst the intrinsic properties and structures of any natural kind there are at least some which are essential to things of that kind; and if there be any such properties or structures which are neither accidental nor essential to the kind in question, then there is a species of that kind which has these additional properties or structures essentially.[8] Hence, in my view, every intrinsic property or structure of a natural kind which is not merely accidental is an essential property or structure of some species of that kind.

I further postulate that the laws of action and interaction concerning natural kinds of things depend entirely upon the causal powers, capacities and propensities that these sorts of things have essentially. That is, I suppose that the properties and structures which constitute the real essence of a kind determine its laws of action and interaction. Hence, the natural kinds that exist in our world could not fail to act or interact as they do, because, for things to behave otherwise, they would have to have different essential properties, and hence be different kinds of things. Consequently, the causal laws, i.e. the laws which truly describe the actions and interactions of the various natural kinds that exist in this world, could not be other than they are. A world with different causal laws would have to consist of different kinds of things.

The laws I am calling 'causal laws' describe certain kinds of causal processes, and are to be distinguished from other kinds of laws, such as equilibrium and conservation laws. The causal laws describe those natural kinds of processes which are essentially the displays of specific causal powers, capacities or propensities. They describe the circumstances in which these processes will be initiated, and what their effects will be. The properties which are displayed by these processes are all dispositional, and most of them are quantitative. The fact that they are dispositional implies that there are circumstances apt for their display, and that if or when these circumstances are brought about, the powers, capacities or propensities will be displayed in the appropriate way, against the background of whatever other events may be occurring. The fact that these dispositional properties are mostly quantitative implies that different things may have them to different degrees, and that their displays in similar circumstances will accordingly be quantitatively different from each other. Most of the laws of action of the various causal powers, capacities and propensities of things will therefore be quantitative, as nearly all causal laws are.

Of course, not all classifications of causal processes are natural kinds classifications. So, not all kinds of processes can be described as being essentially displays of specific causal powers, etc. Causal laws must be descriptive of natural kinds of processes. Moreover, processes of the same natural kinds must be essentially, not just accidentally, the same in some respects. Generalisations describing regular patterns of behaviour will not therefore be causal laws, if these patterns of behaviour do not constitute a natural kind. In my view, then, all causal laws are descriptive of natural kinds of processes which are essentially the displays of the intrinsic causal powers, capacities and propensities of things.

If this is right, then all causal laws must be necessary. For the identity of any quantitative dispositional property depends on the quantitative laws describing its display in the various circumstances in which its display is triggered. The laws of causal interactions are, of course, somewhat more complicated, since, in general, two or more quantitative dispositional properties will be involved. But the same conclusion holds for these laws too. The quantitatively described interactions that occur must be derivable from the quantitative laws describing the displays of the causal powers, capacities and propensities of the interacting things.

In my version of scientific essentialism, the most fundamental kinds of things are all members of natural kinds. The members of these natural kinds are characterisable by their causal powers, capacities and propensities, and have no other intrinsic properties. These properties are their essential properties. The fundamental natural kinds belong in hierarchies. At the lowest, most specific, levels of these hierarchies, the members of the natural kinds are intrinsically identical. But as we ascend the hierarchies, the natural kinds become more general, and permit intrinsic variation. The members of these higher order species, therefore, are not all intrinsically identical. Nevertheless, they have a common real essence, defined by the causal powers, capacities and propensities they share. At higher levels still, there are structures of lower order entities. But the only structures that can exist are those that the lower orders permit, and each distinct structure defines a new higher order kind. The essence of any such higher order kind is then given by the structure, and the entities involved in it. However, the intrinsic causal powers of any such kind must depend ultimately on those of its constituents. Therefore, the causal powers, capacities and propensities of all such structured kinds must be essential and inherent properties of these kinds of things.

If I am right in thinking that the world is structured like this, and, consequently, that the causal powers, capacities and propensities of the various kinds of atoms and molecules (which are obviously structured kinds) are amongst their essential properties, then the causal laws involving such things must also be necessary. Consequently, the laws of chemical combination, action and reaction must all be necessary too. Indeed, we may conclude, without going through the detail of the argument, that all causal laws describing natural kinds of processes are necessary. The Humean conviction that they are all contingent must therefore be an illusion.

3. THE HUMEAN METAPHYSIC

In Hume's philosophy, the only properties of which we can have any knowledge are those which give rise to distinct impressions. These properties include the perceptible qualities of bodies, viz. their shape, size, colour, etc., but do not include any properties of power or agency. That one body operates on and affects another in some way is a judgement we commonly make. But Hume insists that we have no impression of power or agency on which any such judgement could possibly be founded, even in cases where the action is by contact between bodies – as, for example, when billiard balls collide. The qualities of bodies, of which we have distinct impressions, are all passive. One event may be seen to follow another, and if events of one kind are seen to follow contiguous events of another kind often enough, then we may judge that events of the first kind are caused by those of the second. But there is nothing in external nature, he says, or at least nothing of which we have any distinct impression, which corresponds to this idea of causation or necessary connection.[9] The events occurring in external nature seem conjoined, he says, but never connected. The connections, he concludes, are only in our minds.

Modern Humeans accept that Hume is basically right about the properties of things. For they accept his conclusion that objects have no intrinsic causal powers. The natural world is, as Berkeley and Hume believed it to be, a passive world. If one object seems to act upon and affect another, then this is just an illusion. Objects do not have any active properties, and so cannot act on other objects. Even objects in motion are causally powerless, since motion neither is, nor implies, any active property. What happens in nature might well happen according to laws, or universal regularities. But these regularities concern only the patterns of behaviour of objects which are intrinsically powerless. If one billiard ball strikes another and knocks it into a pocket, then what the first ball does might well be said to cause what the second one does. But really this is a misleading way of speaking. What happens is just a case of one event being followed by another in accordance with the laws of nature.

Humeans thus hold that the most fundamental laws of nature all refer to intrinsically powerless objects. In support of their position, they argue that if any of the most basic kinds of things in the world had any intrinsic causal powers, then there would be ways in which these things would be bound to act, viz. in accordance with these powers. Consequently, the laws concerning the behaviour of these things would not all be contingent. But, if there is one thing that Humeans insist upon, it is that the laws of nature are all contingent. Therefore, they are, and must remain, deeply committed to the mechanistic thesis of the essential passivity of matter.

The thesis of the essential passivity of matter is not to be confused with the stronger thesis of categorical realism. For categorical realism has no place for basic dispositions of any kind, not even essentially passive ones, like Locke's impenetrability. Categorical realists hold that no dispositional properties whatever are genuinely occurrent.[10] Dispositions, they say, must all have categorical bases, and these bases cannot be identical to, but only contingently associated with, the dispositions they ground. Thus, in another world, things with exactly the same categorical bases might be disposed to behave quite differently, depending on what the laws of nature happen to be. In another world, for example, human beings might be penetrable, as angels were once supposed to be, and be able to walk through each other, or through walls. For the laws of nature might not preclude such things as these from happening.

But categorical realism is a much stronger thesis than neo-Humeans need, if they only wish to avoid commitment to causal powers, propensities, and other active properties. For they could allow, compatibly with their essential passivity thesis, that there are some basic dispositions of a passive nature which do not have categorical bases. Locke included one basic disposition, viz. impenetrability, in his list of primary qualities. Inertia is another property which might plausibly be included amongst the primary properties of matter, compatibly with the essential passivity thesis. For neo-Humeans only need to insist that there are no *active* dispositions, i.e. no causal powers or propensities, which are not further reducible.

But modern Humeans are chary of admitting any basic dispositional properties into their ontologies, lest they should blow their position apart. If, for example, inertia, i.e. the capacity of a body to resist acceleration under the action of a given force, were admitted as a basic property, then why not also admit gravity, i.e. the power of a body to attract other bodies gravitationally? Why discriminate between active and passive dispositions, if neither has a known categorical basis?[11] Moreover, given the equivalence of gravitational and inertial mass, if anything were plausibly a categorical basis for either, then it must, equally plausibly, be a categorical basis for the other. Most Humeans are therefore categorical realists who take the strong and uncompromising view that *all* dispositional properties must have categorical bases, and, whether we know them or not, there must be laws of nature concerning these unknown, but intrinsically passive, bases which entail the dispositions.[12]

Dispositional properties are properties which dispose things to behave in certain ways, either by necessity, or with some degree of probability. Hence they are often called *modal* properties. Therefore, we may put the point of the previous paragraph by saying that in the Humean metaphysic there are no modal properties which do not depend on categorical properties.

The thesis that the modal properties of things must all depend on what categorical properties they have, and what the laws of nature are, is sometimes called the *Humean Supervenience thesis*. For, if this theory is correct, it implies that no modal properties can exist as properties in their own right. The existence and nature of all such properties must be fully determined, once the categorical properties of things, and the laws of nature, are fixed. For the theory implies that things can have the modal or dispositional properties they do, only if they have the appropriate kinds of categorical bases, and the laws of nature are such as to ensure that things having such bases are disposed to behave in the appropriate ways.

However, the Humean Supervenience thesis is untenable.[13] There are no known kinds of objects characterisable only by the sorts of properties (viz. categorical) on which all of their other properties are supposed to supervene, and there are no known laws of nature describing how objects characterisable in this way would or should behave. But for the Humean Supervenience thesis to be true, such objects and laws must exist, and be fundamental in science or ontology. Therefore, there is no evidence whatsoever that there is anything in nature for the most primitive causal powers, capacities and propensities of things to supervene upon, and no evidence that the kinds of laws required for supervenience exist. On all of the evidence available to science, the most basic properties in nature are dispositional, and the most fundamental kinds of things are characterisable only by their dispositional properties. So, why should we hang on to an outdated metaphysic which is scientifically so implausible? There is a much more plausible alternative, viz. that there are fundamental dispositional properties on which the causal laws all depend.

The Humean Supervenience thesis has other unpalatable consequences. If the Humean Supervenience thesis were true, then modal properties, and hence most of the properties which feature in causal laws, would not exist – at least, not as occurrent properties of things in the actual world. But if modal properties do not exist as properties of things in the actual world, as neo-Humeans believe, then the natural necessities which their attributions would seem to imply, cannot be grounded by anything in this world either. On what then could such necessities depend? It seems that we have just two alternatives: Either we must buy the whole Humean package, and deny the existence of natural necessities altogether, or we must find a basis for them in some reality other than actual reality. Astonishingly, many philosophers have chosen this latter option.[14]

4. HUMEAN AND ESSENTIALIST PERSPECTIVES ON REALITY

The Humean metaphysic is deeply flawed. It implies that the causal powers and capacities of things are properties which depend on what the laws of nature are. If the laws were different, the Humeans would say, then so must their causal powers be different. Moreover, since they suppose the laws to be contingent, it must be a contingent matter how things are disposed to act or interact, and consequently impinge upon our senses, or indeed upon any instruments we might use to observe them. Consequently, in other possible worlds, things might seem to be very different from how they appear to us in this world. Conversely, things with very different structures, and made of different substances, might be disposed to behave in exactly the same way, if the laws of nature were sufficiently different.

The Humean tradition thus embraces the medieval distinction between a substance and its species attributes, which was confirmed at the Council of Trent in 1551 in the doctrine of transubstantiation.

> If anyone shall say that, in the most holy Eucharist, there remains the substance of bread and wine together with the body and blood of our Lord Jesus Christ; and shall deny that wonderful and singular conversion of the whole substance of the bread into the body, and of the whole substance of the wine into the blood, the species of bread and wine alone remaining, which conversion the Catholic Church most fittingly calls Transubstantiation, let him be anathema. (Session 13, Canon 2)

What is supposed to occur in the Eucharist is that the wine and bread literally become the blood and body of Jesus, even though all their species attributes (including all of their physical and chemical properties) remain just those of wine and bread. Same observable properties, different substances. No problem, if you think that what a thing is is logically independent of what it does. Mad if you think otherwise.

For scientific essentialists this doctrine is bizarre. Of course the substances are still bread and wine! Nevertheless, many philosophers would argue that it is not logically impossible for transubstantiation to occur, since there is no formal contradiction involved in the supposition that it does. Therefore, they would say, it is at least logically possible that the bread and the wine should really become flesh and blood, despite all appearances to the contrary. However, it does not follow from the absence of any formal contradiction that there is a possible world in which transubstantiation occurs. For whether it is possible in this broader sense hinges on what the identity of a substance depends on. If it depends on its causal powers, capacities and propensities, as we scientific essentialists would say, then there is no possible world in which flesh and blood could behave in the bizarre way required by the doctrine of transubstantiation. It is only if the identity of a substance is supposed to be independent of anything that might be discovered about it by observation or experiment that the doctrine of transubstantiation makes any sense. On any other assumption, the flesh and blood of Jesus could not manifest itself as bread and wine, in this or in any other possible world. They would have to cease to be flesh and blood in order to become bread and wine.

Although many, perhaps most, Humeans would firmly reject the doctrine of transubstantiation, they nevertheless continue, embarrassingly, to embrace its metaphysics. For every Humean believes that things exactly like those that exist in this world, could exist in worlds in which the laws of nature were different, and so, presumably, even in worlds in which the laws of nature were *very* different. But how these things would behave, and so what effects they would have on us, and on our measuring instruments, would depend on what the laws of nature were in these other worlds. So, Humeans cannot rule out the doctrine of transubstantiation as incoherent.

The alternative, and essentialist, view of the matter is that causal powers, capacities and propensities exist in nature, and the identities of the various kinds of things that exist are dependent on them. What makes things flesh and blood are their physical and chemical properties. Flesh and blood cannot lack these properties, and have those of bread and wine instead. Conversely, if anything has the physical and chemical properties of bread or wine, then it is bread or wine, as the case may be. It is certainly not flesh or blood.

The physical properties of bread and wine are mostly dispositional. For they depend on how the samples would respond to various tests. But plausibly, the chemical constitutions of these substances are categorical properties. For, although certain tests must be carried out in order to discover what the chemical constitutions of bread and wine are, the claim that they have these constitutions is not simply a summary way of stating the results of these tests. However, the claim that flesh and blood might have exactly the same physical and chemical properties as bread and wine, if the laws of nature were sufficiently different, presupposes that the constituents of flesh and blood might behave in exactly the way that the constituents of bread and wine actually behave. It presupposes, therefore, that the identities of the constituents of flesh and blood do not depend on their dispositions to behave physically or chemically as flesh and blood do.

On what, then, do the identities of the constituents of substances like flesh and blood depend? Ultimately, they must depend on the identities of their constituents, and so on; and if this line of dependence is pursued, we must either come to some basic constituents whose identities do not depend on anything else, or we must go on pursuing identity down through ever deeper levels. On present evidence, the categorical properties drop out long before we get to the deepest levels, i.e. the identities of the most basic kinds of things depend *only* on their dispositional properties. That being the case, it is metaphysically impossible that flesh and blood, constituted as they are, should behave as the doctrine of transubstantiation requires.

On the view which is taken here, these essential properties of things are not occult, but discoverable and measurable by the ordinary methods of science. There is nothing especially obscure or hidden about them. They can be found just by observing how the instances of the various natural kinds act on, and interact with, each other. Of course, to discover the effects of any particular property, it may be necessary to observe what happens in circumstances in which this property may reasonably be assumed to be the only relevant cause, or to observe it in circumstances from which we may abstract from the known effects of other causes. But this is not abnormal practice in science. On the contrary, it is standard experimental method to attempt to isolate what we wish to examine from causes whose effects are not the objects of investigation, or, where we cannot do this, to discount for the known effects of the causes we are unable to eliminate.

5. DISCUSSION

As a metaphysical thesis, scientific essentialism has a lot to be said in its favour:

1. It accounts well for the necessity of causal laws. They are necessary because they are the identifying descriptions of the basic dispositional properties of things, i.e. of their causal powers, capacities and propensities. Let P be any such property. Then, necessarily, anything that has P must be disposed to display P in some appropriate circumstances, and the identity of the property displayed must depend on what has to happen in these circumstances. A causal law identifies P by describing both the circumstances C and the display E. The general form of a causal law is therefore: For all *x*, necessarily, if P*x* and C*x*, then E*x*. (Note the position of the necessity operator.)
2. Scientific essentialism explains why laws describing the essential properties and structures of natural kinds of things are necessary. If K is a natural kind of thing, then everything of that kind must have the appropriate structure and properties. The general form of such a law is: For all *x*, if *x* is an instance of K, then P*x*. Furthermore, if P is a causal power, capacity or propensity, then *x* must be bound by the causal law concerning P. That is: For all *x*, necessarily, if *x* is an instance of K and C*x*, then E*x*.
3. The theory explains, as no other theory of laws can, the special status of laws, as both *a posteriori* and necessary. They are *a posteriori* because the essential natures and structures of things have to be discovered by empirical investigation. We do not know *a priori* what properties there are in nature. Therefore, we do not know *a priori* how things will behave in various possible sets of circumstances. Nor do we know *a priori* what constitutes substances such as water or gold. We have to find these things out by the ordinary experimental methods of science.
4. Scientific essentialism also gives a good account of the nature of scientific reasoning – at least in those fields where the objects of study mostly belong to natural kinds. In these areas, scientific reasoning is generally like geometrical reasoning, and, characteristically, it involves reasoning from small well-selected samples to total populations. This point has been fully developed in another paper (Ellis, 1996).
5. A scientific essentialist can also explain the frequent focus in the physical sciences on the ways in which idealised things would have to behave in various idealised kinds of circumstances. The primary reason for this focus is that the objects of our investigations are often obscured. They are often obscured because the kinds of processes we wish to study seldom, if ever, occur in isolation, and the effects of their operation may well be swamped by those of other processes occurring to, or in, the same things at the same time. This point has also been developed more fully elsewhere (Ellis, 1992a).
6. Finally, scientific essentialism is a metaphysic appropriate for modern science. It is not embedded in Seventeenth Century mechanism, or in reactionary Humean analyses of causation. Despite its ancient lineage, it is a thoroughly modern doctrine concerning the existence and potentialities of causal powers, capacities and propensities in the universe.

Nevertheless, there are some serious objections to scientific essentialism which many philosophers will find compelling:

Objection 1. The laws of nature are not necessary, but contingent. There is nothing easier than to imagine their violation. But if the contrary of any law is imaginable, then it is possible. Therefore, the law itself is contingent.

Reply 1. Imaginability is a very poor test of possibility.[15] For what is imaginable is constrained only by our powers of imagination, and determines only epistemic possibility. What is really or metaphysically possible depends on the kinds of things we are concerned with, and the kinds of things that can happen to them in our kind of world. Could this flower-pot grow wings and fly away? Certainly, it is imaginable. But could it happen to this flower-pot? Not if the flower-pot and the world are the kinds of things I think they are. Of course, I might be wrong in thinking that this object is a flower-pot, or I might be wrong about what kind of world it is. At least, it is easy enough for me to envisage this possibility. So, it is epistemically possible that I am wrong about these things. Therefore, it is epistemically possible that what I take to be an impossibility is really a possibility. However, this does not show that it is really a possibility. Epistemic possibility is not, and does not entail, real or metaphysical possibility.

Objection 2. If imaginability does not entail real possibility, then we should expect there to be many belief-contravening suppositions which do not describe real possibilities. They do not have to be bizarre ones, like flower-pots growing wings and flying away. Even quite ordinary counterfactual suppositions might not describe genuine possibilities, if imaginability is not the test of possibility. Therefore, many ordinary counterfactual conditionals may, in fact, be counter-possibles, and therefore, at best, only vacuously true.[16]

Reply 2. As you might expect, the theory of causal conditionals I favour is equivalent to a possible worlds theory of a certain kind, viz. one which refers only to possible worlds which are *essentially similar* to ours, having the same global structure, and the same basic ontology. On this theory of conditionals, the proposition "if A were the case, then B would be the case" will be true if and only if in every *essentially similar* possible world in which 'A' is true, in circumstances as near as possible to those which actually obtained at the relevant time in this world, 'B' is also true. It is thus a necessary condition for the truth of the conditional that its antecedent be really possible in sufficiently similar circumstances.

However, as the objection notes, the assertibility of a conditional does not depend on the real possibility of its antecedent, for in many cases this is simply unknown. Presumably, therefore, there are many conditionals which are assertible, but have impossible antecedents, i.e. antecedents which could not possibly be true. For a conditional to be assertible, its antecedent must plausibly describe some conceivable state of affairs. But, as we have seen, what is conceivable depends on our powers of imagination, which can quite easily go beyond the bounds of real possibility. Our imaginations, being primarily cinematographic, can easily conceive of situations, or happenings, which are really impossible, as every cartoonist knows, and as Escher convincingly demonstrated.

There are at least two ways in which we might think about counter-possibles – literally and imaginatively. If we think about them literally, and hence as referring to the kinds of things and substances they purport to be about, then we have to conclude that they are no more than vacuously true. If we think about them imaginatively, however, we may interpret them sympathetically as saying something about how imaginary objects (having the relevant traits and features of their real counterparts), would be likely to behave in the imagined circumstances.

To evaluate the imaginative constructions of counter-possibles, we must put aside our scruples about what is really possible, and think about them as David Lewis thinks about what he calls 'counterfactuals'. Then, if they would come out true on his analysis, then they are to be counted as assertible. Take the proposition that if people could walk on water, they would have less need for boats. To evaluate the imaginative construction of this proposition, we must imagine worlds in which the counterparts of people can walk on the counterpart of water, but which are, otherwise, as much like our world as we can make them. Then, on Lewis' theory, the proposition is to be judged true iff its consequent, regarded as a proposition about the counterparts of people in the imagined worlds, would be true in all of these worlds. However, either the counterparts of people are not people, or the counterpart of water is not water. Therefore, there is no possible world in which the antecedent of this counter-possible is literally true. Therefore, the proposition itself must be literally vacuous, whatever the imagined counterparts of people and water might be constrained to do in worlds that are superficially like ours.

Objection 3. The essentialist theory of laws collapses an important philosophical distinction, viz. that between what is logically necessary and what is only physically necessary. For, according to essentialist theory, physical necessity, which essentialists call metaphysical, is just a species of logical necessity. If it is a law of nature that like charges repel each other, as it surely is, then this law must hold in all possible worlds. In possible worlds in which like charges do not exist, it holds vacuously. In possible worlds in which like charges do exist it holds in virtue of the dispositions that charged objects must have in being charged. For it is the having of this whole set of dispositions which, essentialists say, *constitutes* their being charged.[17] But if this law holds in all possible worlds, as essentialism requires, then it is logically necessary.

Reply 3. It is true that causal laws are necessary, in the sense of being true in all possible worlds. But I should be giving away too much to agree that they are therefore logically necessary. Truth in all possible worlds is characteristic of all species of necessity, not just logical necessity. Logical truths are necessary because we cannot rationally conceive of a world in which they are not true. They are, therefore, necessary *a priori*. Analytic truths are necessary because their truth depends only on the meanings of words, and does not depend on what the world may be like. They are necessary *de dicto*. Causal laws, and other metaphysically necessary propositions, are necessary because they could only fail to be true in another world by failing to be significant in that world. They are necessary *de re*.

Of the three species of necessity, only metaphysical or real necessity is *a posteriori*. This is because metaphysical necessities have first to be established empirically. But once established empirically as true in this world, then it can be seen that they are necessarily true, i.e. true in all possible worlds. Magic, you say? Not at all. If the Morning Star is the Evening Star, then it is so necessarily. But this identity had first to be discovered empirically. If water is H_2O, then it is so necessarily. But this identity too had to be discovered empirically. However, the fact that metaphysical necessities have first to be established empirically does distinguish them clearly from other kinds of necessity, and this fact explains the reluctance of many philosophers to think of them as necessary at all. For surely the science can go wrong, and chemical substances, and other kinds of things, can be misidentified. Indeed they can. There is always this epistemic possibility. But if we have got it right, and the substance is in fact the kind of thing we think it is, then it is so necessarily. For what makes it this kind of thing could hardly be lacking of the same kind of thing in another world.

Objection 4. The distinction that scientific essentialists require between epistemic and real possibility is unsustainable. For if anything that is epistemically possible is not really possible, then in what sense is it possible? How can anything be possible if it is not true in any possible world?

Reply 4. What is epistemically possible is an epistemic rather than an ontological classification, and may be defined: *p* is epistemically possible iff *p* is not known, or knowable *a priori*, to be impossible. This concept of epistemic possibility meshes with the ontological concepts of necessity and possibility in the following ways: (a) epistemic possibility does not entail real possibility. For what is not known, or knowable *a priori*, to be impossible might nevertheless be impossible. (b) If *p* is possible, then *p* can neither be known, nor be knowable *a priori*, to be impossible. Therefore, real possibility entails epistemic possibility. (c) Then, by contraposition, we have that epistemic necessity entails necessity.

Epistemic necessity is therefore a stronger concept than ordinary necessity, and epistemic possibility is weaker than ordinary possibility. However, these epistemic concepts cannot be used in ontology, e.g. to specify truth conditions, for the class of epistemically possible worlds is a paraconsistent class. So, unless one is willing to be a realist about impossible, as well as about possible, worlds, the class of epistemically possible worlds is not an ontologically relevant class.

La Trobe University and University of Melbourne

NOTES

[1] The case for saying that the most fundamental properties in nature are dispositional is powerfully argued in Thompson (1988). Certainly, the most fundamental properties we know about are causal powers or propensities, and whenever one such causal power is seen to depend on other properties, these other properties always include even more fundamental causal powers. There is no prospect in modern physics of being able to characterise the most fundamental entities by intrinsic categorical (i.e. non-dispositional) properties alone, as Prior's (1985) account requires. Nor do any of the most fundamental laws of nature describe how entities characterised only by their categorical properties (if indeed they have any such properties) must behave.

[2] 'Scientific essentialism' is the name which both Bealer (1987) and I (1992b) have given to the view that explanations in the physical and biological sciences typically involve postulates concerning the essential natures of the fundamental natural kinds of objects and processes occurring in the world. On this view, the causal laws which apply in these areas describe the operation of the essential properties of these natural kinds, and are metaphysically necessary.

[3] Such an ontology of properties has recently been defended by a number of authors, including Harré (1970), Achinstein (1974), Harré and Madden (1973 and 1975), Shoemaker (1980), Swoyer (1982), Fales (1986 and 1990) and Bealer (1987).

[4] Bealer (1987) argues that the only defensible form of scientific essentialism is 'local', not 'global'. The distinction between local and global scientific essentialism depends on a distinction between concepts – between *naturalistic concepts* on the one hand (including the concepts we have of such things as water, heat, gold, arthritis, beech, elm (his list on p. 305), and *content and category concepts* on the other. Scientific essentialism is defensible, he holds, if it is restricted to naturalistic concepts, but not defensible if it is

deployed more widely to disallow the sorts of intuitions of possibility on which philosophical reasoning so often depends. (See also note 16 below.)

5 Iltis (1973) argues that the Leibnizian-Newtonian controversy reflected in the Clarke-Leibniz correspondence was, in fact, "fundamentally a clash of philosophical world views on the nature of God, matter, and force" (p. 343). For the Newtonians, God was the source of all power in the universe, and His constant action was required to sustain it. For the Leibnizians, God created the world perfect from the start with the power to sustain itself. See also Laudan (1968).

6 The question of whether Hume himself believed in the regularity theory of laws is controversial. Costa (1989), for example, argues that "Hume does not reduce causation to regularity in any direct or simple way", and that there is a case for saying that Hume was a causal realist, despite his insistence that nothing more than regularity is truly conceivable by us.

7 The concept of scope is related to that of a reference class. The reference class for the law that As are necessarily Bs is the class of As. It is not the class of non-Bs, since non-Bs do not (usually) constitute a natural kind. Moreover, the probability of an A being a B is generally not the same as that of a non-B being a non-A. Nor is the reference class the same as the class of all things, as the classical extensional analysis of laws implies, since the law concerns what is essentially involved in anything being an A, not just what things happen to be Bs or non-Bs.

8 I endorse the view, defended in Elder (1996), that there are generic as well as specific universals.

9 Fales (1990) disputes this. He argues that our concepts of causation are grounded in the feelings of pressure, impact and exertion which accompany our everyday actions and interactions with things in the world.

10 This is the position taken by Prior, Pargetter and Jackson (1982), for example, and developed and defended in Prior (1985).

11 Averill (1990) argues that the property of inertial mass, which is said to measure a body's capacity to resist acceleration, and which is, therefore, almost paradigmatically dispositional, is not dispositional at all. It is not dispositional, he argues, because it does not entail the required counterfactual conditionals. However, Averill's arguments are unsound, as Reeder (1995) adequately shows.

12 Mumford (1994, 1995a and 1995b), for example, takes the view that such underlying categorical bases must exist, even if they are unknown and unknowable. In Ellis and Lierse (1994) we argued that although the dispositions of things may often depend on how they are structured, they must in all such cases also depend on the causal powers, capacities and propensities of the elements of the structure. Hence the basis of a disposition (on which it depends ontologically) is never going to be purely categorical. However, a dispositional property might well be an occurrent property which is not ontologically dependent on any other property. Presumably, some of the most fundamental properties in nature are of this kind.

13 There are, in fact, several Humean supervenience theses. These are described and discussed in Chapter 3 of Carroll's (1994). Carroll argues convincingly that they are all untenable, including the thesis that is being discussed here.

14 The case against such theories, e.g. Lewis' modal realism, is presented in Elder (1994). Criticising Lewis' account of causal power, Elder says: "It is as if one said that Gambler Joe possesses a power over the cards he plays with, and that his having this power consists in the fact that other gamblers distinct from Joe, who do not actually exist, always just happen to win, but with different cards – non-actual cards, which exist in different worlds" (p. 657). On the Ersatzers, he says: "But if the information that rust has thus far formed on iron exposed to water does not in and of itself justify us in expecting rust next time, can information about what words occur in certain fictions complete the justification" (p. 658).

15 Blanshard (1962) distinguishes between imaginability and conceivability. To conceive of an object is to have a complete knowledge and understanding of it. To imagine an object is much more superficial. It requires no such complete knowledge or understanding. Blanshard argues that imaginability is neither a necessary nor a sufficient condition for possibility. However, if a sequence of events is conceivable, in his sense, then it is possible. In this paper, I am using the words 'conceivable' and 'imaginable' more or less interchangeably.

16 This objection is due to Bealer (1987). 'Global scientific essentialism', he says, 'impinges upon traditional philosophical inquiry not only in specific theoretical issues. It also significantly affects traditional philosophical method. For example, in the course of philosophical dialectic one would often advance hypothetical examples (usually counterexamples). One's *a priori* intuition that the example is possible (i.e. not necessarily not the case) would usually suffice to establish that it is possible. However, global scientific

essentialism invalidates this traditional procedure. For example, suppose that in an argument in metaphysics (or philosophy of biology) you propose the hypothetical example of life forms (i.e. living beings) whose bodies are composed not of hydrocarbons but of something else. With the advent of scientific essentialism, your example might now be dismissed as follows: "You are not justified in accepting that example as genuinely possible. For all you know, you are mistaking a mere epistemic possibility for a genuine metaphysical possibility. Only a genuine metaphysical possibility will win your point in this metaphysical dialectic" (p. 293). To deal with this problem, Bealer urges that we should restrict the use of essentialist strategies to naturalistic concepts. However, it is doubtful whether this restriction, even if justified, would deal with the problem of imaginative counterexamples, or with the associated problem of counterfactual conditionals. A large number of quite assertible counterfactual conditionals are going to turn out to be counterlegals.

[17] McGinn (1974/5) makes a distinction between two kinds of nomic properties, which he labels 'primary' and 'secondary'. The primary nomic properties of a natural kind of substance such as gold are the fundamental ones on which all of the properties of this substance, other than its accidental ones, depends. The secondary nomic properties are those other properties of the substance which it has as a consequence of its having the primary ones. He then asks: What is the nature of the relation between these two kinds of properties? Is it contingent or non-contingent? If it is contingent, he says, then Mackie (1973/4) is right in thinking that the observable physical and chemical properties of a substance such as gold (heavy, yellow, malleable, soluble in *aqua regia*, etc.) are properties of gold only contingently (although he is wrong in ascribing this view to Kripke). If it is necessary, then Kripke (1972) is right, and gold has these properties necessarily. Whether Kripke is right or not, McGinn argues, depends crucially on whether it is the case that to have these secondary properties *just is* to have certain primary properties, i.e. on whether the secondary category can be reduced to the first. McGinn takes the view, which he thinks is also the view of Locke, that it can be reduced. He quotes Locke as saying: "I doubt not but if we could discover the figure, size, texture, and motion of the minute constituent parts of any two bodies, we should know without trial several of their operations one upon another; *as we do now the properties of a square or triangle*" (*An Essay Concerning Human Understanding* Book Four, III, 24. McGinn's italics). In this paper, I follow Kripke in supposing that a natural kind has all of its intrinsic powers, capacities and propensities essentially.

REFERENCES

Achinstein, P. (1974), 'The Identity of Properties', *American Philosophical Quarterly* **11**, 257–76

Averill, E.W. (1990), 'Are Physical Properties Dispositions?', *Philosophy of Science* **57**, 118–32

Bealer, G. (1987), 'The Philosophical Limits of Scientific Essentialism', in J. Tomberlin (ed.), *Philosophical Perspectives 1*, Atascadero: Ridgeway, 289–365

Berkeley, G. (1720), *De Motu*, Abridged and translated as 'Concerning Motion' in T.E. Jessop (ed.), *Berkeley: Philosophical Writings*, Nelson, 1952, 202–16

Bigelow, J.C., Ellis, B.D. and Lierse, C. (1992), 'The World as One of a Kind: Natural Necessity and Laws of Nature', *British Journal for the Philosophy of Science* **43**, 371–88

Blanshard, B. (1962), *Reason and Analysis*, La Salle, Ill: Open Court Publishing Company

Carroll, J.W. (1994), *Laws of Nature*, Cambridge: Cambridge University Press

Costa, M.J. (1989), 'Hume and Causal Realism', *Australasian Journal of Philosophy* **67**, 172–90

Elder, C.L. (1994), 'Laws, Natures and Contingent Necessities', *Philosophy and Phenomenological Research* **54**, 649–67

Elder, C.L. (1996), 'Realism and Determinable Properties', *Philosophy and Phenomenological Research* **56**, 149–60

Ellis, B.D. (1992a), 'Idealization in Science', in C. Dilworth (ed.), *Idealization IV: Intelligibility in Science, Poznan Studies in the Philosophy of the Sciences and the Humanities* 26, Amsterdam, Atlanta GA: Rodopi

Ellis, B.D. (1992b), 'Scientific Essentialism', *Victorian Centre for the History and Philosophy of Science Preprint Series* **1/92**, La Trobe University

Ellis, B.D. and Lierse, C.E. (1994), 'Dispositional Essentialism', *Autralasian Journal of Philosophy* **72**, 27–45

Ellis, B.D. (1996), 'Natural Kinds and Natural Kind Reasoning', in P. Riggs (ed.) *Natural Kinds, Laws of Nature and Scientific Methodology*, Dordrecht: Kluwer Academic Publishers, 11–28

Fales, E. (1986), 'Essentialism and the Elementary Constituents of Matter', *Midwest Studies in Philosophy* **11**, 391–402

Fales, E. (1990), *Causation and Universals*, London: Routledge and Kegan Paul

Harré, R. (1970), 'Powers', *British Journal for the Philosophy of Science* **21**, 81–101

Harré, R. and Madden, E.H. (1973), 'Natural Powers and Powerful Natures', *Philosophy* **48**, 209–30

Harré, R. and Madden, E.H. (1975), *Causal Powers: A Theory of Natural Necessity*, Oxford: Blackwell

Iltis, C. (1973), 'The Leibnizian-Newtonian Debates: Natural Philosophy and Social Psychology', *The British Journal for the History of Sciene* **6**, 343–77

Kripke, S. (1972), 'Naming and Necessity', in D. Davidson and G. Harman (eds.) *Semantics of Natural Language*, Dordrecht: Reidel, 252–355

Kripke, S. (1980), *Naming and Necessity*, Oxford: Basil Blackwell (A revised and enlarged version of Kripke (1972))

Laudan, L.L. (1968), 'The *Vis Viva* Controversy, a Postmortem', *Isis* **59**, 131–43

Lewis, D.K. (1994), 'Humean Supervenience Debugged', *Mind* **103**, 473–90

McGinn, C. (1974/5), 'A Note on the Essence of Natural Kinds', *Analysis* **35**, 177–83

Prior, E.W. (1985), *Dispositions*, Aberdeen University Press

Prior, E.W., Pargetter, R.J. and Jackson, F.C. (1982), 'Three Theses about Dispositions', *American Philosophical Quarterly* **19**, 251–7

Reeder, N. (1995), 'Are Physical Properties Dispositions?', *Philosophy of Science* **62**, 141–9

Shoemaker, S. (1980), 'Causality and Properties', in P. van Inwagen (ed.), *Time and Cause: Essays presented to Richard Taylor*, Dordrecht: Reidel, 109–35

Smart, J.J.C. (1993), 'Laws of Nature as a Species of Regularities', in J. Bacon, K.K. Campbell and L. Reinhardt (eds.) *Ontology, Causality and Mind: Essays in Honour of D.M. Armstrong*, Cambridge: Cambridge University Press, 152–69 Reply by D.M. Armstrong, *op. cit.*, 169–74

Swoyer, C. (1982), 'The Nature of Natural Laws', *Australasian Journal of Philosophy* **60**, 203–23

Swoyer, C. (1996), 'Review of *Laws of Nature* by J.W. Carroll', *British Journal for the Philosophy of Science* **46**, 603–9

Thayer, H.S. (1953), *Newton's Philosophy of Nature*, New York: Hafner Publishing Company

Thompson, I.J. (1988), 'Real Dispositions in the Physical World', *British Journal for the Philosophy of Science* **39**, 67–80

D. M. ARMSTRONG

COMMENT ON ELLIS

This comment is based first upon a letter that I sent to Brian Ellis about his paper, and second upon the quite elaborate reply that he returned to me. The reply, for which I thank him, gave me a much better grip on his position and so assisted me in the present enterprise.

1. *Meinongianism.* The first point I make is that a position such as Ellis' has to embrace a Meinongian metaphysic in respect of unmanifested dispositions and causal powers. Part of the interest of this point is that, if correct, it applies also to the views of Sydney Shoemaker (1980, 1983), Chris Swoyer (1982), C.B. Martin (1993) and George Molnar (unpublished to date). It is to be noticed that this point has nothing to do with a matter raised by Ellis: the degree of unification, or lack of unification, in the structure of the dispositions or powers postulated by particular theories.

Ellis identifies properties of particulars with causal powers. Consider, then, a particular that has a certain property. It is a causal power. But, in general at least, the particular will not *manifest* this power at all times. Because its power constitutes the essence of the property on Ellis' theory, then if it does manifest the power, then it manifests that power *of necessity*. That is ontological gain for the theory. But, as so often, there is ontological pain also. Suppose the manifestation does not occur. The causal power that the particular still has remains essentially one that 'points' to that manifestation, even where the manifestation does not occur. George Molnar speaks of this as 'physical intentionality', and accepts that it is real. Ellis, I maintain, has to accept this also, as should Shoemaker, Swoyer and Martin.

But as Jack Smart has pointed out in discussion, this is Meinongianism. A relation to a non-existent object! Meinong himself did not spot the point with respect to powers. In developing his theory of objects, he started from Brentano's intentionality, which Brentano linked with the mental. Hence Meinong did not consider the powers of non-mental things. But it would have been in the spirit of his theory. (See my 'Reacting to Meinong', 1996.)

I do not regard this as a conclusive refutation of the power theory of properties, if only because conclusive refutations are not to be had in metaphysics. It is perfectly possible to accept a Meinongian position with respect to property-powers. But I myself incline to a *modus tollens* against property-powers because they involve a Meinongian theory.

2. *Swinburne's argument.* The next argument to be developed does *not* directly

H. Sankey (ed.), Causation and Laws of Nature, 35–38.

threaten Ellis' position. Perhaps the only upholder of property-powers that is directly affected is Shoemaker (see Richard Swinburne's 'Reply to Shoemaker', 1983). But the argument is important, because it can only be met by a theory that *exempts* certain properties (or relations) from being reduced to powers.

Suppose that all properties and relations are reducible to mere powers. *Manifestation* of a power, as I suppose everybody would accept, is a matter of an object gaining or losing a new property, non-relational or relational, or of an object having certain properties being created or destroyed. But if all properties and relations reduce to powers without remainder, these manifestations must themselves be analysed in terms of acquiring or losing of powers, and so on to infinity. Putting it into scholastic terms, potency never gets translated into act, because even where the power is manifested its only result is the acquiring or losing of further potencies by particulars. I am not sure that this is a contradiction, but it seems undesirable in the extreme. Upholders of power theories seem to agree, with the exception of Shoemaker. He gives an account of all properties in terms of causal potentialities and says that the account "could be extended to cover relations as well as properties" (Shoemaker, 1983, p. 296). Swinburne's argument seems to catch him.

One theory that completely escapes this difficulty is C.B. Martin's. This is because he holds what might be called a double-aspect theory of properties. All properties have a power 'side', but they also have a categorical 'side'. Hence 'act' is always present when some object acquires a property. Swoyer, Molnar and Ellis, however, do not accept Martin's view. They prefer to meet the Swinburne problem by making certain properties or relations categorical (but powerless) in order to provide a certain amount of 'act'.

All I know of Swoyer's position is a brief personal communication where he says that he does not reduce all properties to powers. Ellis and Molnar take more or less the same position as each other (personal communications again). Ellis exempts relations from the power theory, Molnar exempts spatio-temporal relations (and causality presumably!). They differ about shape and size. Ellis takes these to be relations, and so not to be powers. Molnar holds that shape and size are properties, and so would reduce them to powers. (For myself, I agree with Molnar that they are properties, not relations, but deny that they are mere powers.)

These concessions in the positions of Ellis and Molnar would seem at least to blunt the force of Swinburne's argument. The 'act' side of the world becomes particulars in various sorts of relation, especially spatio-temporal relation. (It is perhaps worth noting that upholders of this sort of view will find it difficult to reduce particulars to bundles of properties, a popular move with quite a few metaphysicians. But it is not a reduction that Ellis or Molnar accept.)

As may be expected, however, these concessions, though apparently necessary, do have a certain cost. An opponent of Ellis such as myself can say "If relations are thus excepted, why may there not be *properties* that are more than powers?" And one argument, in particular, used by Ellis and by his collaborator Caroline Lierse (Lierse, 1996) against 'categorical', non-dispositional, properties seems to be weakened. Those of us who accept categorical properties tend to agree that it is a contingent matter what powers these properties bestow upon the particulars that instantiate them. This is then made a reproach to us by Lierse. All we can know about these properties, she argues, derives from the *effect* that they produce upon further objects and, ultimately, upon our minds. The properties themselves are an unknown somewhat, things whose nature we can never know. Forget about them, then, and substitute properties that are nothing but powers to produce certain effects.

But now we see that, in order to deal with the Swinburne objection, *certain* properties, or at least certain relations, have to be exempted. They are more than mere power-bestowers. But if so, is not sauce for the goose sauce for the gander? Some properties, or relations, are categorical even in a scheme of the sort proposed by Ellis and the others. Why should not *all* properties, then, have at least a categorical *side*, as Martin for instance maintains?

There is a still worse problem, though. If these categorical properties are powerless, how is it possible that we should know anything about them?

Ellis is aware of this second problem and has attempted to answer it. I do not understand this answer, which is rather brief, but he says it depends upon "the multi-dimensionality of the quantitative laws of action of the causal powers of the things they relate". If the causal powers of things had not been of the right nature (and Ellis allows that they might have been different, thus necessitating different outcomes), the categorical relations would have been inaccessible to us.

The causal powers themselves, however, according to Ellis, have built into their essence the spatiotemporal environment in which they operate. For instance, assuming for simplicity the truth of Newtonian gravitation, the gravitational mass of an object is the power to attract other massive bodies according to the product of the masses but in inverse ratio to the square of the distance. To contemplate an inverse cube law is to contemplate a different property, mass* we might call it, for the very same distance. So whatever happens when a certain distance obtains, is already built into the power.

A trouble with this, though, is that, assuming a causal theory of perception, our only epistemic connection with distance is then via the property-powers. The categorial relation of distance becomes a purely theoretical entity, and to say the least, the grounds for postulating it will need to be spelt out very carefully indeed. In any case, some distances appear to be as directly, non-inferentially, perceived as anything is perceived.

It seems that we can, in favourable circumstances, perceive the distance of two things from each other, or at least we can perceive that *a* and *b* are at a greater distance than *a* and *c* are (*a* might be our own body.) For Ellis, the distance is categorical and bestows no power. But *a*, *b*, and *c* will have to have power-properties such that, when the three objects are spatially related thus, they together act on a suitable perceiver and so produce a correct perception of relative distance. This seems strange. The natural thing to say is that the two *distances* make a direct causal contribution to the perception. But if so, then two categorical relations will have to have powers, and have them contingently.

Distance is only one case. Ellis will need to tell a story about the whole spatio-temporal framework of the world, including directions and dimensionality. A mighty thing to be in itself categorical and powerless!

3. *One property, many powers*. If properties are constituted by the powers they bestow, then it will be impossible that one property can bestow wholly distinct powers. Yet the situation is obviously conceivable (which does not settle the matter) and

furthermore seems to be possible. One case here is that of the identity of gravitational and inertial mass. Ellis comments that this distinction is eliminated in general relativity. But he says that an analogous identity, the power of a body to curve spacetime and its power to resist acceleration, is not so easily dealt with. His main suggestion here is that the powers are not really distinct, and that their essential unity will be revealed in some future unifying theory.

I do not object to this sort of defence. Ontological schemes may well have to submit to the discipline of 'our best science', and that science lies in the future. But such a defence should at least discourage any attempt to refute categoricalist theories of properties simply on the basis of *a priori* considerations.

4. *Kinds*. I note finally that one way that Ellis wishes to develop his scheme is not found in the other power-theories briefly mentioned here. Ellis appeals to kinds, taken to be special sorts of universals, in addition to properties and relations. A question does arise whether we do require universals, such as electronhood, in *addition* to particulars having properties and relations. My own inclination is to think that given all the particulars and their true properties and relations (as discerned by total science) then the kinds supervene and are not required as a further ontological postulation. But I will not argue this here.

University of Sydney

REFERENCES

Armstrong, D.M. (1996), 'Reacting to Meinong', in *Meinong und die Gegenstandstheorie*, ed. Rudolf Haller, Amsterdam: Editions Rodopi B.V.

Ellis, Brian (this volume), 'Causal Powers and Laws of Nature', 19–34

Lierse, Caroline (1996), 'The Jerrybuilt House of Humeanism', in *Natural Kinds, Laws of Nature and Scientific Methodology*, ed. P.J. Riggs, Dordrecht: Kluwer, 29–48

Martin, C.B. (1993), 'Power for Realists', in *Ontology, Causality and Mind*, ed. John Bacon, Keith Campbell and Lloyd Reinhardt, Cambridge: Cambridge University Press, 175–94

Shoemaker, Sydney (1980), 'Causality and Properties', in *Time and Cause*, ed. Peter van Inwagen, Dordrecht: Reidel, 109–35; reprinted in *Identity, Cause and Mind*, by Sydney Shoemaker, Cambridge: Cambridge University Press, 206–33, 1984

Shoemaker, Sydney (1983), 'Properties, Causation and Projectibility', in *Aspects of Inductive Logic*, ed. L.J. Cohen and Mary Hesse, Oxford: Oxford University Press, 291–312

Swinburne, Richard (1983), 'Reply to Shoemaker', in *Aspects of Inductive Logic*, ed. L.J. Cohen and Mary Hesse, Oxford: Oxford University Press, 313–20

Swoyer, Chris (1982), 'The Nature of Natural Laws', *Australasian Journal of Philosophy* **60**, 203–23

BRIAN ELLIS

RESPONSE TO DAVID ARMSTRONG

This reply is based on my letter to David Armstrong, which was in response to his earlier criticisms of my paper on Causal Powers and Laws of Nature. I have amended my reply to take account of the objections made in the revised version of his response. I am grateful for the opportunity he has given me to clarify my position.

I. THE MEINONGIAN PROBLEM

In general, we can explain what we think we know, if we can show how it fits in with other things. For explanation proceeds by epistemic location, i.e. by embedding what we think we know in some wider context of knowledge. Something like this, at any rate, is the concept of explanation required for metaphysics. In ontology, the problem is to explain the reasonableness of believing in various kinds of postulated entities. To do this it is necessary to show how entities of these kinds can serve as reasonable and adequate bases for explaining the existence of other things. To be reasonable and adequate, the proposed basis cannot be just an ad hoc construction. The Meinongian problem arises, or at least is seen as a problem, mainly because the proposed basis for explanation appears ad hoc. To explain one's disposition to go to sleep after taking a sleeping draught, an explanation which just refers to the dormitive power of the draught is unsatisfactory, not because it is wrong, but because it does not embed what we already know in any wider context of knowledge. If explanations in terms of causal powers were all like this, then they would all be unsatisfactory for the same reason. The Meinongian problem, it seems to me, arises, at least partly, from the felt inadequacy of such explanations. If we believed that there was nothing in the draught which could be singled out as the ground or source of the dormitive power, (here is the intuition which leads to categorical realism) then the property would indeed be a strange one, and the charge that it was a Meinongian property would be serious.

Scientific explanations which refer to causal powers are not really like this, however. For such explanations are generally quantitative, and the quantitative character of explanations in terms of causal powers makes all the difference. It blocks the charge of ad hocness. A qualitative dormitive power can explain one and only one disposition. But a causal power of the kind likely to feature in a scientific explanation is able to explain any of infinitely many quantitatively distinct dispositions.

If causal powers were Meinongian properties, like dormitive power in the above

H. Sankey (ed.), Causation and Laws of Nature, 39–43.

example, then displays of quantitatively distinct dispositions would have to be displays of quantitatively distinct causal powers. Thus the disposition to E_1 in circumstances C_1 is a different disposition, and hence would have to be considered to be the expression of a different causal power, from a disposition to E_2 in circumstances C_2, even if the differences between E_1 and E_2, and between C_1 and C_2 are just quantitative. But in my view, which I am sure would be the view of the scientific community, the same causal power might well be responsible for a whole spectrum of such specific quantitatively distinct dispositions. The causal power which is responsible for an acceleration a_1 of an object of mass m_1 at a distance d_1 from the body in which this power resides might well be the very same as that which is, or would be, responsible for an acceleration a_2 of an object of mass m_2 at a distance d_2. Thus causal powers may (and usually do) have quantitatively different expressions, depending quantitatively on the circumstances.

Moreover, causal powers not only have quantitatively different expressions, depending on circumstances, or on the quantitative properties of the objects on which they act, they may also differ from each other quantitatively, and so act with greater or less force on the same or identical things. Additionally, causal powers are often directional. For example, they may be centripetal, or centrifugal, or circulatory. All of this complexity adds to the fruitfulness of causal power explanations, and to their embeddedness in the broad context of scientific knowledge. Consequently, it is far from being the case that causal power explanations are ad hoc in the sort of way that the hypothetical dormitive power explanation is. Explanations given in terms of causal powers can thus succeed, as perhaps no other kinds of explanations can, in locating epistemically what is to be explained in a broad network of causes and effects.

Armstrong's Meinongian objection is this: Generally speaking, things do not manifest their causal powers at all times. Some things may even have causal powers which they never manifest. A particular which has such a property then has a property the essence of which involves a relation to an event which never occurs. It "points" to its manifestation, but "the manifestation does not occur". Such a property is a Meinongian property, since its existence entails a relation to a non-existent event. My response is this: A causal power is a particularised relation between natural kinds, viz. between the kind of circumstance (event or state of affairs) which would trigger the causal power of any object which has this power, and the kind of event or process by which this power would then be displayed. For an object to have such a power is for it to be such that it could be triggered in such circumstances to have such effects. The natural kinds of events or processes of which the displays of causal powers are instances exist independently of which, if any, such displays occur in any particular object. They also exist independently of which species of these kinds of events or processes have instances (although it may be necessary for there to be at least one instance of the kind in at least one of its species). For a natural kind of event or process to exist, it is certainly sufficient, whether or not it is also necessary, that some instances of that kind of event or process should occur.

Armstrong's Meinongian objection confuses kinds of events or processes with instances of those kinds. A causal power does not point to some possibly non-existent event (manifestation), although it does imply the existence of some possibly uninstantiated relations between certain kinds of circumstances in which the object may exist, and the kinds of events or processes which may consequently occur. I am not much impressed, therefore, with the Meinongian objection. Causal powers are not infimic dispositions, as the objection seems to presuppose, but the grounds or sources of these dispositions. Therefore, it is not the case that causal powers involve any specific manifestations essentially, as the objection claims. But nor are causal powers properties which have identities which are independent of the dispositions they ground, as the categorialist maintains.

2. SWINBURNE'S REGRESS

The manifestations of the dispositions grounded by causal powers must be either changes in the causal powers (or liabilities or propensities) of things, changes in their circumstances (e.g. of their states of motion, agitation, etc.), or changes in their composition or structure. Some of these changes will result in the extinction of things, or in the creation of new things (when essential properties or structures are changed). Some of these changes will be changes occurring in us, or in our measuring or recording instruments. Some of the changes will be what I should call changes of categorical properties or relations, some of which will be accidental properties of, or relations between, the things affected, and some will be essential properties. Others will be changes of causal powers, capacities, propensities, etc., and again, some of these will be changes in the accidental properties of things, and some of essential properties.

Swinburne's regress depends on the assumption that the only changes that can occur are changes in the causal powers of things, so that causal powers can beget only new or different causal powers, which in turn can beget yet other new or different causal powers, and so on. But causal powers, to be effective, have to be triggered. So, if we grant Swinburne's assumption, the question is whether the triggering of a causal power can be effected by an event which is itself just an alteration, creation or extinguishing of a causal power. I do not see any reason for excluding this possibility, but examples of such actions do not readily spring to mind. Causal powers are generally triggered by changes of circumstances, which are generally not changes in the causal powers of things. They are triggered, for example, by changing the spatial locations of things, or their motions or orientations, or the surrounding fields, and so on – all of which involve changes of relations, and none of which involve changes in causal powers. So I do not see that Swinburne's regress is threatening.

The question for me is whether I can make a satisfactory distinction between (a) the causal powers and propensities of things, and (b) the relations which (typically) characterise their structures or circumstances. To describe a causal power, one has to say how a thing with this power must (in virtue of its possession of this power) act in the various circumstances in the complex range of circumstances in which it would be a causally relevant factor. That is, one must specify its laws of action. The laws of action of any causal power will always, or nearly always, be descriptions of the changes which would occur (in the ideal circumstances in which there are no other relevant causal factors) in the relations between things, or in the structures of things, in response to various possible stimuli (which themselves will normally be just other possible changes in the relations between, or structures of, things).

Now, in my view, the relations involved in causal laws are not causal powers, or propensities or liabilities, or anything of the sort. They have a different role in causal explanation, and are described differently. The relations in question are relations between things of various kinds, and such relations must be described if we wish to describe the circumstances in which causal powers may be exercised. The things related will no doubt have causal powers which will be relevant to their behaviour. But neither the relations themselves, nor any changes in them, are causal powers, even though such changes are the phenomena we usually call 'causes' and 'effects'. (When Hume described causation as a relation between events, he was superficially right. For causal powers are activated by events, and their effects are events. But what is involved is much more than just a relation between events. For the objects involved in these events, and the properties of these objects, have a vital role to play. For they determine, in so far as it is determined, what will happen.)

It is true that knowledge of relations depends on the causal powers of the things related. But it is not true that the relations are themselves causal powers. The relation of being separated by 1m is not a causal power or propensity, although it may be a relevant factor concerning the actions of such properties. Things separated by 1m may be more attracted to each other gravitationally, or more strongly repelled electrically, than similar things separated by 2m. However, it is one thing to be a relevant causal factor, because it is involved essentially in a causal relationship; it is another to be a causal power.

The relations of spatial separation, for example, are independent of the natures of the things separated (once it is settled what are to count as their locations). Therefore, these same relations could exist in a world that is very different from ours, e.g. in which properties such as mass, charge, spin, and the like, did not exist. They could, for example, exist in a classical mechanistic world in which things had no properties at all, other than the passive primary ones. In such a world, the spatial and temporal relations between things would be unknowable. But, conceptually, it is a possible world. On the other hand, the identities of the properties we can know about, because they have the power to affect us or our measuring or observational instruments, depend on their laws of action – laws which involve various spatial and temporal relationships essentially. Therefore, these properties could not exist in worlds without spatial or temporal relationships. The spatial and temporal relationships on which all observable properties depend essentially are therefore more fundamental ontologically than any of the causal powers.

But if spatial, temporal, and other primary properties and relationships are not causal powers, the question arises how we can know about them. We can know about them, I want to say, because of the multidimensionality of the quantitative laws of action of the causal powers of the things they relate. We can, for example, observe that things that are separated by 2m subtend a larger angle from a perpendicular distance of 6m from the mid-point of their line of junction than things that are separated by 1m from a point the same perpendicular distance away. And we can use such observations to aid in the construction of a system of geometry, or theory of spatial relations. But such facts can be observed as directly as this, only because the laws of electromagnetic radiation are, macroscopically at least, as simple as they are.

3. ONE PROPERTY, MANY POWERS

I am happy enough with one power, many quantitatively different manifestations. For the laws describing powers are usually quantitative and hence ground a range of specific dispositions. But I am not so happy with one property, many powers.

We have to distinguish, as before, (in para 2) between powers, and the sorts of properties and relations (categorical ones if you wish) which are essentially involved in causal laws, but are not themselves causal powers (or propensities, or the like), e.g. spatial, temporal and numerical properties or relationships. The latter typically feature in many laws involving causal powers, but their identities do not depend on these laws. The identity of a spatial relationship depends on the system of applied geometry in which it is embedded, that of a temporal relationship depends on the system of chronometry we are using, and that of a numerical relationship depends on our system of applied arithmetic. These concepts are not purely formal; but nor are they straightforwardly empirical. They are theoretical concepts with clearly intended applications.

The question is whether there are causal powers involved in laws which are not laws of action concerning these powers. I should, to begin with interpret laws of action fairly broadly to include laws of reaction or response, so that it might be an entailment of something having the property P, not only that it will act in such and such a way in the circumstances C_1, but that it will react or respond in such and such a way to being placed in the circumstances C_2. That is, I should allow that a causal power might also be a causal liability, these being two sides of the same coin.

However, I would not be happy to go further than this. The example of gravitational mass versus inertial mass comes to mind. Intuitively, these are different concepts: the first is the power of a body to attract other bodies gravitationally, the second is the body's resistance to acceleration. It is true that *this* distinction was eliminated in general relativity theory. But there remains an analogous distinction which, as far as I can see, cannot be removed in the same way, viz. that between the power of a body to curve spacetime, and the power of a body to resist acceleration. So I am inclined to say that there are two properties (powers) here, not just one. But then, what am I to say about the law (for it presumably is a law) that these two powers are always (in suitable units) equal? There are two possibilities which I could entertain: (1) that the powers are not really distinct, and that their essential unity will eventually be revealed in a higher level theory (perhaps a unified field theory), and (2) that this is one of the general features of the kind of world in which we live (as I suppose the conservation laws to be). My inclination is to suppose that the first of these alternatives will ultimately prove to be correct.

La Trobe University and University of Melbourne

[illegible]

6. ONE PROPERTY, MANY POWERS

[illegible]

[illegible]

[illegible]

[illegible]

JOHN BIGELOW

SCIENTIFIC ELLISIANISM

Ellis defends a theory which is of great interest and importance: scientific essentialism. Science seeks the essential properties of things in the world, and the necessary truths which follow from them, argues Ellis. The core doctrine of the Ellis theory, I take to be this:

> Among the *essential* properties of many different *natural kinds* of things, we must include certain important *dispositional* properties.

This is food for thought; and what follows will all be a thorough chewing of this cud.

Harré and Madden (1975), Bhaskar (1975) and others have presented various cases for various theories of very much the same general kind as Ellis's. Personally, I think some theory in this broad family is roughly right. I do not know quite why, but it was only after persuasion by Ellis that I began to appreciate the work of others who had advanced similar theories much earlier, including for instance our own former colleague Michael Levine. Ellis's dogged persistence has something to do with my eventual conversion; but in addition, Ellis has done much to articulate and extend the theory, and even more importantly, to reveal the breadth of its consequences and the depth of the new philosophical vision which it makes possible. Yet I will play the Devil's advocate here, and try to uncover the weakest points in Ellis's case for his version of the theory.

I. INFERENCE TO THE BEST EXPLANATION

One thing Ellis does is to give an exposition of his theory, and of some of its logical consequences. This sheer exposition does serve – by itself – as a kind of argument in support of his theory. If he can present the theory in such a way as to display its merits to full advantage, then that by itself will *invite* something like an inference to the best explanation. I find this "argument-by-display" very persuasive, but I will describe some aspects of Ellis's exposition which trouble me.

If inference-to-the-best-explanation is Ellis's strategy, then it would be good to back up the "exposition-leg" of his argument with a second leg – not necessarily in every paper he writes on the subject, but somewhere – in which the virtues of his theory are contrasted with the vices of alternative theories, in such a way as to persuade us that the Ellis theory is not just plausible and interesting in itself but also *better* than any of its rivals. Limitations of time and space, in a conference or a journal article, make it

H. Sankey (ed.), Causation and Laws of Nature, 45–59.

worth pondering whether it is realistic for Ellis to try to include a refutation of rivals along with an articulation of his own theory.

In his conference paper (this volume), Ellis does embark on a critique of what he sees as his most striking rival: a Humean theory according to which "all things are loose and separate" and according to which matter is "essentially passive". The Humean theory he has in mind is also one according to which every dispositional property has a sufficient categorical basis.

Ellis's target Humean also believes in the existence of a so-called "Hume world" corresponding to any given world. Pick any given world which contains not just correlations but causal connections, and not just regularities but laws of nature. In such a world, various counterfactuals will be true: for instance, that "If the bird had not been nailed to the perch, it would have fallen off". The Hume world corresponding to this cause-and-law-laden world will exactly match it in the distribution of local intrinsic properties, in all correlations and regularities, but will differ from it because it will contain no causal connections, no laws of nature; and counterfactuals true in the former world will be false in its Hume world. Things which are true of some sort of necessity in the one world will be true "by sheer coincidence" in its Hume world.

Ellis's strategy is to support his own theory by undermining the prominent Humean rival which he has selected. That strategy is all right as far as it goes, but it has attendant dangers, one of which is the danger of misleading advertising. One should not take him to be claiming to have refuted all his Humean rivals, just on the strength of his refutation of one Humean package deal. This is especially important to keep in mind when considering those of Ellis's arguments which target *only* those Humean theories which affirm the existence of what is called a *Hume world*.

Belief in a Hume world certainly sounds, at least initially, like a very "Humean" belief; and it generates a rival to Ellis which is well worth the attention Ellis gives it. Jackson (1977) not only impressed many of us with the idea of a Hume world, but also produced some persuasive arguments for the existence of a Hume world. Yet I do put in a word of caution: is it so obvious that *all* Humeans need to accept the existence of a Hume world? I think not. Not if a Hume world is supposed to be one in which there are no causal connections, relevant counterfactuals are all simply *not true*, and so on. And of those who do accept the existence of a Hume world, do all of them need to accept that such a Hume world is a *distinct* world from the actual world? I think not. In contrasting his own theory with his one selected rival, Ellis does cast very useful light on the nature of his own theory; yet this does little—something, but only a little—to show that an inference to *his* explanation is an inference to the "best" explanation currently on the market.

2. ESSENTIAL PROPERTIES OF NATURAL KINDS

Ellis offers a package deal in which things of various important *natural kinds* are said not only to have so-called *dispositional* properties, but also to have these dispositional properties *essentially*.

The Humeans, in contrast, are said to be committed to *the essential passivity* of matter. But take care: there is an ambiguity in the wording which Ellis uses to characterise the Humean position. If you say that something is "essentially passive", this might easily be taken to suggest that: "it is not possible for it to be anything other than passive". Compare Ellis's claim that matter is "essentially passive" with a Cartesian claim that matter is "essentially extended". The Cartesian claim does clearly rule out the possibility of matter which is unextended. So: does Ellis's claim, analogously, rule out the possibility of matter which is not passive? Yet, under that fairly natural construal, the Humean would be saying that things can *never* have dispositional properties. And that, surely, is a ridiculously strong doctrine to foist on the "Humeans". Not *all* Humeans are as extreme as those Quineans who dismiss all modal talk, and so all talk of dispositions, as "second grade discourse".

What Ellis assumes, quite justifiably, is that there is at least a significant coven of Humeans who take it to be a contingent question *which* dispositional properties an individual has. These Humeans deny, of each dispositional property taken separately, that *it* is an essential property of the individual. Yet it is very misleading to re-express this thesis as the doctrine that individuals are "essentially passive". It is no part of this Humean thesis that individuals ever *are* passive. Only that: for each disposition, the individual could have existed without having had *that* disposition. They may indeed be essentially active—it's just that they are not *essentially* active in *any particular way*.

Confusingly, Ellis often seems to agree with the Humeans about this last Humean contention; and yet also, he often protests that his view is diametrically opposed to the Humeans on precisely this issue. An ambiguity arises from the fact that Ellis speaks of essential properties "of natural kinds", rather than essential properties "of individuals".

Ellis sometimes allows that an individual which belongs to one natural kind might continue to exist, yet cease to belong to that natural kind. And that would allow precisely the Humean point: that for each disposition, an individual could exist without having that disposition.

Ellisian essentialism, under his own preferred articulation, consists in the claim that having certain dispositions is an essential property, not of individuals within a natural kind, but of "the natural kind" *itself*. Now, it is not obvious how we are to understand this idea. One suggestion is this: we should say that having certain dispositions is an essential property of an individual *qua* member of a given kind. And this means:

> it would be logically impossible for a thing to *belong* to that natural kind without *having* those dispositions.

However, if that were all that Ellis's essentialism amounted to, it would be impossible to see why any clear-headed Humean would need to disagree with it. Humeans are perfectly happy to allow the existence of a variety of different entailments between predicates. In a sense, therefore, you might say that a Humean has no problem allowing that there may be essential properties *of properties*. "It is an essential property of red that it is a colour": a Humean can make sense of that. Likewise, a Humean can make relatively unproblematic sense of a "natural kind" having essential properties, provided that we take a "natural kind" to be something in the same broad category as *predicables*, or *properties*, or *universals*.

And natural kinds do seem to belong to the same broad category as universals. A natural kind is, after all, a "one over many". Many different things *belong* to a natural kind, just as many different things *instantiate* a universal. There seems to be little more than a verbal difference between saying, "This individual electron belongs to the natural kind, *electron*", and saying "This individual electron has the property of being an electron".

An Ellisian might try distinguishing "natural kind" predicates from other predicates. Perhaps it might be supposed that a "natural kind" predicate is a predicate which plays some specifiable and distinctive role in the formulation of scientific theories. Now, it is perfectly true that natural kind predicates do play distinctive roles in the formulation of scientific theories. Yet that is *not* enough for Ellis's purposes, as Ellis emphasises in various places. Natural kind predicates need to be distinguished in a deeper way, a more *de re* and realist way, than just by their roles in theories. Humeans can easily agree with Ellis that there are "natural kind" predicates *in that sense*, and that the possession of natural kind predicates in that sense will always entail various conditionals.

Ellis is well aware of the need to inject some *de re* modality into his notion of "natural kind" predicates – in order for his theory to have any of the distinctive modal punch which would distinguish that theory from a merely Humean theory. Hence Ellis *does* need to take natural kind properties to be essential properties of *things* and not just of properties.

But what *things* are there, which could be the bearers of these essential properties? The obvious candidates are, of course, the things which *belong* to the natural kinds. Remember, that among the essential properties of a natural kind, like say protons, we are supposed to include such properties as that of the "disposition to resist acceleration". And what sort of thing is the *bearer* of a property like that? Well: a *proton* is the sort of thing which has a disposition to resist acceleration. "Protonhood" certainly doesn't have any disposition to resist acceleration. The natural kind, *proton*, taken "collectively" as it were, might be said to have a disposition to resist acceleration – in the sense that it would be hard to shift the centre of gravity of the aggregate of all protons. But that is obviously irrelevant. Maybe there is some thing – some *res* about which there could be *de re* necessities – called *the natural kind, proton*. But even if there is, it is not *that* thing which has the disposition to resist acceleration. The only sense in which a disposition to resist acceleration is an essential property "of this natural kind", is the sense which entails that the *members* of this natural kind have a disposition to resist acceleration. The bearers of the essential properties of natural kinds must be the things which *belong to* those natural kinds. And, *prima facie*, the things which belong to natural kinds are "individuals".

Casual intuition would suggest that it is possible for something of one kind to *become* something of another kind: for Lot's wife to become a pillar of salt, for instance; or for an atom of one kind to lose a proton and become an atom of another kind. These intuitions are *prima facie* inconsistent with the essentialism I am foisting on Ellis.

Yet these intuitions can be explained away. There are indeed things which at some earlier time *constitute* a thing of one kind, and at a later time these same things (or at least, most of them) may come to *constitute* a thing of another kind. The things which "were" of one kind, and "came to be" of another kind, are however never *numerically identical* either with the thing of the first kind or with the thing of the second kind. And so, for instance, we may say that an aggregate of subatomic particles,

particles which together constitute an atom of one kind, may lose one of its parts, and perhaps "rearrange" the others, and so the members of the resulting new aggregate may come to constitute an atom of another kind. But that does not mean that *the atom* of the first kind is, literally, *the thing* which has come to be an atom of a different kind. This illustrates one way of explaining away the appearance of a thing's changing from one kind to another. There may be other ways as well.

Ellis, after all, should be the last person to be swayed by casual *intuitions* about the *possibility* of an individual's continuing to exist, but changing from one natural kind to another. It is entirely in the spirit of his enterprise to insist that these sorts of appearances of possibility are entirely illusory, and to be explained away. Elsewhere, Ellis has to bite cannon balls, compared to that bullet!

Thus, I strongly urge Ellisians to grasp the nettle here: when it is said that certain dispositional properties are essential properties "for some natural kind", this *should* be construed as meaning that these are essential properties (*de re*) for each of the things *belonging* to that natural kind. Ellis sometimes speaks as though "individuals" are the things which "persist through changes in kind", rather than the things which "belong to those natural kinds". But let the term 'individual' fall where it will. The important point is that it is the things which *belong* to the natural kind which should be taken to have the properties which are said to be essential properties, *de re*, "of that natural kind". This may well be what Ellis has meant all along, by saying that he is speaking of essential properties "of natural kinds" rather than "of individuals": he was speaking of essential properties of the "things which *belong* to natural kinds" rather than of the "things which can persist through changes of natural kind".

3. DISPOSITIONAL PROPERTIES RECONSTRUED

Returning, then, to "dispositional" properties: Ellis maintains, and the Humean allegedly denies, that "dispositional" properties are included among the *essential* properties of natural kinds. The Humean characteristically claims, for any given dispositional property, that it is logically possible (*de re*) for an individual of a given kind to exist, and yet for it not to have that dispositional property. And that means: it is logically possible for an individual of a given kind to exist, and yet for any given conditional to fail to be true. This is the Humean metaphysics which Ellis is rejecting.

Is this characterisation fair to the Humeans? I think many Humeans might indeed, initially, accept the possibility of individuals (of a given kind) existing *without* the conditionals which Ellis speaks of being true. But should a wary Humean really accept this, on mature reflection?

What, more precisely, are these conditionals, that Ellis's Humean is alleged to be saying "could fail to be true"? It is easy to find conditionals which, Humeans will say, "could fail to be true". The trouble is that these easily found examples are all ones over which, on reflection, Ellis would have to agree with the Humean, that indeed they "*could* fail to be true". Ellis would add, of course, that those were not the conditionals which he had in mind. This, then, begs us to ask the question: which, more precisely, *are* the conditionals which Ellis takes to be strictly entailed by the possession of dispositional properties?

It is clear what direction to look, if you want to find examples of very roughly the sort of conditionals which Ellis has in mind: for instance, Ellis takes it to be essential to electrons that they have a propensity to repel one another. And that means that, *under certain conditions*, two electrons will move away from each other. But their tendencies to move away from each other can be *overridden* by other factors. Hence various obvious, tempting, illustrative, but overly simple conditionals will not in fact be true. For instance, it will *not* be true that: "If two electrons are a certain distance apart, then they will move farther apart".

Phenomenalists, following John Stuart Mill, used to try to analyse statements like, "There is a coin in my pocket" as equivalent to a conditional like, "If I had reaching-into-pocket sensations, I would have coin-in-my-hand sensations". It was easy to get the general idea, but almost impossible to display any concrete examples of such conditionals which really were, even approximately, truly *logically entailed* by the sentence which was the target of the phenomenalist's attempted reduction.

Exactly the same pattern recurred over Rylean behaviourism. It was suggested that mentalistic statements like "I feel hungry" entailed behavioural conditionals of something like the flavour of, "If I had food within reach I would eat it". But it was impossible to come up with concrete examples of any such conditionals which anyone would really take as plausibly *entailed* by the mentalistic target of the Rylean reduction.

Exactly the same thing holds for Ellis's thesis that each disposition will entail certain distinctive conditionals. It is not as easy as it might seem, to find a relevant conditional which is really *true*. You need to add a *ceteris paribus* qualification to the conditional, obtaining something like this: "If two electrons are a certain distance apart, then – *other things being equal* – they will move farther apart". These may be true – but they are vague.

Now, Ellis's "Humean" allegedly has to maintain the following: that it is possible for an individual of a certain kind to exist, and yet for conditionals *of the above kind* (complete with appropriate *ceteris paribus* qualifications) to be false. It is far from clear to me that any clear-headed Humean does need to say any such thing. If enough is built into the *ceteris paribus* qualifications, then the conditionals will be true but virtually vacuous; if not enough is built into them, then even Ellis will agree that they are not after all even true, let alone essential to the individuals and kinds in question. When the relevant conditionals are spelled out, a Humean may well agree that the existence of individuals of a given kind *will logically entail* the truth of conditionals of that sort. And that need not in any way compromise their Humeanism.

Lewis, for instance, is a hardcore Humean. Yet he can define what it is for something to be an "essential" property of a thing: it is for this to be a property which is shared by all a thing's counterparts, in all other possible worlds in which it has any counterparts. And something in another possible world is a "counterpart" of some given thing in this actual world just in case it is *sufficiently* similar, and none of its worldmates are *more* similar, to that given thing in the actual world. Lewis is then perfectly free to say that nothing will ever count as "sufficiently" similar to some actual thing unless there are certain rafts of counterfactual conditionals which are true of both these things. (For Lewis, context determines what counts as "sufficient" similarity; and in some contexts, the contextually relevant similarities may well have to be so extensive that they will guarantee the truth of a raft of counterfactuals.) Then, for Lewis, it would be the case that the existence of the actual thing of some given natural kind will *entail* the truth of those counterfactual conditionals. So we would have here a Humean who agrees with Ellis up to this point: that the existence of individuals of a given kind *will logically entail* the truth of a class of relevant (*ceteris paribus*) conditionals.

Hence the distinctive feature of the Ellis theory cannot consist simply in the *entailment thesis*: that the existence of individuals of a given kind will logically entail the truth of certain distinctive conditionals. Clearheaded Humeans may agree with Ellis about the truth of that entailment thesis. Their Humean credentials will be completely uncompromised, so long as – "Euthyphro-style" – they maintain that such conditionals, if true at all, are *not constitutive* of either the individuals or the kinds in question. The key Humean idea, in this territory, is that conditionals are only true *in virtue of* certain occurrent properties and relations of individuals; and that the truth of *no* conditionals is *constitutive* of those properties, relations or individuals.

Let me say a word more about the "Euthyphro" question concerning the "direction of ontological dependence", or of "constitution", between conditionals and dispositional properties. In Plato's *Euthyphro* it is imagined that there is a necessary coextensiveness between the things which are approved by the gods, and the things which are good or just; yet it is suggested that there is a difference between one who says that just things are just "because" they are approved by the gods, and one who says that just things are approved by the gods "because" they are just. Parallel differences can be manifested between the attitudes of different theorists on the relationship between *dispositional* properties on the one hand, and the holding of various *conditionals* on the other.

The believer in so-called "categorical bases" for all dispositions is very like the realist and anti-reductionist about justice. In insisting that "dispositions" need "categorical bases", what is being insisted upon is that properties for which the truth of conditionals *is* constitutive can only be possessed by things which *also* have properties for which the truth of conditionals is *not* constitutive. And it is further insisted, by the "categorialist" whom Ellis has in mind, that the possession of properties for which the truth of conditionals is *not* constitutive will provide a sufficient truthmaker for the possession of any properties for which the truth of conditionals *is* constitutive.

If Ellis's essentialism is to be what it sets out to be, a rival to the Humean metaphysics, it will have to embody not just *an entailment-thesis*, but also a *constitution-thesis*. Ellis will have to maintain more than just that certain occurrent properties and relations of certain natural kinds of individuals will be ones whose possession *entails* the truth of certain (non-vacuous) conditionals. That is the entailment-thesis. Ellis will also have to maintain that the truth of relevant conditionals is *constitutive* of the possession of the occurrent properties, relations, kinds or individuals in question. That is the constitution-thesis.

"Dispositional" properties, then, are properties whose possession is at least partly *constituted* by the holding of certain conditionals. Ellis's metaphysics is distinctive, in affirming not only the existence of dispositional properties in that sense, but in affirming also that dispositional properties of that kind are *essential* properties of members of many important natural kinds. And, Ellis maintains, for the possession of at least some of these dispositional properties, it is impossible to find sufficient truthmakers consisting entirely of the possession of categorical properties.

4. INTRINSIC PROPERTIES

Ellis makes it part of his theory that dispositional properties are among the *intrinsic* properties of things which belong to various important natural kinds. Here is one place where Ellis put forward his theory:

> I postulate that amongst the intrinsic properties and structures of any natural kind there are at least some which are essential to things of that kind. (this volume, p. 22)

This is problematic in several ways. I have already tangled with one of those ways in Section 2, when I considered the question of whether the relevant properties are essential properties "of the kind", or "of the individuals" which belong to that kind. (I have urged Ellisians to say the latter.) Next, I focus on his use of the word 'intrinsic'.

The exposition Ellis gives of his theory is beset with ambiguities whenever he articulates it with the weasel-word 'intrinsic'. Humberstone has written an instructive paper, 'Intrinsic/Extrinsic' (1996), in which he takes the lid off the can of worms which is concealed under that little word 'intrinsic'. There are five worms worth special attention. (a) Sometimes an intrinsic property is said to be one which a thing will share with all its duplicates. (b) Sometimes an intrinsic property is said to be one which is compatible with loneliness, that is, which a thing could still have even if it were the only thing in existence. (c) Sometimes an intrinsic property is thought of as one which does not involve any relations to any other individuals. (d) As Ellis points out, very insightfully and originally, there is a sense in which the intrinsic length of a rubber band is the length it has when it is not being stretched. Note that a rubber band may have a certain "intrinsic shape" even at a time when it does not have that shape. And generally, a thing may have a property "intrinsically", in sense (d), even though it *does not have* that property. At the other extreme, however, we have a fifth sense: (e) in loose common usage, when one says something is an "intrinsic" property of a thing, what is quite often meant is that this is an *essential* property for that thing.

These five senses of 'intrinsic properties' overlap in their extensions, and so it is tempting to equivocate. But there are cases which pull them apart – and that is *especially* so when *dispositions* get into the act.

Lewis and Langton have probed the notion of intrinsicality in a recent paper on "Defining 'Intrinsic'" (1996). In that paper, dispositional properties are noted, several times, as especially contentious cases for a variety of attempts to frame a definition of 'intrinsicality'. One way of framing a definition will ensure that dispositional properties will count as "intrinsic", another way of framing a definition will give the opposite verdict – pick whichever definition you wish, depending on which verdict you prefer on dispositional properties. The dispositional properties, propensities, powers and so forth, of which Ellis speaks, are all especially entangled in *precisely* the areas where any ambiguities in 'intrinsic' will yield differences in extension. For this reason, it is hard to know what to make of an Ellisian claim that dispositional properties are "intrinsic" properties of natural kinds.

It is especially interesting to think about sense (d) intrinsicality. Ellis is original, I think, in attracting attention to this sense. Ask, then, whether dispositional properties are "intrinsic(d)" to important natural kinds of things. For instance, is the disposition of electrons to repel one another "intrinsic" to electrons *in the same way that*, for (say) some given tennis ball, perhaps just one of the shapes it manifests will count as its "intrinsic" shape? That would mean that the relevant disposition could be an "intrinsic" property of electrons, even though quite often electrons are subject to distorting influences, and consequently *do not possess* this disposition. They may have the meta-disposition, to possess the lower-level disposition in the absence of distortions. And they may have the compound property of "intrinsically having the disposition". But they need not have the disposition, at times when they are subject to distorting influences – just as a tennis ball may be "intrinsically(d)" spherical even though at the moment of impact with a racket it is flat as a pancake. This would pull Ellis in the opposite direction from that in which he wants to go, over the transubstantiation argument to be discussed in the next section. But it might rescue him from his relatively severe troubles over counterfactuals, which I discuss in the section after that. And incidentally, if he were to take that route, he would be lead straight to the heterodox Ellisian theory of Leckey, in Leckey and Bigelow (1995).

There is therefore a genuine question of interpretation, whenever Ellis uses the term 'intrinsic' in articulating his own theory: which sense of 'intrinsic' should he be intending in that context? In the interests of clarity, I urge both Ellisians and his critics alike to avoid the weasel-word *intrinsic* when trying to give a canonical exposition of the core doctrines of Ellis's theory.

5. TRANSUBSTANTIATION AND REALISM

One memorable argument Ellis deploys against his rivals is one which evokes historically deep and painful memories of Protestant and Catholic bigotries. Ellis refers us to the doctrine of transubstantiation: in the ritual of communion, bread and wine, "the host", are said by some Christians to become the body and blood of Christ, even though they keep displaying the superficial appearances – and all the physical and chemical properties – of bread and wine.

Ellis then claims that the Humean is committed to the logical possibility of a *global* transubstantiation. *All* empirically accessible properties could remain exactly the same, but the underlying essential properties, and numerical identities, of things might be completely changed. This, Ellis takes to be "mad", and so to furnish a *reductio ad absurdum* of Humean theories. As an expository device, reference to transubstantiation is interesting and helps to make some of the issues more vivid. But as an argument, it is a little irritating.

What exactly is supposed to be involved in the envisaged possibility of global transubstantiation? I presume that a global transubstantiation would involve a change in *some* of the qualities of things, their essential properties and their natural kind properties, without any change whatsoever in those properties which are *empirically, directly accessible* to us. Ellis imagines the Humean to be saying that the connections between the former and the latter properties is *contingent*, and hence that it is logically possible for the properties in these two classes to be uncoupled, in a global transubstantiation.

Under an Ellis global transubstantiation, we are to imagine that things change in their natural kind properties, while changing in neither their dispositional nor their empirically, directly accessible properties. So in particular, we are to imagine a change of *natural kind* from wine to blood, with no change at all in any *dispositions* toward any different displays of *empirically, directly accessible* properties. So the stuff which has been transformed from wine to blood not only continues to look and taste like wine, to pass chemical tests for wine, and so on, but also continues to be *disposed* to look and taste like wine, to pass chemical tests for wine, and so on.

Are all Humeans committed to the logical possibility of a global transubstantiation of that kind? Surely not. Many Humeans are nominalists, and believe only in *de dicto*, and in no *de re*, necessities. They will believe in no property of "being wine" which is in any way *distinguishable* from "deserving to be *called* 'wine' ". Such Humeans will display no hesitation: "If it looks (. . . etc) like wine, then it certainly doesn't deserve to be called 'blood' ". Such Humeans will surely reject the possibility of a global transubstantiation.

Which "Humeans" will be committed to the possibility of a global transubstantiation? Answer: only those who are realists about natural kind properties, and who also take the links between these properties on the one hand, and dispositions on the other, to be purely *contingent*. That is: those who believe both in the possibility of a certain kind of Hume world, and in its distinctness from the actual world.

Is it a weighty objection to that kind of Humean metaphysics, that it commits you to the merely logical possibility of a transubstantiation of that sort? I think not. It is, rather, an index of the strength of such Humeans' *realism* about their posited *natural kind* properties.

Compare this kind of realism about categorical properties with John Locke's representative realism about material things. Locke's theory allowed the logical possibility that our ideas fail to correspond to any real things: our life might all be a dream, as it were, or a merely "virtual reality". Berkeley took that to furnish a weighty objection to Locke's realism. Ellis's transubstantiation argument takes a similar logical possibility, global transubstantiation, to furnish a weighty objection to a certain sort of realism about categorical natural kinds and categorical properties. Yet Ellis is the very last person to side with Berkeley against realism. So it is hard to see any force in his transubstantiation argument, if it were to be directed against a Humean's postulation of the existence of *categorical* properties, and of the existence of merely contingent connections between those properties and directly, empirically accessible dispositional properties.

An Ellisian might, however, launch an argument of a slightly different kind, though still working on the transubstantiation theme. Grant, for the sake of the argument, a realism about natural kind properties of a categorical sort, ones which could indeed conceivably be subject to a global transubstantiation. Ask, however, whether those are the *only* natural kind properties in the world. And ask further, whether *typical examples* of natural kind properties should be construed as natural kind properties of that, categorical, kind.

Take for instance natural kinds like "being wine" or "being blood" – or "being an electron". Plausibly, certain dispositional properties *are* among the essential properties for things of *those* natural kinds. And for those natural kinds and dispositional properties, there would be no logical possibility of any global transubstantiation. It would not be possible to change those natural kinds without changing the dispositions.

This Ellisian argument is very persuasive. But does it refute the Humeans? I do not think so. All that it establishes is that there are *some* natural kinds for which the possession of certain dispositions will be among their essential properties. For those natural kinds, a global transubstantiation would indeed be a logical impossibility.

Yet Ellis's target Humeans can accept that there may indeed be natural kinds of that sort, "dispositional" natural kinds as we might call them. And these Humeans could then easily agree with Ellis that a global transubstantiation would indeed be a logical impossibility, if it required us to tear apart those "dispositional natural kinds" from the dispositions which are essential to them.

The Humeans we are talking about, remember, are ones who believe in "the Hume world" – a world in which things retain their categorical properties and natural kinds but lose their dispositions. But the Hume world is not defined to be a world in which things sustain relevant *dispositional* natural kinds and lose relevant dispositions. It is defined to be a world in which things sustain relevant *categorical* natural kinds and lose relevant dispositions. And no realist should think of *that* logical possibility as a nettle which they should hesitate to grasp.

Thus, Ellis's target Humeans would have to accept the logical possibility of a global transubstantiation which changed categorical natural kinds without changing any dispositions at the directly, empirically accessible level. But Ellis is in no position to point to that as any "mad" absurdity, so long as he is a realist, and so long as he admits the existence of at least some categorical properties.

Ellis's target Humeans could also accept that many natural kinds are, in fact, dispositional natural kinds – natural kinds for which possession of certain important dispositions will count as essential properties. Such Humeans, however, are not vulnerable to Ellis's transubstantiation argument. Their belief in a Hume world did not commit them to a belief in a world which changes *dispositional* natural kinds without changing any directly, empirically accessible dispositions.

Of course, not all Humeans would be happy to accept natural kinds, or essential properties, or any of that sort of metaphysical baggage. But then, those more abstemious Humeans will not believe in the existence and distinctness of a Hume world. So they will certainly not be vulnerable to Ellis's transubstantiation argument.

Hence, I conclude, the global transubstantiation argument works against no Humeans at all. But to end on a more positive note, it should be remembered that the global transubstantiation thought experiment does establish one thing of importance. It shows that it is very plausible to suppose that there are at least some dispositional natural kinds; and indeed it shows more than that. It shows that many of the most plausible examples of natural kinds which come to mind are in fact examples of dispositional natural kinds. Dispositional properties are indeed essential properties of these natural kinds.

6. VACUOUS COUNTERFACTUALS

Ellis concedes that on his theory, a very great many hum-drum, everyday, run-of-the-mill counterfactuals will have to be counted as *vacuous*, even though some of them have the strong appearance of being *true* and others of being *false*, and none of them of being merely vacuously either, or for that matter, of somehow managing to be neither. Ellis's theory tells us that these counterfactuals are not in fact nonvacuously true or false, and that the illusion of nonvacuous truth or falsity is engendered by the fact that they vary in their degrees of *assertibility*.

It is highly assertible, for instance, that: "If we could walk on water, we'd have less need for boats". Yet taken literally, this counterfactual is vacuous – according to Ellis. It is vacuous because the antecedent, "If we could walk on water . . .", puts forward a counterfactual supposition which is metaphysically impossible. There is *no* possible world in which you walk on water, because (to put it in Irish) if you were to do so in any possible world at all, then that would show either that you were not a person, or that it was not water that you were walking on, or else that walking is not what you were doing on it.

Do not be distracted, however, by the far-fetchedness of the water-walking which features in this illustration. Ellis's difficulty is not restricted to such explicitly miraculous counterfactual suppositions as that of walking on water. Lewis has shown convincingly that minor miracles, small exceptions to the laws of nature, may very often be required by almost any counterfactual supposition. To clarify further: Lewis rightly insists that our theory should allow that a great many everyday counterfactuals would still have been true, even if the world had been a simple and deterministic one. And in a world like that, these straightforward counterfactuals could not be true without requiring the logical possibility of small exceptions to the laws of nature. (See especially Lewis (1979).)

Ellis, however, will be compelled to say that, though assertible, it is in fact vacuous to say even, as I sit exhausted in front of the television: "If I had a beer here within reach, I'd drink it". Or at least, Ellis will be compelled to say that this counterfactual *would* be vacuous, if it were said in a deterministic world, in which you could make a change in the present only at the cost of changing the whole of history. Imagine a world in which, in order to insert a beer here now, without creating any exceptions to the laws of nature, you would have to change preceding conditions; and then you would have to change the conditions preceding those preceding conditions, and so forth. And it may be that such changes in preceding conditions will push back like dominoes, right back to the big bang. Once you get back to the big bang, you may find that there is no way of making a little change which would *both* result in a beer here now, *and* result in me still being thirsty. Indeed, there will be possible worlds in which even a slight change in initial conditions at the big bang would, under the laws of that world, result in the failure of human beings to ever evolve in the first place, let alone for me to find myself here now, still thirsty but with a welcome beer within reach. Yet even if our world were a world like that, I say, it would still be true that if I had a beer here now, I would drink it.

I recall this line of argumentation from Jonathan Bennett, in seminars as long ago as 1970. In that context, the argumentation was directed against Nelson Goodman's (1955) *metalinguistic* theory of counterfactuals, not Lewis or Stalnaker's *possible worlds* semantics for counterfactuals. The Ellis theory, that laws are logical necessities, creates as much trouble in that theoretical environment as it does under Lewis's possible world semantics for counterfactuals. Add the antecedent of a counterfactual to some set of propositions, "background assumptions". Adjust this set of propositions, but in ways which fall short of changing the whole of history, back to the big bang. Under the Ellis theory that laws of nature are logically necessary, the set we are envisaging will in fact be *logically inconsistent*. And anything follows from a logical inconsistency. So there is no way of explaining why one, rather than another, consequent should be taken to follow counterfactually from the antecedent we were examining.

It is easy to underestimate how bad it is to have to construe *so much* counterfactual talk as mere illusion. And Ellis does have to say that it is an illusion to think that such a counterfactual is straightforwardly true, in such a world. He cannot say that such a counterfactual is straightforwardly true, because he cannot make sense of the counterfactual supposition of there being a beer here now. In order both to keep me thirsty and also insert a beer into my world, within reach, you would have to introduce an exception to the laws of nature: and that, according to the Ellis theory, means that it is not a beer that has been inserted into the world, and it is not even a human being, let alone me, who drinks it.

Admittedly, almost any theory will have to allow for the existence of *some* vacuous counterfactuals which sound true. But Ellis comes close to saying that *all* counterfactuals are vacuous but assertible. That is a doctrine which has had its adherents – as has the extreme modal nihilism of some Quineans. But I am not alone in thinking that it is wrongheaded to treat all, or nearly all, counterfactuals as either vacuously true, or vacuously false, or else as somehow not in the business of being either true or false. Ellis's paper does not, I think, evince sufficient awareness of how high a cost he has to pay, if he has to treat almost all counterfactuals as vacuous. Especially when he rests so much of his theory on dispositional properties, *and* on their *essential* linkages to counterfactual conditionals!

It is important to see this objection to Ellis in context. On the one hand there may be new things to say about counterfactuals, which will remove his problems. And on the other hand there may be ways in which Ellis could alter his theory, so that standard treatments of counterfactuals would no longer give him any problems.

Ellis gets into trouble because he takes a very particular line on *which* properties are essential properties of natural kinds. He says that *having a certain disposition* is an essential property of a thing like an electron. If he said, rather, that the essential property were that of *having a disposition "intrinsically"* (*sense d*) – then the dialectical situation would be completely different. It would no longer be so difficult to find possible worlds in which antecedents of ordinary counterfactuals would be satisfied. These would just be worlds in which interfering factors put things into unnatural states, in which they lose the dispositions which they have "intrinsically(d)" – as when external forces distort the shape which a ball has "intrinsically(d)". The theory at which Ellis would arrive, by exploiting more fully his own originally discovered sense of "intrinsicality(d)", would be a theory which differs very little more than verbally from the Leckey theory, in Leckey and Bigelow (1995).

7. CONCLUSION

The highest priority for scientific essentialism is, I urge, to cleanse its exposition of ambiguities, and to get clear about exactly what it is saying. When you have explained the theory clearly, I reckon that it will be pretty obvious that it is true. Clear exposition of the theory is a hard enough task in itself. Mixing in pot-shots at the opposition may sometimes help warm up the debate, or to make the theory stand out more brightly by painting in some of the darkest parts of the background. But that is of secondary importance. And there is a risk attendant on giving too much prominence to arguments against rival theories. It directs too much attention onto those rivals and away from the Ellis theory.

The highest priority, for both Ellisian supporters and critics alike, should be to focus on what I call the "argument by display". It is important to get clearer, I urge, about the following things especially.

What should Ellis be interpreted as meaning, when he says that things "of a natural kind" have "an essential property"? I urge the answer: that it is impossible (*de re*) for things which belong to that natural kind to fail to have that property.

What is meant by a 'dispositional property'? I urge in answer, that: "dispositional properties" cannot be explained entirely in terms of *entailments* between, on the one hand, a thing's having the property, and on the other hand, the truth of certain conditionals. Rather, it must be explained, Euthyphro-style, in terms of whether a thing possesses the relevant property *in virtue of* the truth of certain conditionals. A dispositional property is one which is partly *constituted by* the holding of certain conditionals.

In what senses should dispositional properties be said to be "intrinsic"? Answer: it is best to avoid ever saying they are "intrinsic", unless you make it very clear which sense you have in mind. But, having taken that on board, Ellis would be very well advised to make appeal to his own newly-discovered notion of "intrinsicality(d)": dispositions are "intrinsic" properties of natural kinds in the sense that things belonging to these natural kinds would not be the things they are unless they have these dispositions *whenever* they are free of *distorting influences*.

Should the Ellis theory celebrate or mock the logical possibility of a global transubstantiation? I say: as a scientific realist, Ellis should embrace and celebrate the logical possibility of global transubstantiation.

What consequences does scientific essentialism have for the semantics of counterfactuals? I do not know, it's a problem: and this gives us a very good reason for looking much more closely at Leckey's heterodox essentialist theory (in Leckey and Bigelow (1995)), which (among various other virtues) can be seen to have much less of a problem than Ellis does in construing some everyday counterfactuals as nonvacuously true, and others as nonvacuously false.

Monash University

REFERENCES

Bhaskar, Roy (1975), *A Realist Theory of Science*, Atlantic Highlands, NJ: Humanities Press

Ellis, B. (this volume), 'Causal powers and laws of nature', 19–34

Goodman, N. (1955), *Fact, Fiction, and Forecast*, New York: Bobbs-Merrill

Harré, R. and Madden, E.H. (1975), *Causal Powers: A Theory of Natural Necessity*, Oxford: Blackwell

Humberstone, I.L. (1996), 'Intrinsic/extrinsic', *Synthese* **108**, 205–67

Jackson, F. (1977), 'A causal theory of counterfactuals', *Australasian Journal of Philosophy* **55**, 3–21

Kripke, S. (1980), *Naming and Necessity*, Cambridge, Mass.: Harvard University Press

Leckey, M. and J. Bigelow (1995), 'The Necessitarian Perspective: Laws as Natural Entailments', in *Laws of Nature: Essays on the Philosophical, Scientific and Historical Dimensions*, ed. F. Weinert, Berlin and New York: Walter de Gruyter, 92–119 (*principal author: Martin Leckey*)

Lewis, D. (this volume), 'Finkish dispositions', pp. 143–55

Lewis, D. (1979), 'Counterfactual dependence and time's arrow', *Nous* **13**, 455–76

Lewis, D., and Langton, R. (1998), "Defining 'Intrinsic' ", *Philosophy and Phenomenological Research* **58**, 333–45

Martin, C.B. (1994), 'Dispositions and conditionals', *Philosophical Quarterly* **44**, 1–8

Mellor, D.H. (1974), 'In defence of dispositions', *Philosophical Review* **83**, 157–81

Prior, E., Pargetter, R., and Jackson, F. (1982), 'Three theses on dispositions', *American Philosophical Quarterly* **19**, 251–7

BRIAN ELLIS

BIGELOW'S WORRIES ABOUT SCIENTIFIC ESSENTIALISM

John Bigelow's paper 'Scientific Ellisianism' is a more or less sympathetic critique of my development of the metaphysics of scientific essentialism – a programme on which I have been working for some years (much of the time with Caroline Lierse). The critique is focussed mainly on my paper 'Causal Powers and Laws of Nature', but many apparent references are made to other papers, both published and unpublished, in which aspects of the programme are discussed.[1] Bigelow's critique has made me unhappy with my original presentation, but rather than revise the paper and present a shifting target, I have chosen to leave it as it was, and try to explain my position more adequately and clearly in this reply. I think I have learnt quite a lot from this exchange, and I wish to thank John Bigelow for helping me to improve my understanding of the issues.

I. AIMS AND METHODOLOGY

The point of 'Causal Powers and Laws of Nature' was to argue that the scientific essentialist can offer a better perspective on causal laws than a Humean theorist can, and to defend the essentialist theory against likely objections. Specifically, my aim was to show that an essentialist can explain (a) what the causal laws of nature are (descriptions of natural kinds of processes), (b) what their truthmakers are (the causal powers, capacities and propensities of the objects necessarily involved in these processes), (c) what makes these laws necessary (because things of the kinds involved must, in appropriate circumstances, display the causal powers, etc. on which their identities depend), and to argue that (d) these theses can all be defended adequately against the most common objections. This, I should have thought, was enough for one paper.

My reason for choosing a Humean regularity theory to contrast with mine is that I see the Humeanism which informs it as the dominant metaphysic of current Anglo-American philosophy.[2] And this is my real target. The theory that I chose to represent Humeanism is one that is widely accepted.[3] I did not invent it; and it is not, as Bigelow suggests, "an extreme" form of Humeanism, but a fairly standard one. In any case, even if it were an extreme position, it would still have been pertinent for me to consider it. For any theory which concedes that the identity of a thing might depend, even partly, on how it is disposed to behave would not be truly anti-essentialist, and therefore not one which stands in sharp contrast with the kind of theory I want to defend.

H. Sankey (ed.), Causation and Laws of Nature, 61–75.

As the dominant metaphysic, Humeanism has broad implications for many areas of philosophy. It is not just a theory of science, or language, or of what exists; but it gives shape to contemporary theories in these and in many other areas. What I wish to argue for is also a metaphysic. For scientific essentialism, like Humeanism, has implications right across the board in philosophy, not only in the theories of causation, and laws of nature. However, the broader aim of elaborating and defending a new metaphysic is long-term. For a metaphysic cannot be judged as a more specific philosophical theory might be: because it is so wide-ranging, it has to be argued for in a different way. One cannot say: "Here are the problems; here is my solution"; and then argue that this solution is better than any other. For one thing, the problems shift from one metaphysical position to another. Thus, for a Humean, there are two serious problems about laws of nature: the necessity problem (i.e. to explain the nature, or the illusion, of physical necessity);[4] and the idealisation problem (i.e. to explain the existence of laws which appear to range over idealised systems, e.g. closed and isolated ones).[5] For an essentialist these are not serious problems; although there are problems for essentialism arising from illusions of possibility, some of which were considered in the paper.

Because of its basic role in shaping philosophical theories, a metaphysic has to be argued for, and defended, on many different fronts at the same time. This cannot be done satisfactorily in a single paper, as Bigelow says; what is needed is either a book-length study, or a series of papers. The only way I know of arguing for a metaphysical position, such as the one we are in the process of developing, is to set it out carefully, displaying its range and overall coherence, contrast it with other positions, show how it deals with philosophical issues in various fields, and answer specific objections. The form of the argument is, therefore, as Bigelow correctly notes, an argument by display. You show your wares, and invite people to buy them. If your system strikes your audience as being simpler, more coherent, or more promising than any alternative for dealing with the recalcitrant difficulties of other systems, then this may be a good enough reason to buy it.

In 'Causal Laws and Laws of Nature', my aims were somewhat more limited. I sought to show that there were serious problems with Humeanism (my ultimate target), which anyone defending a Humean theory of causality would have to deal with. I did not focus on any particular Humean regularity theory, such as Jack Smart's (1993) theory. For there is not much to be gained by showing that a particular Humean theory lacks credibility. What I sought to do was attack the underlying metaphysic.[6] Many others have shown, to my mind conclusively, that Humean regularity theories of causation, and of causal laws, are unsatisfactory. Rom Harré and Edward Madden (1975), Roy Bhaskar (1975) and David Armstrong (1983) all did so in their own ways. But their message is still largely unheeded, not because their arguments were not good, but because Humeanism is not just a theory about laws of nature, or causal connections. And the alternatives they proposed, viz. an essentialist theory of causation, and a theory of causal laws as relations between universals, could not easily be accommodated within a metaphysical system which was otherwise basically Humean.[7] Consequently, the message that the Humean theory of laws of nature is radically unsatisfactory still does not get through. Hume's theory of causation, and of causal laws, is seen as being in trouble, perhaps, but still, it is supposed, it must ultimately be defensible.

2. HUMEANISM, PASSIVITY AND THE TRANSUBSTANTIATION ARGUMENT

I take Humeanism to be a group of related theses: (a) that matter is essentially passive, (b) that the laws of nature are behavioural regularities of some kind, (c) that the laws of nature are contingent, (d) that causal relations hold between logically independent events, (e) that the identities of objects are independent of the laws of nature, (f) that the dispositional properties of things are not intrinsic, but depend on what the laws of nature happen to be. I do not say that every Humean holds all of these theses, or even that they all originate with Hume. But they do constitute what most philosophers would consider to be a Humean package.

Against the Humean, the scientific essentialist holds: (a) that matter is essentially active or interactive, (b) that the basic causal laws of nature are not descriptions of behavioural regularities, but of the ways in which things belonging to natural kinds are naturally disposed to act or interact, (c) that the laws of nature are strictly necessary, (d) that causal relations involve necessary connections between events, (e) that what makes a thing something of a given kind may depend on how it is intrinsically disposed to interact with other things, whether of the same or another kind, (f) that there are intrinsic dispositional properties of things which are the causal bases of their behaviour. It is, of course, not possible to argue for the whole of this alternative package here. I will focus just on those that are most relevant to the paper being discussed, and to Bigelow's reply.

The doctrine of the essential passivity of matter is older than Hume. But it is one of the core theses of seventeenth and eighteenth century mechanism. Most of the other theses of Humeanism can be seen to stem from it. The doctrine of passivity amounts to the claim that inanimate or dead matter can have no causal powers. Material objects can be acted upon by motive forces (*vis mortua*) which are external to them (or if internal, then external to the parts on which they act). But causal power could not reside in the objects themselves. *Vis viva*, which was Leibniz's measure of the causal power of an object, was considered by most natural philosophers of the eighteenth century to be a mathematical construct of no physical significance.[8] The real forces which restrain, or compel, bodies to move were thought to have their origins in the continuing and consistent actions of God (or perhaps in the occasional actions of other agents). Consequently, the way that a material body actually behaves in nature must depend on the laws of action, distribution and combination of the forces acting. And these laws, of course, are all contingent. God could have ordained that the same material things should move or be restrained by other forces acting on them in other ways.

However different their philosophies may otherwise have been, the belief that matter is essentially passive is one that was shared by Descartes, Locke and Newton, and by most of their philosophical contemporaries.[9] The main exceptions were the Leibnizians who argued for the reality of causal powers. But the Leibnizians were a minority, and the mechanistic theory of the externality of all forces prevailed for most of the eighteenth century. Hume went one step further than the mechanists, however, and denied not only the existence of causal powers, but also the reality of motive forces, and hence of all genuine causal connections.

Today's Humeans have followed Hume and the Newtonians in thinking that the world consists of ultimate parts (whatever they may be) which exist and have their being independently of the laws of nature. They also think of these ultimate material things as being essentially passive, just as the eighteenth century natural philosophers did. They are divided over whether the laws of nature are just regularities concerning the behaviour of these ultimate things, or whether these regularities depend on the existence of laws of action and distribution of forces, as the Newtonians believed. Those who believe that they depend on forces have, by and large, rejected the Newtonian belief that the forces that exist in nature must depend ultimately on divine agency. So the position they hold is a kind of secularised Newtonianism.

My use of the term 'Humeanism' in the paper was intended to include both classical Humeanism and secular Newtonianism. For both are committed to the essential passivity of matter, which I take to be the seminal doctrine of Humeanism. And this is the doctrine which I claim has come to dominate modern Anglo-American philosophy. It is implied, for example, in Bigelow and Pargetter's (1991) own conception of a Hume world. For, as they define it, the Hume world, *vis a vis* any given world, is a world exactly like the given world, *containing the same things with the same first order properties, and standing in the same first order relations to each other*, but without the causal connections, or other relations of natural necessity which distinguish laws from accidental generalisations (my paraphrase, pp. 279–80). This definition clearly presupposes that the identities of things are independent of their causal powers, since the possibility of there being such a world depends on the possibility of the *same* things with the *same* (first order) properties existing without any of their causal powers. It is not clear what properties Bigelow and Pargetter think could exist in a Hume world, but whatever they are, it must be possible for them to exist in different worlds governed by different laws (since they hold that the laws of nature are contingent).

Let us call those properties which can exist in a Hume world 'categorical properties'.[10] Then, since the properties of things in a Hume world can only be categorical, the things that exist in such worlds must all be without causal power, and so essentially passive, i.e. their identities must be independent of anything that they might be disposed to do. But remember, that for Bigelow and Pargetter, there is a Hume world corresponding to the actual world which contains the same things with the same first order properties. Therefore, everything that exists in the actual world is essentially passive.

Yet this doctrine of the essential passivity of things presents the modern Humean with some very nasty problems. Specifically, a classical Humean, for whom the world *just is* a Hume world, must say that causal laws are nothing but regular patterns of behaviour which happen to occur amongst things whose identities are independent of these laws. A Newtonian can believe that these regularities are brought about by the actions of God, or of forces whose distribution laws and laws of action are apt for the production of these specific regularities. But this commits modern defenders of essential passivity to some bizarre theses about reality, much more bizarre than most philosophers realise. For example, they must countenance the possibility of a wholly *transubstantiated world*. We are familiar with the concept of a Hume world, which is bizarre enough. But, to my knowledge, no one has pointed out that a Humean, or a secular Newtonian (it matters not which), is also committed to the possibility of *global transubstantiation*, i.e. to a world which, by every test that we could possibly make, is identical to this world, but which contains none of the same kinds of things. Such a world would be a law-governed world, and, as far as we were able to tell, the laws would be identical to the laws in this world. But it would still be a radically different kind of world from ours – one which contains *none* of the same substances, and is governed by *completely different* causal laws.

Bigelow misses the point of the transubstantiation example, and thinks my rejection of the possibility of transubstantiation is just Protestant bias. It is not, of course. Let the transubstantiation be global to make it perfectly clear that the point has nothing to do with Catholicism. The example of the Holy Eucharist was chosen, (a) because it is striking and memorable, (b) because it suggests, what I believe to be true, that the origin of the doctrine of the essential passivity of matter has more to do with religion than with science, and (c) because the quotation from the official records of the Council of Trent was just too good to pass over. The point concerns the relationship between the passive parts of things, on which their identities must ultimately depend, and the manifest ways that things behave, which depend on what the laws of nature are. The doctrine of the essential passivity of matter clearly makes it possible to preserve the manifest ways that things actually behave, while changing both the natures of the things on which forces act, and making compensating changes in the laws of distribution or action of these forces. For a classical Humean, who does not believe in forces or causal connections, it should be even easier to change the natures of things without changing how they behave. Essential passivity thus implies the possibility of global transubstantiation. Yet, this doctrine, which probably arose in the first place from the need to have a metaphysical system which can accommodate miracles, has, astonishingly, become the dominant metaphysic of the secular Anglo-American tradition.

Now some will say that my worries about Hume and transubstantiated worlds are just remnants of my positivist past, and that any healthy Humean who believes in the essential passivity of matter will resist this verificationist argument. Bigelow himself suggests as much. But here again I think he misunderstands me. One does not have to be a verificationist to appreciate that these are bizarre consequences of Humeanism. If you had a choice between two theories, one that had such consequences, and one that did not, then, other things being equal, which would you choose? I say that the rational person will choose the theory which provides the best explanation, and that the best explanation is one that does not have such bizarre consequences. This is not verificationism at work; it is a commonsense application of the epistemic value system involved in all theory choice. If such reasoning is disallowed, on the ground that it is verificationist, then science is disallowed, and so is scientific realism. 'Verificationism' is a naughty word, only if it is used to exclude meaningful discourse by declaring it to be meaningless. If, on the other hand, it is just the demand that we should prefer plausible theories for which there is decent evidence, to implausible theories for which there is no possible evidence, then I am all for verificationism. For without such a principle, science would be impossible, and scientific realism would be untenable. But this is the only sense in which my argument against Humeanism based on its bizarre consequences is verificationist. I have not argued that the Humean thesis of the essential passivity of matter is meaningless. I have only argued that it is *a very bad theory*, because it requires us to postulate the existence of unknown and unknowable entities (the passive realities) for which there is no good evidence in the first place. The essentialist theory that I advocate is just a much better theory, even if, according to the theory, certain miracles which are supposed to have occurred, could not possibly have occurred.

3. INTRINSIC AND ESSENTIAL PROPERTIES

There are at least three concepts necessary for an understanding of the essentialist perspective that I propose as an alternative to Humeanism. They are the concepts of intrinsicality, of natural kind, and of kind-essence. I will not say much here about natural kinds, because this concept is probably well enough understood for present purposes. It is important to note only that scientific essentialism is based on an ontology of natural kinds of objects, properties and processes. That is, it depends on the assumption that the most fundamental objects, properties and processes occurring in the world occur as members of natural kinds, and that the natural kinds that exist in each category are related to each other as the infimic species of a natural kind hierarchy.[11] Crucial for the purposes of this paper are the concepts of intrinsicality and kind-essence.

Bigelow complains about my use of the 'weasel' word 'intrinsic' in formulating scientific essentialism. However, the theory of intrinsic properties that I am using is clearly explained, although not as well as it might have been, in my 1992 paper on scientific essentialism (Ellis, 1992c). I am not responsible for the fact that there are many different, and mostly unsatisfactory, theories of intrinsicality around. In my 1992 paper I argued that what science needs is a *causal theory of intrinsicality*, as opposed to a logical one. In most of the unsatisfactory theories it is assumed that intrinsicality is a *formal property of properties*, and that we should therefore be able to distinguish *a priori* between intrinsic and extrinsic properties.[12] What is needed for science, however, is not an *a priori* distinction, but an *a posteriori* one which reflects the structure of causal explanations. Roughly, the distinction we need is between what is due to internal forces alone, and what is due, wholly or partly, to external forces or circumstances. This distinction is a fundamental one in the physical sciences, because abstract theoretical models of physical systems are usually designed specifically to represent the properties and structures of systems in causal isolation from their surroundings.[13]

For most philosophers, shape is paradigmatically intrinsic. For the shape of an object is logically independent of anything external to it. However, the shape of an object is not necessarily the shape it has *naturally*. Nor is its actual shape a very important property from a scientific point of view. It is much more important to know what the *natural shape* of the object is, what forces are acting upon it, and what its resistance co-efficients are to these various forces. The natural shape of the body, and the resistance co-efficients, are amongst what I should call the intrinsic properties of the body. For these are the properties of the body we need to know about, if we wish to understand how it will respond to external influences. So, from my point of view, the actual shape of the body is not an intrinsic property of it. Like a resultant force, it is really a superposition of shapes, of a distortion (produced by the action of the forces) on to the natural shape of the body. From the point of view of a causal theorist, therefore, the actual shape of the body is not intrinsic.

In general, the properties that I would count as intrinsic to a body are those that it has, or would have, independently of external influences. These properties are often not the manifest properties which the body has or displays. Rather, they are the properties which are the causal bases of the manifest properties. That is, they are the properties which, along with the external forces acting upon the body, are determinative of its manifest properties (e.g. its actual shape) and its manifest behaviour. The properties which constitute the intrinsic causal bases of an object's behaviour are its intrinsic dispositional properties. These include all of the body's intrinsic causal powers, capacities and propensities, and all of its intrinsic structural properties.

Some bodies have intrinsic properties which can change over time. Metals can suffer fatigue, structures can become altered, things can become more elastic, people can acquire beliefs they did not have before, their skills can deteriorate, their concerns for their fellows can diminish, and so on. So, not all bodies hold all of their intrinsic dispositional or structural properties necessarily. Some intrinsic properties are held contingently, and therefore, as we say, accidentally. However, there are some things, viz. those which belong to natural kinds, which hold some or all of their intrinsic properties necessarily. The sets of intrinsic properties which these things hold necessarily are the sets of properties on which their kind-identities depend. That is, they could not lose any of these properties without ceasing to be things of the kinds they are, and nothing could acquire any set of kind-identifying properties without becoming a thing of this kind. These kind-identifying sets of intrinsic properties are the ones I call "the real essences of the natural kinds", although they might, perhaps more correctly, be designated the 'kind-essences' of the individuals which possess them, since the bearers of these properties are undoubtedly individuals.

Besides having a kind-identity, an individual thing may also have certain properties in virtue of which it is the individual it is. That is, it may have certain properties which it could not lose without ceasing to exist, and which nothing could have acquired, except in the process of becoming that very individual. And often, when people speak of the real essence of a thing, it is the individual essence they have in mind. Saul Kripke, for example, talks of individual essences rather than kind-essences (Kripke, 1972).

Mostly it is assumed that the individual essence of a thing belonging to a natural kind includes its kind-essence. That is, the identity of an individual is supposed to depend on its being just the natural kind of thing it is. If this is right, then an individual of one kind could not possibly be transformed into something of another kind, although it might cease to exist and be replaced by something else. However, I am not at all sure that this is right. Certainly, there are limits to the possibilities of interspecific transformations. A horse could not become a cow. But when an atom loses an electron by b-emission to become an atom one greater in atomic number, it is not at all obvious that the former atom has just ceased to exist, or that a new atom has come into being at precisely the place where the first atom was. Most plausibly, the former atom has just lost a nuclear electron, and thereby changed its nature. For there is a powerful continuity argument to suggest that, as an individual, the former atom still exists, but now as an atom of another kind.

For these reasons, I am reluctant to accept, as Bigelow urges me to, that the individual essence of a thing belonging to a natural kind includes its kind-essence. Nor do I think that I have to accept this thesis if I am to provide a sound basis for scientific essentialism. Bigelow has picked up on my distaste for individual essences. As he says, "they set my teeth on edge". They do so, because individual essences would seem to have very little to do with kind-essences. The identity of something as an individual seems to depend primarily on its temporal and causal history, and therefore on its extrinsic, not its intrinsic, properties. Therefore, a separate argument would be needed to show that an individual cannot change its essential nature. If, as a matter of fact, things cannot change natures, or cannot do so except within very narrow limits, then this is presumably a fact about the kinds of things that exist in our world, rather than a necessary condition for individual identity.

Be that as it may, it is not necessary for me to suppose that the individual essence of a thing belonging to a natural kind necessarily includes its kind-essence. For the theory of natural necessity that I propose does not depend on it. To deduce the causal laws we need to know only what the kind-essences of the various natural kinds of things are, and *what the essential properties of the properties which constitute these kind-essences* are. For remember that the properties which constitute the kind-essences are intrinsic properties of the things which have them. And these intrinsic properties must include all of those intrinsic dispositions of the things of these kinds which they have solely in virtue of being things of these kinds. Therefore, the intrinsic dispositions of these things to interact with each other must be entailed by the fact of their being the kinds of things they are. Now, the causal laws concerning two natural kinds of things A and B is the set of ways in which things of the kinds A and B are intrinsically disposed to interact with each other, in virtue of their being the kinds of things they are. Of course, as Bigelow points out, things may not interact as they are intrinsically disposed to interact. For other forces may come into play. But then the laws of nature which we call 'causal laws' allow for this. The causal laws are not universal generalisations about how things actually behave, but necessary truths about how they are intrinsically disposed to behave.

Bigelow's mistake is to think that if the laws of nature do not derive from the individual essences of things, then they cannot be necessary *de re*. At best, they can only be necessary *de dicto*, if indeed they are necessary at all. But this is not so. The laws of nature that I call causal laws are all necessary *de re*. The laws in question are straightforward descriptions of the essential properties of the intrinsic dispositional properties which fundamental things must have in virtue of being things of the kinds they are. Let K_1 and K_2 be natural kinds, '$\Rightarrow$' be the connective 'if . . . then . . . ', 'C_i' the predicate 'are in the specific circumstances C_i within a range of possible circumstances C', and 'IE_i' the predicate 'are intrinsically disposed to interact in the manner E_i within the range of possibilities E'. Then the general form of a causal law involving two natural kinds of things K_1 and K_2 is:

L1. For all x, y and i, necessarily $[x \in K_1 \ \& \ y \in K_2 \Rightarrow (C_i x,y \Rightarrow IE_i x,y)]$

For example, let K_1 be the kind which consists of samples of pure hydrogen, K_2 the kind which consists of samples of pure oxygen, and i an index of circumstances in which such pure samples of hydrogen and oxygen are disposed to interact chemically. Then the causal law would be something like:

For all x and y, it is necessarily true that if x is oxygen and y is hydrogen, then if x is ignited in y, then x and y are intrinsically disposed to combine chemically to form water.

Note that the necessity operator which occurs in a law such as this, which spells out the essential properties of the intrinsic dispositional properties of the kind-essences of things, occurs within the scopes of the universal quantifiers, that is, in a *de re* position. If, as Bigelow insists, individual essences include kind-essences, then the causal law is certainly necessary *de re*. On Bigelow's thesis what we have is:

L2. For all x, y and i, [$x \in K_1$ & $y \in K_2 \Rightarrow$ necessarily ($C_i x,y \Rightarrow IE_i x,y$)]

Consequently, if we knew that $a \in K_1$ & $b \in K_2$ we could instantiate to a and b and detach the consequent to obtain 'necessarily ($C_i a,b \Rightarrow IE_i a,b$)'. However, such a strong essentialist claim entails that individuals which belong to natural kinds cannot, in any circumstances, change their natures. This is the thesis which Bigelow urges me strongly to accept. However, for the reasons given, I am not sure that I have to, or that I want to. I remain to be convinced.

The weaker thesis, represented by **L1**, has one clear advantage over **L2**. If **L1** is accepted, then it will be accepted on the basis that causal laws derive not from *the essential properties of individuals*, but from *the essential properties of the intrinsic properties* which individuals belonging to natural kinds must have, in virtue of being members of these kinds. The beauty of this is that it allows us to generalise the account of natural necessity, and to explain the necessity of those laws of nature which can be stated without reference to natural kinds of objects.

In our (1994c), Caroline Lierse and I argued for an ontology which includes natural kinds of properties as well as natural kinds of objects. Gravitational mass, for example, is a generic kind of intrinsic causal power. The infimic species of this generic kind are the specific gravitational masses which individuals may possess. Let M_1 and M_2 be the kinds consisting of individual objects of masses m_1 and m_2 respectively, and r(x,y) be the proposition that x and y are separated by a distance r. Then, analogous to **L1**, we have:

M1. For all x,y and r, necessarily [$x \in M_1$ & $y \in M_2 \Rightarrow$ ($r(x,y) \Rightarrow IE_r(x,y)$)]

where $IE_r(x,y)$ is the statement that x and y are intrinsically disposed to accelerate towards each other with accelerations proportional to m_2 and m_1 respectively, and inversely proportional to r^2. However, the law analogous to **L2**, viz.:

M2. For all x,y and r, [$x \in M_1$ & $y \in M_2 \Rightarrow$ necessarily ($r(x,y) \Rightarrow IE_r(x,y)$)]

is fairly evidently untenable, unless x and y happen to belong to natural kinds. For **M2** requires that the members of M_1 and M_2 have their specific masses essentially. They may do so; but fairly clearly they need not do so.

4. DISPOSITIONS AND CATEGORICAL REALISM

Bigelow argues that I misrepresent the position of categorical realists on dispositions, and claims that I exaggerate the distance between their theory of dispositions and the theory defended by Caroline Lierse and me (Ellis and Lierse, 1994). The categorical realist's position that I had in mind when writing the current paper was actually that held by David Armstrong (in Armstrong, 1978). However, as Bigelow says, it is arguable that the position taken later by Elizabeth Prior, Robert Pargetter and Frank Jackson (1982 [hereafter PPJ]), is not as far removed from our 1994 position as Armstrong's earlier one. I agree with this. Nevertheless, there is still a considerable distance between PPJ and us, and they are still basically in the categorical realist's camp.

PPJ argue that every behavioural disposition must have a causal basis. That is, they hold that there must be an intrinsic property or structure in virtue of which a given thing has the behavioural disposition in question. They are non-committal, however, about the nature of the causal basis, and they explicitly deny that it must be categorical (PPJ, 1982, p. 253). We have no reason to disagree with them so far about any of this. For it is no part of our theory that behavioural dispositions do, or could, just hang in the air, without being rooted in occurrent properties or structures. However, some (but not all) behavioural dispositions are grounded occurrent properties of the kind we call 'dispositional properties', i.e. properties whose identities depend on the behavioural dispositions they support.

Dispositional properties support behavioural dispositions in the following sense: If an object x has a dispositional property P, and x exists in any of the circumstances C_i which belong to its triggering range C, then it must, in virtue of having P, be intrinsically disposed to behave in the manner E_i, depending on C_i. Of course, an object x having P in the circumstances C_i need not *actually* behave in the manner E_i. Other forces could come into play which would frustrate the disposition. But if there are no other such forces, then the behavioural disposition to E_i in circumstances C_i must be displayed. Thus, on our view, the disposition is displayed because the object has the dispositional property, and exists in circumstances apt for its display. It does not have the dispositional property because the behaviour E_i happens regularly to be displayed in circumstances C_i.[14] On the other hand, a behavioural disposition such as this, need not, in our view, always have some other (non-dispositional) kind of causal basis.

Dispositional properties, like all other genuine occurrent properties must be the same in all possible worlds in which they exist. Fragility, which is assumed uncritically by PPJ to represent all behavioural dispositions, is not even a genuine property. There is a predicate 'is fragile' which is true of some things. But there is no unique *property* in virtue of which things are fragile. Certainly, there are occurrent properties and structures, indeed a great many of them, which make for fragility. But the fragile/non-fragile distinction is a social construct, useful for the purposes of warning people to take care, but of no use whatever in scientific explanation. There simply is no natural property of fragility. Long-stemmed glasses, ancient parchments, ecosystems and spiders' webs are all in their own ways fragile; but they have nothing intrinsic or structural in common to which the name 'fragility' could be attached. These things are grouped together just because they can all be broken or damaged easily.

Nor is fragility a typical behavioural disposition; for it is quite unlike any of the thousands of behavioural dispositions investigated in the natural sciences which are grounded in occurrent dispositional properties. The causal powers, capacities and propensities which feature in scientific explanations, for example, are all dispositional properties, in this sense. For each is supportive of a wide range of specific behavioural dispositions, often an infinite number of them; and for each of them it is true that their identities depend on the sets of behavioural dispositions they support. The refractivity of a certain kind of glass, for example, is a dispositional property of the glass which grounds infinitely many behavioural dispositions. The dispositions it grounds are characterised by a range of quantitatively different circumstances (different angles of incidence, different frequencies of incident light, different refractivity of the medium, etc.) resulting in a range of quantitatively different effects (different angles of refraction, total internal reflection, and so on). This is the sort of thing we have in mind when we speak of a dispositional property. It is a causal power, capacity or propensity which underlies a range of behavioural dispositions.

Dispositional properties such as these are clearly not categorical properties. For they could not survive changes to the laws of refraction, or to the laws of quantum mechanics from which they derive. However, PPJ claim not to be committed to the view that all behavioural dispositions must have categorical properties or structures as their causal bases. Is there, then, any reason why they should not just accept our analysis, and agree that the causal basis of a behavioural disposition might simply be a causal power? I think there is. The principal difficulty *for them* lies in their acceptance of the distinctness thesis. For this would commit them to saying that, in another possible world, the same causal power might ground a different set of behavioural dispositions, or conversely, that the same set of behavioural dispositions might be grounded in a different causal power. For, they insist, *the connections between behavioural dispositions and their causal bases are contingent.* They are contingent, they suppose, because the laws of nature are contingent. To accept that there are necessary connections between behavioural dispositions and their causal bases is precisely what they, as Humeans, cannot do. Therefore, despite their reticence about saying that the causal bases of behavioural dispositions must be categorical, they are nevertheless committed to the view that they are. For, given their distinctness thesis, the causal basis of a disposition must be able to survive changes in the relevant laws of nature.

In reply, PPJ might be willing to accept the argument so far, but argue that my so-called dispositional properties *are not the real underlying causal bases for the behavioural dispositions they ground.* For refractivity is itself a disposition which plausibly requires a causal basis. What then might PPJ have in mind if they were to speak of the causal basis of a disposition such as refractivity? What they would say, I think, is that the causal basis must lie in the molecular structure of the glass, a structure which, they might say, can be characterised without reference to the laws of refractivity or of quantum mechanics. However, what we might say by way of characterisation is one thing, what the reality is, is another. For the supposed causal basis of refractivity is not categorical in the sense that it could survive even the slightest changes to the laws of nature relevant to refractivity. In reality, the molecular structure, indeed the molecules themselves, depend ontologically on these laws. So, the molecular structure that is undoubtedly the causal basis of refractivity just could not exist at all in a world in which the relevant laws of nature were different.

5. VACUOUS CONDITIONALS

Vacuous conditionals are admittedly a problem for scientific essentialism. But the problem may not be as serious as Bigelow suggests, and perhaps it is only a problem if one thinks that our intuitive judgements of logical possibility are reliable. As a scientific essentialist I have to argue that such judgements are generally unreliable. They are so for at least two reasons: first, because we have been systematically trained as philosophers to think that whatever is imaginable is possible; and second, because it is so easy to confuse epistemic possibility with real, or metaphysical, possibility.

Take almost any paper written in the Anglo-American tradition of philosophy, and in it you will find examples of allegedly possible states of affairs. These supposedly possible states of affairs, which are drawn from the imagination, are the examples and counterexamples used standardly by philosophers in their arguments for the philosophical positions they hold. But the imagination works without scientific constraint, and hence without having to consider the sorts of limitations on possibility that science imposes. Consequently, the examples used by philosophers often seem absurd to scientists, and to others who are not well trained in philosophy, and it has to be explained patiently that, as philosophers, we are concerned, not with *physical* possibility, but only with *logical* possibility, the test for which, apart from there being no obvious contradiction in describing the case, is just imaginability or conceivability. At any rate, what is possible is supposed to have nothing to do with what is possible from a scientific point of view.

The methodology of testing philosophical positions by using imaginatively constructed, but scientifically implausible, or even physically impossible, counterexamples is perhaps a reasonable one if you are a Humean, and believe that all things are loose and separate. For, in a Humean world, anything that is consistently imaginable is possible, because the laws of nature in such a world can easily be changed to make it conform to the way we imagine it to be. But, in the kind of world that science has revealed to us, we can no longer be so sanguine about the looseness or separateness of things; and our imaginations, which are necessarily superficial and cartoon-like in their representations, can no longer be considered to be reliable sources of information about real possibility. A scientific essentialist has to be much more cautious than a Humean when making judgements about what is really possible.

A scientific essentialist cannot rely on arguments from ignorance to establish that anything is really possible. For the fact that something is not known to be impossible does not imply that it is possible in the required metaphysical sense. It only shows that it is 'epistemically' possible (i.e. 'possible for all we know'). But, as I argued in the paper, epistemic possibility is a different concept from real possibility, even though it is related to it. Nevertheless, epistemic possibility is a concept that is widely used, and is undoubtedly one that is important in connection with the theory of conditionals.

It is in connection with the theory of conditionals that Bigelow sees trouble for scientific essentialism. He is not worried that some counterfactual conditionals with antecedents which are obviously physically impossible, like "If we could walk on water, then we should have less need for boats", should turn out to be vacuous. What concerns him is that many quite ordinary-looking counterfactuals, like "If I had a cold beer in front of me, then I should drink it" might well turn out to be vacuous. For, on Lewis's theory of counterfactuals (Lewis, 1979), the truth of this proposition requires that there be a possible world with the same laws, and with a past that is at least very like ours, except that, miraculously, there is now a cold beer in front of my counterpart in this world. The miracle required might only be a small one, but it is a miracle nevertheless, and its occurrence would require that some causal laws be violated in a possible world which is retrospectively otherwise much like ours.

There are two ways of dealing with this objection. The first is to accept that most counterfactual, and perhaps even most indicative, conditionals might turn out to be vacuous, and to argue that what matters for conditionals is not whether they are true, but whether they are assertible. For example, we might argue that conditionals are assertible if the information they provide indirectly about the world, in particular about the dispositional properties of things, is both true and appropriate. The second is to provide realistic truth semantics for conditionals along the lines of Lewis's possible worlds semantics, but substituting genuinely possible worlds for the merely imagined worlds of Lewis's theory. However, I am persuaded that this strategy does not work, and that it is not generally possible to specify adequate, realistic, possible worlds semantics for conditionals. Consequently, I think we shall just have to settle for assertibility conditions for conditionals, and forget about truth conditions.[15]

The attempt to provide adequate real possible world semantics for conditionals fails because the evaluation of conditionals does not require us to consider what exists in other really possible worlds. A knowledge of the dispositional properties of things, of the actual circumstances, and a good imagination, are all that are needed. The causal conditional '$C_i x,y \Rightarrow E_i x,y$' is assertible, for example, iff

(1) in the imagined circumstances, arrived at by making whatever changes to the actual circumstances are minimally required to accommodate the antecedent supposition $C_i x,y$, a dispositional property triggered in these circumstances, and having the consequent $E_i x,y$ as its outcome, exists,
(2) there are no countervailing dispositional properties possessed by any of the individuals in the imagined circumstances, and also triggered in these circumstances, which are strong enough to overcome or swamp the display of the dispositional property having the outcome $E_i x,y$.

But what about truth conditions? We can easily imagine things to be the case which could not possibly be so. Therefore, we cannot just assume that the above assertibility conditions for causal conditionals are also truth conditions. The truths relating to causal conditionals are the underlying ones on which their assertibility depends. That I am thirsty, for example, is a fact about me. It is also a fact that I like beer, and that there is nothing in the world that I would like better at the moment. Consequently, I would say, "If there were a beer in front of me I would drink it." It simply does not matter whether it is really possible for there to be a beer in front of me. And if this makes it vacuous, then it does not matter whether it is vacuous. By asserting the conditional I tell you graphically what my desires are at the present time, and what you could do to satisfy them. It is better than saying "I am thirsty", because you might then offer me water, which is not what I want most. It is better than saying "I am thirsty, and I like beer", because this is compatible with my not wanting a beer at the moment.

La Trobe University and University of Melbourne

NOTES

[1] Others include Bigelow, Ellis and Lierse (1992), Ellis (1992a, 1992b, 1992c and 1996) and Ellis and Lierse (1994a, 1994b and 1994c), Lierse (1996), and a number of unpublished papers.

[2] I am, of course, aware that there are other theories about laws of nature than the two that are contrasted here. There is, for example, the conventionalist theory, according to which the laws of nature are conventions adopted for their utility in organising or structuring our knowledge. On this theory, laws of nature are, strictly speaking, neither true nor false, and if, as a façon de parler, we should say that they are true, then we can mean only that they are the best for this purpose. My colleague, Caroline Lierse, and I have discussed this, and also David Armstrong's (1983), theory about laws of nature, in an unpublished paper entitled 'Three Problems about Laws of Nature' (1994b).

[3] For example, by J.J.C. Smart (1993) in his paper on 'Laws of Nature as a Species of Regularities'.

[4] Bas van Fraassen (1989) calls this 'the identification problem', because for him the problem is to identify the relationship in virtue of which a universal generalisation acquires nomic status. However, this is like looking for a relationship which will convert a material conditional into a subjunctive conditional, and this is just the wrong way of looking at it. Subjunctive conditionals are essentially modal; they are not material conditionals which have somehow acquired modality.

[5] These two problems are discussed in Ellis and Lierse (1994b).

[6] The criticisms I have to make of Humeanism were developed in other ways in Caroline Lierse's (1996) paper.

[7] First order logic, for example, is a logical theory arising out of a Humean world-view. It is not a logical theory which can readily be adapted to reasoning about universals, or about necessary connections in nature. I do not say that more elaborate logical systems that could do the job adequately cannot be constructed. It is just that a lot of work would need to be done in logical theory to accommodate either of these non-Humean theories of laws of nature.

[8] See Larry Laudan (1968) for an account of the eighteenth century attitudes to *vis viva*. Today's Humeans are not, of course, so reluctant to admit the physical significance of modern descendants of Leibnizian force concepts, such as energy. Nevertheless, they still think about causal interactions in a neo-Newtonian way as being due to forces acting externally on the objects involved, rather than as transformations involving several kinds of energy (including mass-energy) occurring within a system of objects according to their intrinsic dispositional properties to undergo such transformations.

[9] For a scholarly and informed account of the metaphysics of eighteenth century science in Europe, and a clear exposition of the doctrine of the essential passivity of matter, see Carolyn Iltis's excellent paper (Iltis, 1973). She writes: "Newton, in sharp contrast with Leibniz, was concerned with demonstrating the passivity and non-activity of matter. The active principles which provide the source of motion in bodies were external to matter. He held that matter depended ultimately on the will of God for its existence and motion" (p. 351).

[10] Categorial properties have never been clearly defined in the literature, and the concept of a categorical property as any property that could exist in a Hume world is as good as any. The properties that Caroline Lierse and I (Ellis and Lierse, 1994a) have called dispositional are not categorical in this sense, because they are, by definition, the very sorts of properties that could not exist in a Hume world, viz. causal powers, capacities, propensities, and the like. Bare dispositions could indeed exist in a Hume world, because bare regularities could exist in such a world. But bare dispositions are not occurrent properties.

[11] This ontology is explained and developed in Ellis and Lierse (1994c).

[12] The best known theory of this kind is probably Jaegwon Kim's theory of internal properties (Kim, 1982). Kim's definition of an internal property is equivalent to the following:

G is intrinsic =df possibly some object x has G at t, although no contingent object wholly distinct from x at t exists.

[13] As Bhaskar (1975) argues so persuasively.

[14] This is my answer to Bigelow's worries stemming from the Euthyphro.

[15] This does not worry me as much as it might worry someone like John Bigelow, because the logic of conditionals that I have developed elsewhere (Ellis, 1979) is included in a theory of rational belief systems which does not require an underlying truth semantics. Nor should it worry Frank Jackson who bit a similar bullet some time ago when he argued that conditional probability does not measure the probability of truth

of an indicative conditional, but the degree to which it is assertible (Jackson, 1979). The probability of its truth, he said, is the probability of the corresponding material conditional – a claim which can be shown to have some strongly counter-intuitive consequences (Ellis, 1984). But see also Jackson (1984).

REFERENCES

Armstrong, D.M. (1978), *Universals and Scientific Realism*, 2 Vols. Cambridge: Cambridge University Press

Armstrong, D.M. (1983), *What is a Law of Nature?*, Cambridge: Cambridge University Press

Bhaskar, R. (1975), *A Realist Theory of Science*, Atlantic Highlands, NJ: Humanities Press

Bigelow, J.C. (this volume), 'Scientific Ellisianism', 45–59

Bigelow, J.C., Ellis, B.D. and Lierse, C.E. (1992), 'The World as One of a Kind: Natural Necessities and Laws of Nature', *British Journal for the Philosophy of Science* **43**, 371–88

Bigelow, J.C. and Pargetter, R.J. (1991), *Science and Necessity*, Cambridge: Cambridge University Press

Ellis, B.D. (1979), *Rational Belief Systems*, Oxford: Basil Blackwell

Ellis, B.D. (1984), 'Two Theories of Indicative Conditionals', Part 1 of a symposium on indicative conditionals, *Australasian Journal of Philosophy* **62**, 50–66

Ellis, B.D. (1992a), 'Idealisation in Science', in *Intelligibility in Science*, edited by Craig Dilworth, *Poznan Studies in the Philosophy of the Sciences and the Humanities* **26**, Amsterdam-Atlanta, GA, 265–82

Ellis, B.D. (1992b), 'Scientific Platonism', an essay review of *Science and Necessity* by John Bigelow and Robert Pargetter, *Studies in the History and Philosophy of Science* **23**, 665–79

Ellis, B.D. (1992c), 'Scientific Essentialism', La Trobe University: Victorian Centre for the History and Philosophy of Science Preprint Series **3/92**. Paper read at the 1988 Annual Conference of the Australasian Association of History, Philosophy and Social Studies of Science, Perth, Western Australia

Ellis, B.D. (1996), 'Natural Kinds and Natural Kind Reasoning', in P.J. Riggs (ed.) *Natural Kinds, Laws of Nature and Scientific Methodology*, Dordrecht: Kluwer Academic Publishers, 11–28

Ellis, B.D. and Lierse, C.E. (1994a), 'Dispositional Essentialism', *Australasian Journal of Philosophy* **72**, 27–45

Ellis, B.D. and Lierse, C.E. (1994b), 'Three Problems about Laws of Nature', Unpublished manuscript

Ellis, B.D. and Lierse, C.E. (1994c), 'The Fundamental Importance of Natural Kinds', La Trobe University: Victorian Centre for the History and Philosophy of Science Preprint Series **3/94**

Harré, R. and Madden, E.H. (1975), *Causal Powers: A Theory of Natural Necessity*, Oxford: Blackwell

Iltis, C. (1973), 'The Leibnizian-Newtonian Debates: Natural Philosophy and Social Psychology', *The British Journal for the History of Science*, **6**, No. 24, 343–77

Jackson, F.C. (1979), 'On Assertion and Indicative Conditionals', *Philosophical Review* 88, 565–79

Jackson, F.C. (1984), 'Two Theories of Indicative Conditionals: A Reply to Brian Ellis', Part 2 of a symposium on indicative conditionals, *Australasian Journal of Philosophy* **61**, 67–76

Kim, J. (1982), 'Psychophysical Supervenience', *Philosophical Studies* **41**, 51–70

Laudan, L.L. (1968), 'The *Vis viva* Controversy, a Post-Mortem', *Isis* **59**, 131–43

Lewis, D.K. (1979), 'Counterfactual Dependence and Time's Arrow', *Nous* **13**, 455–76

Lierse, C.E. (1996), 'The Jerrybuilt House of Humeanism', in P.J. Riggs (ed.) *Natural Kinds, Laws of Nature and Scientific Methodology*, Dordrecht: Kluwer Academic Publishers, 29–48

Prior, E.B., Pargetter, R.J. and Jackson, F.C. (1982), 'Three Theses about Dispositions', *American Philosophical Quarterly* **19**, 251–7

Smart, J.J.C. (1993), 'Laws of Nature as a Species of Regularities', in J. Bacon, K.K.Campbell and L. Reinhardt (eds.), *Ontology, Causality and Mind: Essays in Honour of D.M. Armstrong*, Cambridge: Cambridge University Press, 152–69. Reply by D.M. Armstrong, *op. cit.*, 169–74

van Fraassen, B.C. (1989), *Laws and Symmetry*, Oxford: Clarendon Press

MARTIN LECKEY

THE NATURALNESS THEORY OF LAWS

If I were to drop an apple, then it would fall. It is not possible that it would fly upwards. It is necessary that it would fall. I think these statements are true, but at the same time I believe that it is not logically necessary that the apple would fall. I believe that there is in nature a kind of necessity weaker than logical necessity: natural necessity.

This view contrasts with the views of Ellis and Lierse. According to Ellis and Lierse (Ellis, this volume [a]; Ellis and Lierse (1994)), things behave the way they do because of the kinds of things they are. Furthermore the identity of those things depends rigidly on their behavioural dispositions: Ellis and Lierse suppose that the behavioural dispositions are essential to the kinds of things they are so that anything of that kind could not fail to behave according to these dispositions without either ceasing to be something of that kind or ceasing to exist altogether. Thus, they say, the laws of its behaviour must be necessarily true. According to this view, which they label a variety of "scientific essentialism", the way things behave is grounded in the essential natures of the fundamental natural kinds of objects and processes occurring in the world. On this view, the causal laws describe the operation of the essential properties of these natural kinds and are logically necessary.[1] This is quite a radical view. The more usual view, in both everyday usage and in philosophical accounts of laws, is that laws of nature are contingent rather than necessary: on the usual view, in many other possible worlds the laws differ from the laws in our world. Not only must the laws be the same in all possible worlds, according to Ellis and Lierse, but also exceptions to the laws of nature are logically impossible. For instance, it is logically impossible for water to become wine in, say, a half a minute: if something which seemed watery were to become winey in half a minute then it could not have been water that we started with, and it would not be wine that it had become.

The theory I will present here, which I will call the naturalness theory, is more complicated than the Ellis and Lierse theory. It allows for the contingency of laws, and the logical possibility of miracles. For me, the laws are not products of the essences of physical things or physical properties on their own, although I have sympathies with this general approach, of grounding laws in the propensities and dispositions of the world as a whole, as well as things in the world. At the 1993 Australasian Association of Philosophy conference, John Bigelow and I presented a paper defending a version of the Ellis/Lierse strong theory, and David Lewis impressed us with an objection. Lewis wanted to know if our theory left him any space to make sense of ordinary,

H. Sankey (ed.), Causation and Laws of Nature, 77–82.

run-of-the-mill counterfactual suppositions: such as, "If she could have walked on water, then she would not be so wet". We became convinced that Lewis was right to think that it was at least logically possible for things to break the laws of nature. This was one of the driving forces behind my development of the naturalness theory, a variant on the theory we had developed up until then.

One way to put my idea is to say that although things can break the laws of nature, if and when they do then there is something very different, something "unnatural", going on. Sometimes things are behaving entirely naturally: and in those contexts everything will obey the laws of nature. As long as things are in an entirely *natural* state, it follows conditionally on that naturalness that they will obey the laws of nature – if they did not obey the laws, then they could not consistently be supposed to be in an entirely natural state.

According to this idea, the laws are not grounded on the essences of physical things and properties in themselves, as Ellis and Lierse maintain, but on the essence of an extra property, one which is shared in common by all and only those things which are in a "natural" state. Things obey the laws of nature only when they have this extra property in common.

I assume that the property of naturalness is not an essential property of things. Even if things are, in the actual world, always in natural states, they would not cease to be the things they are if they became involved in a miracle. Naturalness would be an essential property for some given thing just in case this thing could not be the thing that it is if it were not in a natural state, if the only way this thing could lose the naturalness property is by ceasing to exist altogether. However, naturalness is not, I think, an essential property of any of the things I can think of – or any of the actual things which are, at least for most of their careers, governed by the laws of nature. Though naturalness could be an essential property of some things, it is not to be laid down in advance that it would have to be an essential property of anything which had it. Laws should be defined in such a way as to allow the logical possibility of things breaking the laws. If it were logically impossible for actual things to break the laws, that impossibility would have to be grounded on a very strong metaphysical thesis concerning the essences of actual things; it should not be grounded on just the definition of laws alone.

I assume that although it is not an essential property of things to have the naturalness property, things at least have a very strong propensity to take on the property of naturalness. It is part of what it is to be a natural property for our world, an essential property of it, that all things, or at least things in an appropriate realm, including the realm of most things we are aware of, have a strong propensity to have this property. It is because of this propensity that we are right to be confident that things we experience will continue to have this property of naturalness, and would have the property of naturalness in various counterfactual circumstances. The modal force of the laws of nature comes from this primitive propensity for the naturalness property to be instantiated by things in our world.

Ellis supposes that the behavioural dispositions are essential to the identity of kinds and properties. I agree that certain behavioural dispositions are essential to the identity of many properties. However, I do not accept that this position will lead us to suppose that laws of nature are logically necessary. This is because the kinds of necessary truths about properties that Ellis and Lierse identify have no direct consequences for whether individuals in our world instantiate those properties or not, and I argue that in order for something to count as a law of nature for our world it must have consequences for what properties are instantiated by individuals in our world.

Consider for example Newton's law of gravitation. Ellis (this volume [b]) claims that the inverse square law of gravitation can be grounded in the essential properties of the property "gravitational mass". Ellis supposes that it is essential to the notion of gravitational mass that things with gravitational mass are intrinsically disposed, in certain circumstances, to accelerate towards each other with accelerations proportional to their masses and inversely proportional to the square of the distance between them. This claim could certainly be questioned. In physics text books 'gravitational mass' is defined fairly narrowly in terms of gravitational force, so that while it is (arguably) essential to the concept of gravitational mass that the objects' accelerations toward each other be proportional to their mass, under the appropriate circumstances, it is not essential that their accelerations be proportional to the inverse square of the distance. On the text book definition, something could still retain the same gravitational mass if it obeyed an inverse cube law of gravitation rather than an inverse square law. However suppose I agree, for the sake of argument, to identify the concept mass, including gravitational mass, by its role in a large network of causal roles that it enters into in our world, including the inverse square law of gravitation. In that case I would be willing to accept that it is essential to the concept of mass that things with mass are disposed to obey the inverse square law under appropriate circumstances. I would agree that it is necessarily true, true in all possible worlds, that "things with mass, in certain circumstances, are intrinsically disposed to accelerate towards each other with accelerations proportional to their mass and inversely proportional to the square of the distance between them". However I do not accept that this necessary statement captures a law of nature. The problem as I see it is that this statement is true in worlds in which there is no mass, or only some objects with mass, the rest of the objects having "shmass", where shmass is a property like mass except that it follows an inverse cube law of gravitation rather than an inverse square law. On the Ellis view, as presented in this case, it is a purely accidental fact, independent of the law of gravitation, that mass is instantiated both here and in the Andromeda galaxy, and everywhere else in the universe. The fact that only objects with mass are found in our world – the fact that all things in our universe obey the inverse square law of gravitation (as far as we know) – is normally considered to follow from the law of gravitation rather than being independent of it.

I will make the condition that for something to qualify as a law of gravitation of our world, or any other law of our world, it must follow from it that all objects in our world obey that law (in the range of circumstances in which the law is applicable), or at least that there is a very strong propensity for things in a certain realm to obey that law, including most of the things we are aware of. (This qualification allows for the logical possibility of miracles, and for the possibility that things in certain regions, or certain types of things, are not disposed to obey that law.) The Ellis account of the law of gravitation fails this test. On the other hand, the naturalness theory of laws does meet this condition. On the naturalness account of the law of gravitation, it is a natural state of things in our world for those things, under the appropriate circumstances, to accelerate towards each other with accelerations proportional to their masses and inversely proportional to the square of the distance between them. There is a very great propensity for natural states to be instantiated by things in our world, and this explains why most things we are aware of, and perhaps all things, obey the law of gravitation.

The naturalness theory holds that there exist natural states, and when a thing is in a natural state then it follows, as a matter of logical necessity that it will obey the laws of nature. The laws of nature are all bound up in the essence of the property of naturalness.

What kind of a thing could this property of "naturalness" be? It is a primitive of the theory, but some of the properties of the property of naturalness have already been indicated: individuals have this property when they instantiate only natural states, and it is an essential property of natural states that they have a very great propensity to be instantiated by many things in our world, including most of the things we are aware of, so that most things we are aware of share the naturalness property. Other properties of the naturalness property will depend on the kind of world we are in. It is important to keep a wide range of possibilities in mind, if we are to get a grip on the concept of laws of nature – I wish to be able to include, as possibilities for the way laws are, many of the conceptions of laws of nature that have been thought of throughout history. One possible reading for the property of naturalness that goes at least some way to elucidating it is that the naturalness property is the property of instantiating only those states that can be derived from the natural states of the fundamental particles and fields. This reading of the property of naturalness might be appropriate in the case that all laws supervene on the laws of fundamental physics. Keep in mind the possibility that the property of naturalness could be a *relational* property. One thing which could be shared in common by many things is, for instance, the property of *being created by God*. All things created by God share the relational property of having been brought into existence by an all-powerful being with such-and-such a plan.

There are two ways a common origin in God could bestow a common property on things – one top-down and the other bottom-up. God could cause each thing he creates to have a nature which interacts harmoniously with all of his other creations so that they behave according to his plan – this is a bottom-up picture, with God as common cause underlying the natures of things. Another picture is that of the Cartesian occasionalists like Malebranche and Berkeley, according to which physical things are inert, passive, and lacking in any intrinsic natures (apart from their passivity). On this view things fall into regular patterns because of the continuous intervention of God, who causes everything to happen according to his general plan. This is a top-down conception par excellence. It can be subsumed as a special case of the naturalness theory. I remain officially agnostic between top-down and bottom up versions of the naturalness theory.

If there were a hierarchy of naturalness properties, there would be an ambiguity in the term 'laws of nature'. There could be, for instance, a common nature to all the things created by God, and there could be various rough-and-ready regularities which emerge as a consequence of that. But sometimes, we may suppose, the propensities of these things could be overridden by God. Yet, we may suppose, God does not share in the same nature as the things he created. So when God intervenes and causes a miracle, there is one sense in which this event is unnatural, and the laws of nature are broken. Yet in another sense it is perfectly natural for God to do this. And in a sense it is perfectly natural for the things created by God to behave as God wills. What is unnatural with reference to one intrinsic nature of things, the nature which generates such regularities as we do see in the natural world, is perfectly natural with reference to another intrinsic nature, their natural propensity to do as God wills.

It is not only a theological context which can give rise to distinct natures in the world. As another example, consider the concept of the supernatural. Suppose there are supernatural powers such as human psychic powers or the powers of other agents such as spirits. Then we might suppose that those ordinary processes that occur in the absence of intervention by these powers share some common nature, and from the point of view of this nature the supernatural powers break the laws of nature, giving rise to non-natural states. However, it is not obvious that the supernatural must always fail to be law-governed. If these powers are exercised with any kind of pattern there may well exist law or laws to encompass them. Thus on our view this would simply be to discover more laws of nature, and some former laws may need to have their circumstances of application limited to those where the powers are not exercised. This means that so-called supernatural events and processes would also have some common nature which they would share with more ordinary events, at a deeper level, and from the point of view of this nature, these processes and events are perfectly natural. Under this view of the natural, the only truly non-natural states of affairs are ones where laws that previously held were broken by an "anomaly" – something, whether "natural" or "supernatural", that did not share the property, if there is one, which underlies all law-abiding phenomena.

Laws of nature exist only to the extent that things in the world share in common a fundamental nature, at some level. If these laws are to be exceptionless and universal, then this universal, this common nature, must be instantiated by everything which exists, without exception. If it is right, then we can expect there to be non-trivial exceptionless regularities in the world. However, our conception of laws should be distinguished from this supposition.

The naturalness conception of laws leaves space for magic and miracles, sheer randomness and anomalies, events which break the laws. It leaves space for different laws to operate in different parts of the world: things may have different propensities, and different pecking orders of propensities, in different parts of the world. For similar reasons the laws of nature might change with the passage of time. There is a big gulf between the laws of nature and the regularities which will emerge as a consequence of those laws.

Let me conclude by briefly illustrating the formal representation of the laws of nature. I have been convinced by the arguments of Dretske, Tooley, and Armstrong that a law should take the form of a relation between universals. The formal account presented below expresses laws as relations between universals, while at the same time capturing the idea of basing laws on the property of naturalness. The formal account of laws follows that presented in Leckey and Bigelow (1995), where it is described in more detail. As will be shown, the formal account can be recast in the form of a relation of "natural necessitation" between universals, so that it resembles

Tooley's (1987) formal account of laws, although our formulation has the advantage that we can give some analysis of that relation.

The general form of a law can be written as follows:

$\hat{x}(Cx \;\&\; N_{@}x)$ entails $\hat{x}(Rx)$.

When this entailment holds between properties of these kinds, then it is a law that things in circumstances C manifest state R. The condition $N_{@}x$ is the naturalness condition, which can be read "x is natural" or more fully "x instantiates only natural states for the actual world."

This could be written

$\hat{x}(Cx)$ naturally entails $\hat{x}(Rx)$,

where natural entailment is defined as logical entailment conditional on the supposition that the system x involved in the law is natural. Alternatively put, it is a matter of natural necessity that state R is instantiated given that state C is instantiated, or C naturally necessitates R. This provides an analysis of a relation of natural necessitation in terms of the naturalness condition.

From the above account of laws it can be seen that the laws of our world can be characterized by the states that are natural for our world and the states that are non-natural for our world. For if the state $\hat{x}(Cx \;\&\; Rx)$ is natural and the state $\hat{x}(Cx \;\&\; \sim Rx)$ is non-natural, then the general form of the law given above follows. Similarly, the laws of other possible worlds can be characterized by the natural states and non-natural states for those worlds. Any world that has the same natural and non-natural states as our world will have the same laws of nature as our world, and any world with different natural or non-natural states will have different laws of nature.

Monash University

NOTES

[1] Ellis uses the expression 'metaphysically necessary' rather than 'logically necessary'. In this paper I will be following the practice of taking 'logically necessary' to be synonomous with 'true in all possible worlds'. According to Ellis, if something is metaphysically necessary then it is true in all possible worlds, so this means that it is logically necessary, under the definition used here.

REFERENCES

Ellis, B.D. (this volume [a]), 'Causal Powers and Laws of Nature'

Ellis, B.D. (this volume [b]), 'Bigelow's Worries About Scientific Essentialism', 61–75

Ellis, B.D., and C.E. Lierse (1994), 'Dispositional Essentialism', *Australasian Journal of Philosophy* **72**, 27–44

Leckey, M.J., and J.C. Bigelow (1995), 'The Necessitarian Perspective: Laws as Natural Entailments', in *Laws of Nature*, edited by F. Weinert, Berlin: de Gruyter, 92–119

Tooley, M. (1987), *Causation: A Realist Approach*, Oxford: Clarendon Press

CAROLINE LIERSE

NOMIC NECESSITY AND NATURAL STATES: COMMENT ON THE LECKEY–BIGELOW THEORY OF LAWS

It is a widely held belief that the laws of nature, although necessary in a restricted sense of the word, could have been different from what they are in this world. This view that the laws may have been otherwise is known as the *contingency theory of laws*. In recent years, however, the domination of this popular doctrine has been challenged by a new analysis of lawhood that denies the contingency thesis. This new analysis, known as Essentialism, rebuts the claim that the laws of nature are contingent. According to the essentialist doctrine, laws are *necessary* in that they are grounded in, and thus dependent on the essential natures of the fundamental natural kinds of objects and processes that occur in the world.

Leckey and Bigelow (1995; Leckey, this volume) have recently offered a new theory of lawhood, which they call the "naturalness" theory. Although they claim to sympathise with the essentialist analysis, they reject the central tenet of an essentialist theory which grounds the necessity of laws in the intrinsic natures of properties, on the grounds that it is, in principle, logically possible for things to break the laws of nature. A decent theory of laws, they maintain, should be able to accommodate strange events that violate the laws of nature in this world, such as, for instance, water instantaneously turning into wine. Their commitment to the view that properties do not necessarily have to behave in the way that laws in our world dictate, renders their theory a contingency theory of laws.

As far as I can discern, the Leckey–Bigelow theory is built on a conventional ontology of categorical and dispositional properties. And like the traditional necessitarian contingency theories of Dretske, Tooley and Armstrong, the necessity is "top-down". That is, there is nothing in the properties *themselves* that compel them to behave in a particular kind of way. The laws are extrinsic to them; imposed from outside, or "above" so to speak.

How then, does the Leckey–Bigelow theory inject the requisite necessity required for a satisfactory analysis of lawhood? The necessity is supplied by a special ingredient; the property of "naturalness". According to their theory, this property works in the following way: things obey the laws of nature *only when* they have this extra property in common. Thus, they claim that laws are not grounded in the *essences* of properties per se, as the essentialist theories have it, but rather, they are grounded in the essence of an *extra* property – one that is shared and is common to all and only those things which are in a "natural state". Hence this property of naturalness (which

H. Sankey (ed.), Causation and Laws of Nature, 83–88.

they assert is an accidental property) when present, grants laws their nomic status. Regularities have to occur when this property is present, because its presence means that the properties *have* to behave in that way.

The property of naturalness is central to their theory, and yet Leckey and Bigelow are somewhat coy when it comes to the issue of cashing out exactly what this mysterious property consists of, and how it influences law-like behaviour. However, despite this reticence regarding the nature of this enigmatic yet influential property, from the outline of their theory it is possible to deduce what it can and cannot be. We are told that its presence accounts for the regularity of the laws of nature, and yet it is accidental. Thus it is, in some way, responsible for the kinds of causal interactions properties engage in, but is extrinsically related to them.

Now, how does this property of naturalness work? The most plausible interpretation is that the property of naturalness behaves like some sort of cosmic police officer, in that it restricts the range and frequencies of behaviour that properties can exhibit. Hence without the property of naturalness, things in the world are free to do what they want; it is only through its presence that they behave in an orderly fashion.

However, I see some difficulties in this approach. Whilst prima facie, it may seem a plausible analysis to apply to the actual world, it encounters problems when we extend the analysis to include other worlds. Take, for example, two worlds, both containing the same set of properties and both containing the property of naturalness. Now, according to their theory, is it possible for each world (which are exact duplicates of each other) to obey a different set of laws? If their theory is genuinely a contingency theory of lawhood, I claim that the answer must be "yes". However, if they do uphold the claim that these otherwise intrinsically identical worlds can have different laws, there is nothing in their theory that can explain this difference. The worlds are the same, yet somehow they are inexplicably different in their behaviour. Thus whilst their property of naturalness can explain the *presence* of regularities in a given world, it cannot explain *what kinds* of regularities exist in a given world.

Perhaps Leckey and Bigelow could amend their theory by stipulating that there is a different naturalness property for each world that has different laws. This way, the difference between the two worlds could be attributed to the fact that each world possesses a different *kind* of naturalness property. This approach would, of course, necessitate a multiplicity of naturalness properties, perhaps infinitely many if we were to have a different kind for each different law-abiding world. But exactly how the one type of naturalness property in a given world manages to influence the vast variety of dispositional behaviours associated with all the different properties in a world remains to be explained.

One major objection that Leckey (this volume) raises against essentialist theories such as the kind proposed by Brian Ellis and myself, is that it makes the laws of nature *logically necessary* (meaning true in all possible worlds). Whilst Leckey concedes (somewhat paradoxically, I will later argue) that behavioural dispositions, such as those governed by laws, may be essential to the identity of certain properties, he rejects the claim that this entails that the laws of nature are true *in all worlds*. This assertion is based on the fact that the necessitation relation that holds between properties and their associated dispositional behaviour does not necessitate, or have any bearing on, which of the unaccountably infinite set of possible properties are actually *instantiated* by any of the entities in a given world. Thus, he argues, whilst an essentialist can assert the conditional statement that *if* an entity possesses a particular property then it is logically necessary (and hence true in *all* worlds) that the entity also possesses the behavioural disposition associated with that property, it is not possible to infer from this statement *which* properties and *which* behaviours (and therefore *which* laws) are instantiated *in this world.* And yet, Leckey claims, it is incumbent on any theory of laws that the properties *instantiated in this world* can be derived from the laws which are *true in this world.*

However, I disagree with this claim, for it is generally accepted that not all law statements which are both true and have considerable explanatory utility, refer to states of affairs that actually exist in this world. We have ideal gas laws, and yet notoriously, we have no ideal gases. Similarly, we have ultra-heavy elements that possibly never have been and never will be instantiated, but we have laws that would govern their behaviour if such extreme trans-uranic elements were ever produced. Thus, it does not follow from the truth of a law statement that the entities which it describes exist in this world.

One other key objection that Leckey levels against essentialist theories is that they rob laws of their contingent aspect. Laws of nature, Leckey claims, should be defined in such a way as to allow for the logical possibility of things breaking the laws. Interestingly, however, Leckey offers no direct argument for this assertion. And yet I see no strong argument in favour of it, and indeed, as I shall shortly argue, believe that there is strong motivation for abandoning it.

The central hallmark of a contingency theory of laws is that there exists a sharp distinction between a property and its associated dispositions in that there is no necessary connection between the properties that form the basis of an ontology, and the behaviour that the individuals possessing these properties display.

This metaphysical wedge between a property and its dispositional nature is often taken as axiomatic in that it is rarely argued for, and the ontological consequences of such a sharp division have rarely been questioned. However, as I have argued elsewhere (Lierse, 1996), such a distinction yields an ontology unfriendly and inappropriate for science, in that it furnishes an ontology of properties whose natures are *ex hypothesi* inscrutable.

This problem arises because of the empirical nature of science itself, and consequently, the central role that causal powers assume. The properties that form the basis of a scientific ontology are discovered *a posteriori*; that is, they are discovered in virtue of causally interacting with our scientific apparatus in some way. Furthermore, we use the different causal characteristics as distinguishing features to *qualitatively differentiate* the different properties in our ontology. Properties which possess identical dispositional natures are deemed to be the same type of property, whereas properties which possess different dispositional natures, are deemed to be different in kind.[1] A property that is causally inefficacious generally requires strong arguments for inclusion in our ontology. At the very least, a causally impotent property would generally be regarded as failing to qualify as a "physical" property.

However, a contingency theory of laws cannot, on pain of contradiction, employ a causal criterion to qualitatively distinguish the different properties, or indeed, use it to determine which properties are "physical" and which properties are not. For if the causal natures are only contingently associated with properties, then they cannot be used as distinguishing features to identify the different kinds of properties, for the same kinds of properties are deemed to possess different causal powers in different possible worlds. Moreover, properties that are causally efficacious in this world, and hence are classed as physical properties in this world, may totally lack causal powers in another, and thus, may not qualify as physical properties in other worlds.

Given that a causal criterion of qualitative identity is incompatible with a contingency theory of laws, then some other method for distinguishing the different types of properties is required. Unfortunately, however, those who embrace a contingency theory either erroneously utilise a causal criterion, or else fail to offer an alternative.[2] And this, I argue, is a serious defect in their analysis. For how can a theory which purports to offer an exposition of lawhood offer an adequate theory, if it fails to even offer a method for distinguishing and identifying the properties that form the theory's very foundation, or else employs a criterion which contradicts its central tenet?

Unfortunately, the Leckey–Bigelow theory is not immune to this defect. Whilst they claim that one of their theory's virtues is that it can allow for the contingent status of laws, in formulating an objection to the essentialist theory, Leckey inadvertently employs a causal theory of identity. Leckey (this volume) writes:

> In physics text books 'gravitational mass' is defined fairly narrowly in terms of gravitational force, so that while it is (arguably) essential to the concept of gravitational mass that the objects' accelerations toward each other be proportional to their mass, under the appropriate circumstances, it is not essential that their accelerations be proportional to the inverse square of the distance.

I will not debate here whether it is essential or otherwise that the concept of mass obeys an inverse square law. What is of interest here is that Leckey is even willing to concede the *first* part of the definition. Granting even *this* much, I argue, undermines the contingent aspect of their theory, as the contingency theory requires that there be no necessary connection between a property and its behaviour. This being the case, then why, in the framework of a contingency theory of laws, should the property of gravitational mass be defined in terms of an object's acceleration? Why should it not be defined in terms of an object's predisposition to flash pink then orange, or any other kind of behaviour?

Perhaps it could be argued that Leckey is referring to the property of gravitational mass as it is defined *in this world*. However, if this is the case, then adding this caveat only serves to undermine his own alleged counterexample. Leckey mentions gravitational mass, not to illustrate the way laws are in this world, but as a vehicle for contemplating the possibility that the laws could be different from what they are, and hence *contingent*. But if Leckey wants to defend a contingency theory, then to be consistent, he ought to reject the first part of his definition of gravitational mass as well.

If the idea that gravitational mass could, somewhere, in some possible world, be nomically associated with an entity's propensity to flash green then orange grates with our intuitions and sounds "just wrong", it grates, I argue, because deep down, there is something fundamentally very wrong with the contingency theory of laws. What gives weight to our intuitions about the truth of the contingency theory is that many of the examples (initially) used to justify it, seem terribly plausible, and at least, appear doxastically possible. For instance, there seems to be nothing in the nature of billiard balls that suggests that they must behave the way they do. We can quite easily imagine them behaving in some other way. And something such as the transformation of water into wine does not appear *too* far fetched. In fact, it is almost possible to visualise it, and the resulting image does not seem to contain any apparent contradiction.

But where such examples begin to break down is when we descend into the realm of the sub-atomic. At the most fundamental level of science, properties and their associated dispositions are so tightly interconnected that it is difficult, or indeed, nonsensical to disentangle properties from the behaviour they are disposed to instantiate. And it is here that confusion abounds. For instance, is negative charge a property, or a contingently associated disposition? If it is a disposition, then can an electron lack charge and still be an electron? And what about its other attributes such as "spin"? If these are only contingent dispositions, then how can we characterise an electron? What are its defining features? What criterion can we use if we cannot use a causal criterion? And how can we even begin to make sense of electron talk in worlds with laws other than ours?

It is no accident that science chooses to study and codify the behaviour of these fundamental properties. For it is in this realm that the essential relationship between properties and their behaviours are the most "pure" and less complex in that they form the basis of our ontology. The more complex structures, and their associated laws, supervene on this ontological base. Thus it is really not surprising that the more complex and less pure an object is, the less likely it is that our text books will contain simple, elegant laws to explain the connection between the nature of an object and its behaviour. This is why text books on physics are filled with laws, and why there are far, far fewer laws in the biological sciences.

Part of the reason why supporters of the contingent nature of laws have largely eschewed couching their discussions regarding properties and their associated dispositions in the realm of the sub-atomic, is that our intuitions grounding our belief in this doctrine derive their force from us observing and contemplating our everyday world. It seems possible when imagining the macro-world that things *could* have been different. We can imagine billiard balls suddenly coming to a halt, or jumping up off the table for no apparent reason. And we can seemingly consistently imagine water turning into wine right before our eyes. But we should be careful to ask ourselves what exactly are we imagining. Are we merely picturing a decanter full of a transparent liquid slowly turning red? How do we know it is water in the decanter? How can we distinguish it from, say, methylated spirits? And what is it that we are imagining when we picture the decanter of wine? And what is it that we are imagining when we visualise this transformation occurring?

The point is we cannot specify the stuff we know as water without, at some point, referring to its dispositional properties, *as we know them in our world, according to our laws*. And here is the rub. To speak of a property obeying *different* laws requires us to first identify the property by specifying it in terms of the dispositional behaviours associated with it in this world, and then attach the name of the property to a different set of behaviours. But this yields absurd situations such as talk about electrons without charge, or photons with mass, or indeed, worlds where protons possess the dispositions of electrons and vice versa, or worlds where protons and electrons are causally inefficacious, and thus, do not count as physical properties. The contingency theory of laws allows (and in some versions, even demands) the existence of these possibilities, along with many other circumstances, much, much more bizarre.

This is why I argue that, ultimately, all explications of a contingency theory of laws embody a deep inconsistency. For when proponents of this doctrine refer to entities contained in law statements, they refer to an ontology whose properties are specified and identified in terms of our laws, i.e., in terms of dispositional behaviours or causal natures that are associated with the properties in our world. But what they fail to acknowledge is that they cannot, on pain of violating the contingency thesis, define properties in terms of their causal natures, or regard their causal natures as essential attributes.

By appealing to the property of naturalness, the Leckey–Bigelow theory tries to capture something which we think is central to our understanding of why things behave the way they do. That is, it tries to capture our deep felt intuition that there is something very right and "natural" about the order that we observe in our world. However, although their theory is still only in its infancy, I fail to see how the property of naturalness married to a contingency theory of laws offers any advancement on the more conventional contingency theories offered by Dretske, Tooley and Armstrong. Not only does it appear to possess the pitfalls associated with these contingency theories, the addition of the elusive property of "naturalness" merely serves to supply a few more mysteries of its own.

La Trobe University

NOTES

[1] For example, see Armstrong (1978, vol. 1, p. 133 and vol. 2, p. 11). Although I argue in Lierse (1996) that these identity conditions are in conflict with his contingency theory of laws.

[2] See Lierse (1996) for a detailed discussion of this.

REFERENCES

Armstrong, D.M. (1978), *Universals and Scientific Realism* (2 vols.), Cambridge: Cambridge University Press

Leckey, M. (this volume), 'The Naturalness Theory of Laws', 77–82

Leckey, M., and Bigelow J. (1995), 'The Necessitarian Perspective: Laws as Natural Entailments', in F. Weinert (ed.), *Laws of Nature: Essays on the Philosophical, Scientific and Historical Dimensions*, Berlin: de Gruyter, 92–119

Lierse, Caroline (1996), 'The Jerrybuilt House of Humeanism', in P.J. Riggs (ed.) *Natural Kinds, Laws of Nature and Scientific Methodology*, Dordrecht: Kluwer, 29–48

III.

LAWS, QUANTITIES AND DISPOSITIONS

ALAN BAKER

ARE THE LAWS OF NATURE DEDUCTIVELY CLOSED?

INTRODUCTION

If a proposition is entailed by a set of laws of nature, does that make it a law? Call this the *Entailment Question*. There are two intuitively appealing ways of thinking about laws which produce sharply conflicting answers to this question.

One view of laws is that they are the bare, fundamental principles which describe the nature of the physical world. Laws are like axioms. The set of laws is the smallest possible set that is sufficient to derive all the general truths about the world. On this minimal view of laws, the answer to the Entailment Question is a resounding no: *no* distinct consequence of a set of laws is itself a law.[1]

A second view of laws is that they are the general truths about the nature of the physical world. Laws are like theorems, not axioms, and the set of laws is deductively closed. On this maximal view of laws, the answer to the Entailment Question is a resounding yes: *every* consequence of a set of laws is itself a law.[2]

Both the maximal and the minimal view of laws presuppose that there is a single, unambiguous answer to the question of whether laws are closed under entailment. Some philosophers disagree. They argue that a division should be made between basic laws and derived laws. There is a core set of basic laws which are not closed under entailment, and a larger set of derived laws which contains all consequences of the basic laws. The answer to the Entailment Question depends then on which of these sets we are referring to when we use the term 'law', and this is not fully determined by ordinary usage. Thus the only way to answer the Entailment Question is by stipulation.

So we have on the table three general views of laws of nature – the minimal view, the maximal view, and the stipulational view – each of which yields a clear, simple answer to the Entailment Question. My aim in this paper is show that none of these three simple answers can be correct. Each view captures some of the philosophically distinctive aspects of laws, but each also conflicts with other aspects. If I am right then the answer to the Entailment Question is more complicated. Some consequences of laws are themselves laws, and others are not.

APPEAL TO SCIENTIFIC PRACTICE

It would be helpful if scientists' use of the term 'law' were to yield a clear answer to the Entailment Question. However, deferring to the actual practice and terminology of science in order to settle a philosophical question can be a double-edged sword. On

H. Sankey (ed.), Causation and Laws of Nature, 91–109.

the one hand, the practice of scientists ought to provide ideal raw data for the naturalistically-minded philosopher of science. On the other hand, the practice itself is often unsystematic and inconclusive; what seems like an important question to the philosopher may be dismissed or ignored by working scientists.

Scientists use a variety of labels – seemingly haphazardly – for the basic tenets of their theories. They talk about 'Snell's Law', and 'Boyle's Law', but 'Schrödinger's *Equation*', 'Heisenberg's Uncertainty *Principle*', and 'Rayleigh's *Criterion*'.[3] It is hard to believe that there is some substantive difference here which grounds these fine-grained terminological variations, though it is possible that scientists' use of lawlike terms considered as a whole does have a more systematic character.[4]

Both the minimal and the maximal views of laws conflict with aspects of scientists' terminological practice. On the minimal view, laws are independent axioms. But some of what scientists call laws are consequences of other laws. For example, Newton's Second Law – that the acceleration of an object is directly proportional to the net force applied – entails Newton's First Law – that a body subject to no net force continues in its state of rest or uniform motion. Similarly, the principles of Special Relativity are entailed by the principles of General Relativity; the former apply in the special case where the reference frame is not accelerating.[5] On the maximal view of laws, every consequence of a law is itself a law. But scientists don't in fact say that every trivial consequence of a law is itself a law. Scientific terminology reflects the widespread view of scientists that laws are *fundamental* principles, and hence that trivial consequences don't count as laws.[6]

It would seem, then, that parts of scientific practice support the minimal view of laws and other parts support the maximal view; we are not going to get a clear answer to the Entailment Question solely by focusing on the terminological practice of scientists. Thus the supporter of the stipulational view is right insofar as our actual use of the term 'law' is ambiguous. However, this does not mean that we are forced to answer the Entailment Question by stipulation.

LAWS AND CLOSURE

I propose to take a more direct approach to answering the Entailment Question. Instead of trying to codify the vagaries of scientific practice, I want to look at various features that philosophers commonly attribute to laws of nature – features such as objectivity, necessity, universality, and explanatory role – and examine whether or not each such feature is preserved under entailment. Some features may be more important than others, and it may be the case that a proposition can be a law without possessing all of these features. In other words, our concept of law of nature may be a *cluster concept*.[7] If all or most of this cluster of central features of laws is preserved under entailment then this would provide strong support for the maximal view. Conversely, if few or none of these features are preserved under entailment then this would favour the minimal view.

SEMANTIC FEATURES OF LAWS

The first sub-cluster of features I want to examine are what might be termed logical, or semantic features of laws. This includes features such as truth, necessity, and lack of existential commitment, all of which turn out to be straightforwardly preserved under entailment.

One basic semantic feature of laws is truth. If a proposition, L, is a law of nature, then L is true.[8] But if L is true, then any logical consequence of L is also true.[9] Thus truth is (trivially) preserved under entailment.

A second semantic feature of laws is necessity. Laws are supposed to be necessary not in the logical sense, but rather in some weaker sense of physical, or nomological, necessity (whose precise character will not concern me here).[10] Necessity is not in general preserved under material implication. The material conditional, 'If Schrödinger's Equation holds then I have three coins in my pocket' may be true, and Schrödinger's Equation may be physically necessary, but that does not mean that it is physically necessary that I have three coins in my pocket. However, entailment is stronger than material implication, and modal features are in general preserved under entailment. All that is required are the axioms of the very weak modal system, K, to show that physical necessity is preserved under entailment.[11]

A third feature of laws which is straightforwardly preserved under entailment is lack of existential commitment. It is often claimed that laws of nature, considered in isolation, do not entail the existence of any particular kind of object. I would argue that this is a genuine feature of laws; laws are typically expressed as universal conditional statements, and such statements are vacuously true in a world with no objects.[12] It is clear enough that if a law, L, lacks existential commitment then any proposition, P, entailed by L also lacks such commitment. Otherwise P would be false, and L true, in a world with no objects, and this contradicts the assumption that P is a logical consequence of L.[13] Hence lack of existential commitment is preserved under entailment.

Truth, necessity, and lack of existential commitment are precisely definable, semantic features of laws which cannot be disrupted by logical weakening. The fact that there are several important features of laws which are preserved under entailment lends important initial support to the maximal view.

UNIVERSALITY

A second important aspect of laws is *universality*.[14] Laws are general, unrestricted, and not tied to particular times, places, or individuals. It is difficult to give a precise definition of universality because several distinct features of laws fall under this general heading. The first point to make is that universality is not a mere syntactic feature of laws. It is true that laws are typically expressed as universally quantified statements; however, it is easy to see that this criterion is not sufficient to mark off the universal propositions from the specific. Consider some specific proposition, for example, "Fred is in the house". This is logically equivalent to the universally quantified statement, "For all x, if x is identical to Fred then x is in the house". This sort of logical reformulation makes a purely syntactic criterion of universality unlikely.

There are at least a couple of senses in which laws might be said to be universal. One aspect of universality is unrestrictedness. Laws are universal in part because their spatial and temporal quantifiers are unrestricted. Philosophers tend to gloss this idea with certain standard sorts of remarks: laws state what happens "always and everywhere";[15] laws are "omnitemporally and omnispatially true".[16] Unfortunately, these motivational remarks gesture at a distinction with no real content. "Everyone in Princeton on May 8th 1996 is under 8 feet tall" is a true claim, and it involves restricted quantifiers. What is the difference between this statement being true and it being 'omnitemporally and omnispatially' true? It is a confusion to think that this claim could be true at some places and times, false at others.[17] It is therefore a mistake to identify unrestrictedness of quantifiers with being true "always and everywhere". But, once this misleading rhetoric is peeled away, it is unclear what substance there is to the claim that laws are unrestricted.[18]

A second aspect of universality is unlocalization. Laws are universal in part because they are *unlocalised*;[19] they are expressible using no local predicates, where a local predicate is one which refers to a specific time, place or object – for example '_ is medieval', '_ is polar', and '_ is within 3 miles of the Eiffel tower'.[20] Is unlocalization a genuine feature of laws? It is true that most of the propositions which scientists call laws are in fact unlocalised, but there are exceptions. Kepler's laws of planetary motion, for example, refer to specific celestial objects such as the sun. And there may be cosmological laws which refer to the specific time at which the universe was created by the 'big bang'.

Neither unrestrictedness nor unlocalization is preserved under entailment. Unrestricted propositions entail various restricted propositions. For example, the unrestricted proposition, 'All ravens are black', entails the restricted proposition, 'All ravens within 100 yards of a purple object are black'. Also, unlocalised propositions entail various localised propositions. The unlocalised proposition, 'All metals conduct electricity', entails the localised proposition, 'All metals conduct electricity during June 1996'. Features that are linked to universality are unlikely to be preserved under entailment, because entailment tends to proceed from the general to the specific. If universality is a genuine, substantive feature of laws then the maximal view is in trouble. I am doubtful, however, whether there is any interesting sense in which laws of nature are universal. If this is right, then the maximal view of laws is still tenable.

NON-VACUOUSNESS

So much for universality. I want to move on now to consider another putative feature of laws which is not preserved under entailment, namely *non-vacuousness*. A problem for naive regularity analyses of lawhood is that vacuous generalisations are automatically counted as laws. It seems counterintuitive that, 'All purple ravens conduct electricity', should be a law of nature simply because there are no purple ravens. One response is to insist that genuine laws are non-vacuous; 'All A's are B's' expresses a law only if there exist some A's.

Non-vacuousness is not preserved under entailment. A non-vacuous generalisation of the form, 'All A's are B's', entails various generalisations whose antecedents are more restricted. If enough restrictions are packed into the antecedent then it will eventually yield a vacuous generalisation. 'All ravens are black' is a non-vacuous generalisation. But it entails the (presumably) vacuous generalisation, 'All cubic ravens are black'. Hence non-vacuousness fails to be preserved under entailment.

Why think that laws of nature must be non-vacuous? There are plenty of examples of vacuous laws in science, including 'limiting case' laws such as the ideal gas laws, or the law that at absolute zero there is no molecular motion. It might be objected that such laws are non-vacuous when considered as approximations; things are sometimes very close to being at absolute zero, but nothing is close to being a cubic raven. Whether this distinction between idealising principles and trivially vacuous principles can be made to work is unclear. In any case, the requirement that laws be non-vacuous seems to be an *ad hoc* manoeuvre designed to patch up regularity accounts of lawhood, and the failure of non-vacuousness to be preserved under entailment is therefore no real threat to the maximal view.

What conclusions can be drawn from our examination of these features of universality and non-vacuousness? It is clear that neither feature is preserved under entailment, for there can be consequences of laws which are vacuous and consequences of laws which are not universal. If universality and non-vacuousness are genuine features of laws then the maximal view is in trouble, but I have serious doubts about each of them. Their status is too unclear for their lack of deductive closure to force the rejection of the maximal view.

SCIENTIFIC ROLE

The third and final sub-cluster of lawlike features I want to consider are those which pertain to the role played by laws in scientific practice. The role played by laws in confirmation and explanation is frequently cited as a feature which distinguishes laws from accidental regularities. It is often claimed, for instance, that laws – but not regularities – may be cited in explanations of particular phenomena. The law that all ravens are black helps explain why the raven I saw in my garden this morning was black. But the accidental regularity that everyone in the room is wearing a wristwatch does not explain why I am wearing a wristwatch.

Another popular view is that confirmation and explanation are *converses* of one another, at least in the case of causal laws: an observation, O, confirms an hypothesis, H, if and only if H explains O.[21] I doubt whether the connection between confirmation and explanation is quite this straightforward, but since the two topics raise much the same sorts of issues, I shall set the issue of explanation aside and focus principally on confirmation.

How are laws related to confirmation? It is sometimes said that laws are confirmed by their instances; the law that all ravens are black is confirmed by the observation of a black raven. Discussions of confirmation by instances tend to oscillate between taking an instance to be an *object* – for example a black raven – which satisfies the antecedent and consequent of the law, and taking an instance to be the act of *observation* of such an object. There are also well-known paradoxes, discovered by Hempel, which plague attempts to specify the principles which govern this notion of confirmation by instances.[22] In order to avoid getting needlessly bogged down in this side-issue I shall leave open what exactly a law is confirmed *by* and shall focus instead on the more neutral claim that laws are *confirmable*.

To say that an hypothesis is confirmable has certain epistemological implications. It implies that observing an object which satisfies the hypothesis gives good grounds for thinking that other objects, not yet observed, also satisfy the hypothesis. The hypothesis that all ravens are black is confirmable because observing several black ravens gives good grounds for thinking that the next raven I observe will be black. An hypothesis is confirmable, in other words, if and only if it is capable of receiving non-trivial inductive support.[23]

Since the notion of confirmability is going to play a key role in what follows, I want to say a little more about it. Remember that I am taking confirmation to be the converse of explanation, and as part of what distinguishes laws from accidental regularities. A proposition is confirmable if and only if it is capable of playing a role in the explanation of phenomena. I am also concerned with confirmation as it relates in particular to *scientific* practice, including the gathering of observational evidence and the testing and modification of hypotheses. Hence I am not interested in ways that hypotheses might be confirmed that have little connection to normal scientific practice – for example through testimony, divine revelation, oracles or whatever.

Is this feature of confirmability preserved under entailment? If confirmation satisfies what Hempel calls the 'Special Consequence Condition' then the answer to this question is a straightforward yes. The condition is as follows: if E confirms H, and H' is a logical consequence of H, then E confirms H'.[24] If this condition holds then any logical consequence of a set of laws is confirmable, since it is confirmed by the same evidence which confirms the individual members of that set. However, it is easy to construct counter-examples to the Special Consequence Condition which undermine its status as a general principle of confirmation. Let H be the hypothesis that a concealed die comes up 2, and let H' be the hypothesis that it comes up even. So H entails H'. Let E be the observation that the number rolled is prime. Then E confirms H (in the sense of raising its probability) but fails to confirm H'.[25]

Thus confirmation by a particular body of evidence is not in general preserved under entailment. This shows not that statements derived from laws are unconfirmable, but merely that they are not necessarily confirmed by the evidence that confirms the laws which entail them. Confirmation, then, is not closed under entailment, but what about *confirmability*? I shall argue that confirmability also fails to be preserved under entailment. I shall support this thesis by focusing on a particular class of propositions – namely *disjunctive generalisations* – which are entailed by laws yet are not confirmable.

My strategy in the next few sections will be as follows. First I will define the notion of a disjunctive generalisation, and give some examples to show how they can be entailed by laws. Then I will consider a pair of arguments – one due to William Seager, and one due to David Owens – that certain generalisations of this sort are not confirmable. I will offer a diagnosis of why this is so, focusing on the concept of pairs of predicates being *disparate*. Finally I will go on to consider the role that disjunctive generalisations can play in scientific practice.

DISJUNCTIVE GENERALISATIONS

I shall use the term 'disjunctive generalisation' to refer to any universal conditional with a disjunctive antecedent. A set of laws will entail many such disjunctive generalisations. Consider two arbitrary laws, for example,

(1) All ravens are black.
(2) All metals conduct electricity.

This pair of laws entails the disjunctive generalisation,

(3) If X is a raven or a metal then X is black or conducts electricity.

(3) is a disjunctive generalisation, since its antecedent contains the disjunctive predicate '_ is a raven or a metal'. I shall argue that any disjunctive generalisation formed in this way from two unconnected laws is unconfirmable. There is no doubt that (3) is an ugly and unwieldy proposition, but why should these aesthetic shortcomings render it unconfirmable? In the sparse literature on disjunctive laws, a couple of arguments have been offered for the thesis that disjunctive generalisations are unconfirmable, one due to William Seager and one due to David Owens. Let us examine these two arguments in turn.

Seager's argument proceeds roughly as follows.[26] First, he equates confirmation with probabilistic support. Evidence, E, supports an hypothesis, H, if and only if E raises the probability of H. Next Seager notes that (3) is logically equivalent to the conjunction

(3′) [All ravens are either black or conduct electricity] and [all metals are either black or conduct electricity].

(3) and (3′) are logically equivalent, hence they have the same probability, and a piece of evidence confirms (3) if and only if it confirms (3′). Moreover, the *relevant* evidence divides into two classes – ravens which are black or conduct electricity, and metals which are black or conduct electricity. However, the observation of a raven can support only the left-hand conjunct of (3′), while the observation of a metal can support only the right-hand conjunct of (3′). Seager concludes that no evidence can confirm (3′) in its entirety, and hence no evidence can confirm (3). Hence (3) is unconfirmable.[27]

Seager's argument strikes me as flawed for two reasons. First, it is unclear why confirmation of one conjunct of (3′) is not enough to confirm (3′) as a whole. It seems plausible that the behaviour of ravens is irrelevant to the behaviour of metals. But if this is so, then the observation of a black raven raises the probability of the left-hand conjunct, and has no effect on the probability of the right-hand conjunct. The probability of the whole statement is thereby raised, so this evidence confirms (3′) in its entirety. The inference from confirmation of one conjunct to confirmation of the whole is certainly not valid in general, but it seems unobjectionable if the predicates involved are sufficiently unrelated to each other (as is the case with ravens and metals).

A second problem with Seager's argument is that it proves too much, for a precisely parallel argument can be given to the conclusion that *non*-disjunctive generalisations are also unconfirmable. Consider a simple non-disjunctive law, for example

(1) All ravens are black.

We can perform Seager's 'division of cases' trick on (1) by noting that it is logically equivalent to the conjunction

(1′) [All square ravens are black] and [all non-square ravens are black]

But nothing can be both a square raven and a non-square raven, so no evidence can confirm both conjuncts of (1′). Hence (1′) is unconfirmable. So (1) is also unconfirmable. This absurd conclusion casts additional doubt on the soundness of Seager's argument.

The second argument is due to David Owens.[28] Owens' argument for the unconfirmability of disjunctive generalisations is less formal than Seager's argument but more persuasive. Owens argues that the problem with (3) – and the reason why it is not confirmable – is that it fails to support inductive inferences from observed to unobserved cases. Say I observe a raven that is black. This satisfies both the antecedent and consequent of (3). Does it give grounds for inferring that other objects which are either ravens or metals are either black or conduct electricity? No, because the observation of a black *raven* gives no grounds for thinking that other objects which are *metals* will either be black or conduct electricity. The law makes claims about ravens and about metals. But observations of ravens do not inductively support the claims about metals. And observations of metals do not inductively support the claims about ravens. Hence (3) is unconfirmable.

This version of the argument avoids the objections which Seager's argument faced, since it does not rely on claims about probability or division of evidence. Owens focuses instead on the inductive impotence of disjunctive generalisations. The 'grue'-like nature of (3) means that it cannot function as part of a good inductive argument. Owens, I think, gets much closer to the real problem with disjunctive generalisations, and to why they cannot play a role in ordinary scientific practice. How watertight is this argument?

Consider the following objection to Owens' argument. It may be agreed that the disjunctive generalisation

(3) If X is a raven or a metal then X is black or conducts electricity.

is not confirmed by the observation of a black raven, and that it is not confirmed by the observation of a metal which conducts electricity. In each case the 'cross-induction' – from ravens to metals, or from metals to ravens – fails to hold. But what about the compound observation

(O) r is a black raven and m is a metal which conducts electricity.

Doesn't this confirm (3)? If so, then (3) is confirmable by some evidence, and the Owens argument is undermined.

There are a number of ways one might respond to this objection. One response is to deny that non-atomic propositions can function as evidence in the confirmation relation. However this seems *ad hoc*, and it fails to mesh with other putative cases of confirmation. For example, the hypothesis that all ravens are black seems intuitively to be confirmed by the compound observation 'r is black and r is a raven'. A second response is to take 'unconfirmable' to mean 'unconfirmable by a single, atomic observation', and then to argue that the fact that (O) confirms (3) does not undermine the *atomic* unconfirmability of (3). A third response is to argue that the failure of evidence to confirm an hypothesis is additive. If two distinct observations of distinct and independent objects each fails to confirm H, then the conjunction of the two observations also fails to confirm H.

Maybe one of these three responses can be made to work, but my strategy for defending Owens' argument will be different. Recall that our interest in the issue of confirmability arose out of the more general claim that laws play a distinctive role in scientific practice. My argument is that conjunctions of evidence such as (O) are irrelevant to this issue because they are not the sorts of evidence claims that scientists would ever actually make. Combining mutually irrelevant observations in this way is not part of normal scientific practice. Thus the fact that (O) may confirm (3) does not show that (3) can play a non-pathological role in science. In the next section I shall try to make this notion of mutual irrelevance more precise.

DISPARATENESS

The above discussion has relied at various points on the intuitive notion of two properties being *irrelevant* to one another. The idea that there are objective facts concerning relevance, or concerning similarity between objects, is one that swings in and out of philosophical fashion. Those who favour this view have taken various approaches to cashing out its details. One option is to bring to bear heavyweight metaphysical notions such as universals, classes, or natural kinds in order to underpin the analysis. A second option is to appeal to a primitive unanalysable distinction between genuine respects of similarity and spurious ones. The choice is largely a matter of philosophical taste, and in what follows I shall help myself to a primitive distinction between natural and non-natural properties without stopping to analyse or defend it.

The basic reason why (3) is unconfirmable is because the properties of being a raven and of being a metal are mutually irrelevant. In what follows I shall use the term 'disparate' to describe pairs of predicates which are sufficiently unrelated to make disjunctive generalisations which involve them unconfirmable. In this example, '_ is a raven' and '_ is a metal' are a disparate pair of predicates.

Asked to explain precisely why these two predicates are disparate, a natural response would be to point out that ravens and metals have very little in common. This lack of shared features is, I think, the key to analysing this concept of disparateness. Let us restrict attention here to natural properties. It is not true that ravens and metals have *no* natural properties in common, for they share properties such as mass, solidity, and opacity. Notice, however, that these are all very general, non-specific properties that are also shared by numerous things which are neither ravens nor metals. Compare this with the case of two closely related predicates – for example, '_ is a dog' and '_ is a wolf'. Dogs and wolves share many quite specific properties, such as being quadrupedal, being mammals, and being carnivorous. These are properties which, taken together, are shared by relatively few things which are neither dogs nor wolves. Is there a way of making precise this intuitive distinction between sharing specific properties and sharing only general properties? Consider the following definition, due (in a slightly modified form) to Schlesinger: [29]

> *Definition*: a pair of (natural) predicates, P, Q, are *disparate* iff there is some natural predicate, R, such that each R-thing has every property common to P-things and Q-things, but no R-thing is a P-thing or a Q-thing.

This definition makes precise the intuition that disparate classes of things have no features in common that are peculiar just to them. It is not that ravens and metals have no properties in common, but that there is nothing *peculiar* just to ravens and metals, and this is why we think of the pair of predicates as disparate. A problem with Schlesinger's definition is that it makes the disparateness of predicates dependent on accidental facts about what things there are. Thus the predicates '_ is a dog' and '_ is a wolf' would be disparate predicates if there happened to have evolved a closely related species which shared all the properties common to dogs and wolves. A second problem is that the disparateness of predicates looks like it should be a matter of degree, but this definition makes disparateness an all or nothing thing. So this definition fails to capture perfectly the required notion of disparateness. Nonetheless I think the idea that disparateness is tied to lack of shared individuating properties is a useful one, and that some notion of disparateness is what distinguishes confirmable from unconfirmable disjunctive generalisations.

To see how disparateness is tied to unconfirmability, consider again the example of the disjunctive generalisation

(3) If X is a raven or a metal then X is black or conducts electricity.

If '_ is a raven' and '_ is a metal' are disparate predicates then ravens and metals have no set of uniquely individuating properties in common. No observational evidence could establish that some object is either a raven or a metal without establishing which, for if the evidence left open both possibilities then it would also leave open the possibility that the object in question was something else. It may be possible to construct far-fetched scenarios involving 'raven-or-metal' detectors, computers with disjunctive outputs, and the like. But once again, this all seems to be far enough removed from normal, non-pathological scientific practice to have little bearing on the issue of whether disjunctions such as (3) can play a suitable role in science.

To summarise: my claim is that no disjunctive generalisation involving disparate predicates can be a law. This is because there is no purely observational evidence which could confirm – or be explained by – such a generalisation. This is true even if the disjunctive generalisation is entailed by a set of laws, and this is why (3) is not a genuine law. The maximal view of laws is therefore untenable because according to this view *every* consequence of a set of laws is itself a law. In particular the disparate disjunctive generalisation, (3), counts as a law because it is entailed by the laws (1) and (2). If disjunctive generalisations of this sort cannot be laws then the maximal view must be rejected.

OBJECTIVITY AND THE MINIMAL VIEW OF LAWS

Thus far I have been focussing on undermining the maximal view of laws. I want to turn now to examine the implications of our discussion for the minimal view – the view that *no* consequence of a set of laws is itself a law. Our survey has revealed several important features of laws that are preserved under entailment and other important features that are not, but this in itself neither supports nor undermines the minimal view. Against the background of this general approach, there are at least two distinct strategies for establishing the minimal view. The first strategy is to argue that in no case is *every* important feature of laws preserved under entailment. This claim looks implausible since none of the law-like features we have considered are automatically destroyed under entailment. Moreover – as we have seen – there are a number of cases of derived principles which scientists do refer to as laws. It is hard to see, for instance, on what general grounds the principles of Special Relativity ought not to be counted as laws, despite the fact that they are straightforwardly derivable from the laws of General Relativity.

The second strategy for establishing the minimal view is to argue that it follows from an important feature of laws missing from our survey, namely *logical independence*. This feature is captured by the condition that L is a law only if L is not derivable from any set of laws which does not include L. Following this strategy amounts to establishing the minimal view by stipulation, since logical independence is clearly a property that is automatically destroyed under entailment, and this brings the minimal view suspiciously close to the 'stipulational view' mentioned in the Introduction. The stipulational view is premised on a distinction between 'basic laws' and 'derived laws', whereby the answer to the Entailment Question depends on which notion of law is in play. On this view, the only way to get a determinate answer to the Entailment Question is to stipulate which type of law is being referred to. The minimal view cuts through this ambiguity by insisting that there are no genuine derived laws; basic laws are the only laws that there are. This may be a defensible position, but the onus is on the supporter of the minimal view to demonstrate why there can be no derived laws. Since it appears that neither scientific practice nor philosophical analysis of the behavior under entailment of various law-like features bears out this negative claim, it is difficult to see what further grounds there could be – other than arbitrary fiat – for preferring the more committal minimal view over the stipulational view.

Although the above argument casts doubt on the motivation for the minimal view of laws, it does not undermine the view itself. A more direct problem with the minimal view is that it conflicts with an important feature of laws that we have not yet considered, namely objectivity.

It is natural to think that laws of nature are objective, in the sense of being independent of us and our mental states. This objectivity operates on two separate levels. Consider the law that all ravens are black. This expresses the objective fact that all ravens are black, a fact we think would be true even if we had never observed a raven and never conceived of the law. Also, the fact that this generalisation is a law is also an objective fact.[30] The distinction here is between the objectivity of laws and the objectivity of *lawhood*. It is natural to think that both the truth and the lawlike status of the laws of nature are independent of us and our cognitive apparatus.

According to the minimal view, laws function as axioms rather than as theorems. However, the choice of axioms for the theories we normally deal with is often quite arbitrary. Take the case of classical first-order logic; each system of first-order logic aims to capture all and only the logical truths as theorems, thus the *theorems* of each system are the same. But which particular logical truths are taken as *axioms* varies from system to system.[31] There is no guarantee, of course, that an analogous situation will arise in the case of the ultimate best theory (or theories) of which the laws of nature are axioms. However, extrapolating from these familiar cases raises some serious questions concerning the legitimacy of the minimal view.

Firstly, the fact that the choice of axiomatization is rarely uniquely constrained in the sorts of theories we are familiar with lends force to the worry that the choice of axiomatization for our ultimate best scientific theories will also not be uniquely constrained. The possibility of multiple equally-good axiomatizations, when combined with the minimal view, does not directly force the conclusion that what the laws are is not objective. Objectivity would be undermined if the only option was to make an arbitrary – and hence subjective – choice between the competing sets of axioms. However, there are other options for the supporter of the minimal view. One option is to identify the laws with the *intersection* of all the best sets of axioms. A second option is to identify the laws with the *union* of all the best sets of axioms. In each case the set of laws of nature is objectively determined. However, this objectivity is secured at a cost since there are specific problems with both the intersection and union views. A problem with the intersection view is that it leaves open the real possibility that there are no laws at all; this would be the case if there is no axiom that is common to all the best sets of axioms. A problem with the union view is that it undermines the very feature of logical independence on which the minimal view is premised. Let $A = \{A_1, \ldots, A_m\}$ and $B = \{B_1, \ldots, B_n\}$ be two (distinct) optimal axiomatizations of our best theory T. According to the union view, all of $\{A_1, \ldots, A_m, B_1, \ldots, B_n\}$ are laws of nature. Pick some B_i that is not in A.[32] B_i is a theorem of T, and all the theorems of T are consequences of A, hence B_i is a consequence of A. But both B_i and the members of A are laws of nature. Hence the laws of nature are not logically independent and the minimal view is undermined.

To summarise, the supporter of the minimal view can maintain the objectivity of lawhood in the face of multiple axiomatizations, but only by embracing something like the extreme intersection view described above. The true threat to objectivity lies not with the multiplicity of axioms *per se*, but with the criteria for choosing between competing axiomatizations of a given theory. Even once the theory has been fixed there remains a further issue of selecting the axioms for the theory. In doing so we tend to appeal to considerations such as syntactic simplicity and aesthetic appeal which – it seems reasonable to suppose – depend to a significant degree on the peculiarities of our cognitive apparatus. That there is an element of subjectivity in the choice of axioms for a theory is borne out by consideration of specific examples – such as first-order classical logic – in which choice between competing axiomatizations does seem to be largely a matter of taste.[33]

The issue here is far from clear-cut, and I am not claiming that the above line of argument provides a direct refutation of the minimal view. What it does show, I think, is that the minimal view – in regarding laws of nature as logically independent axioms – takes on the extra burden of showing that what the laws are is an objective fact.

The defender of the minimal view does have a partial response to these objections which implies that the non-subjective constraints on laws of nature are stronger than for other types of axioms. In the case of laws, the various law-like features we have considered – such as universality, non-vacuousness and confirmability – help to further constrain the class of alternative axioms that could function as laws. However, the presence of these extra constraints does not imply that more subjective considerations do not also play a part in selecting some axioms and ruling out others. For example, if 'P' is an axiom (of our best theory T) which satisfies all the objective constraints on lawhood, then '$P \wedge P$' will also satisfy these constraints. Yet we would not want to include '$P \wedge P$' in any best axiomatization of T.

To conclude, my argument against the minimal view falls into two parts. The first part of my argument questions the motivation of the minimal view – I do not see what *a priori* reason the supporter of the minimal view has for insisting that the laws of nature are logically independent of one another. Neither scientific practice nor philosophical reflection forces this conclusion, and without some such reason the minimal view seems to involve a risky empirical prediction about how our best scientific theories are going to turn out. The second part of my argument is that the minimal view, in regarding laws as axioms, makes lawhood only as objective as axiomhood is in general. And since choice of axioms is typically somewhat subjective this casts corresponding doubt on the correctness of the minimal view.

I hope to have shown that there are persuasive reasons for rejecting both the minimal and the maximal views of laws. The minimal view should be rejected because the axiom-like status of laws threatens to undermine their objectivity. The maximal view should be rejected because the theorem-like status of laws conflicts with the fact that there are some disjunctive generalisations that are entailed by laws but are not themselves laws.

THE ROLE OF DISJUNCTIVE GENERALISATIONS IN SCIENCE

In this final section I want to look at whether there is any role for disjunctive laws in science. Thus far I have said nothing to suggest that there are (or could be) any genuine disjunctive laws. I have argued that disjunctive generalisations cannot play an appropriate role in scientific practice if they contain disparate predicates, and hence cannot be genuine laws even if they are entailed by laws. What about disjunctive generalisations that do not contain disparate predicates? These have the potential to be confirmed by evidence and to underpin explanations and predictions, and therefore they may be laws. But do disjunctive laws and disjunctive generalisations ever actually get used by scientists? Are there ever situations in which a disjunctive law is appealed to in order to explain some result, or to predict some future observation? I think that the answer is yes, and that there are at least three sorts of situation in which scientists may need to make use of disjunctive generalisations.

One situation where disjunctive generalisations have a role to play is when the observational evidence is *inconclusive*, and this prevents any simple law from being applied. Imagine the following scenario. A chemist is given an unidentified grey powder and asked to determine whether or not it is toxic to humans. Not wanting to risk testing the toxicity of the powder directly, she resolves instead to try to determine what she can about its chemical composition. The only test which she has the resources to perform involves mixing the powder with nitric acid and then observing the colour of the flame when it is ignited. She performs the test and the powder burns with a green flame. Using her background theoretical knowledge, she infers from this flame test that the powder is either a copper salt or a barium salt. Armed with this evidence, she turns to her medical encyclopedia and consults the toxicology section. There she finds the following two claims;

(4) All copper salts are toxic to humans.
(5) All barium salts are toxic to humans.

Armed with this information, together with the evidence that it is either a copper salt or a barium salt, the chemist concludes that the powder is toxic to humans.

Question: on what law did the chemist base her prediction? Not on either of (4) or (5) taken individually, since neither of these laws is sufficient to derive the prediction from the evidence. It is clear – I think – that what underpins the chemist's prediction of toxicity is the disjunctive law

(6) If X is a copper salt or a barium salt then X is toxic to humans.

This disjunctive law is required because the observational evidence is inconclusive. (4) cannot be directly appealed to because the powder might be a barium salt. And (5) cannot be directly appealed to because the powder might be a copper salt. The evidence from the flame test yields a disjunctive conclusion, that the powder is *either* a copper salt *or* a barium salt. Hence the only way to underpin a prediction based on this evidence is to use a disjunctive law.

This point holds not just for prediction but also for confirmation and explanation. Imagine that a colleague walks into the laboratory and decides to test the chemist's conclusion by eating some of the powder. A few minutes later he collapses, overcome by nausea. The colleague's collapse supports the hypothesis that the powder is toxic, and thereby confirms the disjunctive law (6). Conversely, the disjunctive law can be cited to explain why the colleague collapsed. This web of confirmation and explanation connects the disjunctive law to the evidence in a way in which neither of the simple laws is connected.

This phenomenon of inconclusive evidence also bears on our previous discussion of disparateness; the predicates '_ is a barium salt' and '_ is a copper salt' are not disparate predicates because barium salts and copper salts have an individuating property in common, namely burning green in the flame test. One reason why disjunctive generalisations involving disparate predicates are not confirmable is that in the case of disparate disjunctions the phenomenon of inconclusive evidence does not apply. It is hard to imagine how there could be evidence which indicates that some object, X, is either a raven or a metal, but does not indicate which.[34]

In the case of inconclusive evidence it seems clear that the disjunctive generalisation involved is a genuine law. A second kind of scenario in which disjunctive generalisations may be used in science is when a putative connection is unexpectedly discovered between two properties which previously seemed to be disparate. The disjunctive generalisation postulates a *surprising correlation* between two properties, and in this case it is considerably less clear whether to count the disjunctive generalisation involved as a law. Let me illustrate this with a quasi-historical example.

The hyrax is a small hamster-like mammal which is indigenous to east Africa. It was recently discovered that the hyrax is in fact the closest living relative of the elephant. This came as some surprise, since hyraxes and elephants look like very different species. The evidence for this hypothesis comes from examination of the fossil record and on the anatomical structure of present-day hyraxes and elephants. This same evidence also indicates that the elephant and the hyrax are the only descendants of the now extinct woolly mammoth. In other words, it supports the disjunctive hypothesis

(7) If X is a hyrax or an elephant then X is a descendant of the woolly mammoth.

Assume that (7) is indeed a true generalisation. Is it also a law? To answer this we need to know whether (7) is confirmable, and this in turn depends upon whether '_ is a hyrax' and '_ is an elephant' are disparate predicates. Before the evidence from palaeontology and comparative anatomy came in, the features which hyraxes and elephants were known to have in common would be insufficient to individuate them uniquely. Properties such as being a mammal, not being extinct, and being indigenous to Africa are not sufficient to distinguish hyraxes and elephants from other species, such as lions, or giraffes. Hence before this new evidence came to light, the predicates '_ is a hyrax' and '_ is an elephant' would have counted as disparate.

Adding (7) to our theory overturns the verdict that ' _ is a hyrax' and '_ is an elephant' are disparate predicates. The empirical discovery that hyraxes and elephants have a specific shared ancestry shows that objects falling under these predicates do have something significant in common, and hence that the predicates are not in fact disparate. (7) implies that there is a *surprising correlation* between hyraxhood and elephanthood.

How does the fact that (7) establishes this surprising correlation affect its confirmability? At the point in the scientific inquiry at which (7) is being entertained as a hypothesis, there are no specific features which are known to be shared by hyraxes and elephants. It is therefore hard to see how (7) could ever be directly confirmed by an observation. The only way to confirm (7) is indirectly, by collecting evidence – separately – for the two hypotheses,

(8) All hyraxes are descendants of the woolly mammoth.
and
(9) All elephants are descendants of the woolly mammoth.

If enough evidence is collected to individually support these claims, then we may come to assert them as laws. (8) and (9) together entail (7) so we would be thereby justified in asserting (7), if not as a disjunctive law then at least as a true disjunctive generalisation. Propositions such as (7), which postulate surprising correlations between seemingly disparate pairs of predicates, have a curious intermediate status between laws and non-laws. But whatever their status, they provide an interesting example of how disjunctive generalisations can play a role in science.

A third potential role for disjunctive generalisations in scientific practice is in cases of what I shall call *evidence transfer*. Consider the following (fictitious) sequel to the hyrax/elephant story.

Let us assume that it has been established that hyraxes and elephants share a common ancestry. Some time after this discovery, biochemists begin to investigate the chemicals secreted from the pancreases of various different animals, hoping that there might be useful medical applications. They discover, after studying various other species, that common ancestry is an excellent indicator that the similar chemicals are secreted. If two species are descended from a common (and not too distant) ancestor then it is highly likely that they secrete the same chemical from their pancreases. A number of elephants are tested and each is found to secrete chemical Y from its pancreas. The testing of elephants is straightforward because an elephant's pancreas is so large that copious amounts of the secretion can be collected for testing. Hyraxes, on the other hand, pose insuperable problems for the biochemists' testing procedure. The small size of the hyrax means that its pancreas produces only minuscule amounts of secretion, and the quantity is too small for its chemical composition to be determined.

Thus the biochemists have direct evidence for the law,

(10) All elephants secrete chemical Y from the pancreas.

but they have no way of obtaining direct evidence for the law

(11) All hyraxes secrete chemical Y from the pancreas.

In this situation, the only way for the biochemists to make progress in establishing (11) is to consider the disjunctive generalisation,

(12) If X is a hyrax or an elephant then X secretes chemical Y from its pancreas.

I claim that (12) is confirmable, hence it is a disjunctive law. (12) is confirmable because evidence about the pancreatic secretions of elephants is relevant to claims about the pancreatic secretions of hyraxes. This relevance is a consequence of the background biological theory, which implies, first, that hyraxes and elephants have a common ancestry, and, second, that species with a common ancestry tend to secrete the same chemicals from the pancreas. (12) allows the *transfer* of evidence about elephants to support claims about hyraxes. Direct evidence can be gathered for the disjunctive law, which entails the law about hyraxes. Thus the disjunctive law allows the law about hyraxes to be established, even though no direct evidence can be gathered for this latter law.[35]

I have looked at three kinds of scenario in which disjunctive generalisations may play a role in scientists' actual practices of confirmation, explanation, and prediction.[36] I conclude that disjunctive generalisations and disjunctive laws can sometimes play a useful role in scientific inquiry, provided that the disjoined predicates are appropriately connected to one another.

CONCLUSIONS

In this paper I have attempted to show that there is no simple answer to the Entailment Question. I have offered no substantive positive theory of laws, but have instead approached the question by looking at how various characteristic features of laws behave under entailment. My aim has been to use this method to undermine both the minimal and the maximal views of laws, and I have devoted particular attention to the maximal view only because I think that it has greater intuitive appeal.

There are several sorts of reasons why one *might* want to reject the maximal view. Philosophers who stress features such as universality, unrestrictedness, and non-vacuousness will tend to reject the maximal view, because many propositions entailed by laws are specific, restricted, and vacuous.[37] Philosophers who tie laws to connections between simple universals, or natural classes, will tend to reject the maximal view because not every Boolean compound of universals is a universal, and not every subclass of a natural class is natural.[38] I think that the maximal view should be rejected, but I do not find either of these sorts of reasons compelling. I know of no way of making the concepts of universality and non-vacuousness precise. And I do not have sufficiently well-developed intuitions concerning universals and natural classes to know where the relevant lines should be drawn. We should reject the maximal view of laws, I have argued, because there are propositions entailed by laws which cannot play any appropriate role in normal scientific practice; such propositions are not confirmable and cannot be used for prediction and explanation. In particular, disjunctive generalisations involving disparate predicates are not directly confirmable by observation, nor can they function in genuine explanations of observed phenomena.

My final point concerns the provisional nature of our claims about lawhood. The identification of laws of nature at a given point in the development of science can at best be provisional. There are numerous historical examples of 'laws' which have had to be abandoned (or modified) in the face of later evidence because they have turned out to be false. However, if my analysis of the Entailment Question is correct then there is a second important sense in which the identification of laws is provisional, a sense which has nothing directly to do with a proposition's truth or falsity. Recall that whether a disjunctive generalisation counts as a law depends partly upon whether the predicates involved are disparate. But the disparateness of a pair of predicates is an open empirical question. Two predicates are disparate if members of the classes which they define have no individuating properties in common. As scientists discover more connections between different classes of things, their verdicts concerning disparateness may change. We saw one example of how this might happen in the hyrax and elephant story. Different theories sometimes give conflicting verdicts concerning disparateness, and this may lead to predicates being grouped together in different ways. The predicates '_ is diamond' and '_ is graphite' would not count as disparate from the point of view of chemistry, because diamond and graphite are both lattice forms of pure carbon. But from the point of view of commonsense, the wildly divergent physical properties of diamond and graphite might well lead to the predicates being counted as disparate. As scientific theories develop, apparently disparate predicates may be unified, and apparently related predicates may be disconnected. In the latter case, what seemed at one point to be a law may no longer be judged to be a law, even though the truth of the proposition has at no point been undermined.[39]

Princeton University

NOTES

1 See e.g. Mill (1846, Book III, Chapter IV, Section 1) who asks, concerning laws of nature; "What are the fewest general propositions from which all the uniformities which exist in the universe might be deductively inferred?"

2 See e.g. Hempel (1965, p. 346), or Ramsey (1978, p. 138), who argues that laws of nature are "consequences of those propositions which we should take as axioms if we knew everything and organised it as simply as possible in a deductive system".

3 Cf. van Fraassen (1989, pp. 36–7), where a similar point is made.

4 R.S. Walters makes essentially this point when he writes: "The term 'law' is used by scientists with no consistent meaning. It is not one of the technical terms defined by any science, and the scope of its application has varied. Once a statement is accepted in a science, it has not much mattered whether it is called a law, or principle, or theory, or anything else" (1967, p. 410). For a contrasting view, see Carroll (1994,

p. 45), who argues that "[s]cientists presumably do have reasons for accepting as laws only the propositions that they do".

5 Two more examples are the derivability of Kepler's Laws of planetary motion from Newton's Laws, and the logical equivalence of Ohm's Law [V = IR], Ampere's Law [I = V/R] and the Resistance Law [R = V/I].

6 Physicist-turned-cosmologist Paul Davies writes: "the laws of physics stand at the base of a rational explanatory chain, in the same way that the axioms of Euclid stand at the base of the logical scheme we call geometry" (1996, p. 35).

7 Van Fraassen argues for a cluster-concept view of lawhood in his (1989, p. 26).

8 Van Fraassen (1989, p. 29) refers to this condition as the 'Inference' criterion for lawhood.

9 And, more generally, any logical consequence of a set of true propositions {L, L′, L″, . . . } is true.

10 If, however, it turns out that physical necessity can itself only be explained in terms of the non-violation of laws of nature then the circle of explanation will be a very tight one.

11 The proof is straightforward, and relies on the axiom $\Box_p (x \supset y) \supset (\Box_p x \supset \Box_p y)$.

12 I am concerned here only with existential commitment to concrete objects. Whether laws of nature entail the existence of abstract objects is a murkier and more contentious issue.

13 Could a set of laws, each lacking individual existential commitment, together entail the existence of objects of a certain sort? This could happen only if the laws were jointly inconsistent, but this contradicts the assumption that laws are true propositions.

14 Cf. van Fraassen (1989, p. 26): "The laws of nature are universal laws, and universality is a mark of lawhood." Also, Carroll (1994, p. 25): "[T]he thesis that the laws of nature are in some sense general or universal is very compelling."

15 Van Fraassen (1989, p. 26).

16 Molnar (1969), quoted in Armstrong (1983, p. 12).

17 Such statements are 0-place open sentences, to use Tarskian terminology, and hence are satisfied by every object or by no object.

18 Even if the notion of unrestrictedness could be made precise, I am sceptical as to whether laws need have their quantifiers range unrestrictedly. It may well be a law, for example, that any object within the horizon of a black hole can never cross that horizon. This is spatially restricted; why can't it be a law?

19 Being unlocalised and being unrestricted are distinct properties. The black hole law just discussed is an example of an unlocalised yet restricted proposition. Whether there could also be propositions which are localised yet unrestricted is less clear.

20 Other examples may be more subtle; for example, '_ is a typhoon' is a localized predicate since a typhoon is defined as "any violent tropical cyclone originating in the Western Pacific". Perhaps predicates describing biological species are also localised, insofar as part of what it is to be a dog, or a canary, is to be of terrestrial origin; if so this would have ramifications for the possibility of higher-level biological laws of nature.

21 Dretske (1977, p. 261).

22 Hempel (1965, pp. 3–49).

23 Goodman makes essentially this point in his (1954, p. 26).

24 Hempel (1965, p. 31).

25 This example is taken from Seager (1991, p. 97).

26 Seager (1991). I have modified Seager's argument slightly to fit the present example.

27 The style of this argument is reminiscent of Hempel's argument for his 'paradoxes' of confirmation. In each case a logically equivalent hypothesis is considered and the conclusions applied back to the original hypothesis.

28 Owens (1989).

29 Cf. Schlesinger (1985, p. 15).

30 Van Fraassen summarises this higher-level constraint as follows: "Whether or not something is a law is entirely independent of our knowledge, belief, state of opinion, interests, or any other sort of epistemological or pragmatic factor" (1989, p. 35).

31 Paul Davies, who supports a minimal view of laws, concedes that this leads to some arbitrariness; "[o]f course, one cannot prove that the laws of physics have to be the starting point of an explanatory scheme, but any attempt to explain the world rationally has to have some starting point, and for most scientists the laws of physics seem a very satisfactory one. In the same way, one need not accept Euclid's axioms as the

starting point of geometry; a set of theorems like Pythagoras's would do equally well" (1996, p. 35).

[32] There must be some such B_i because A and B are by hypothesis distinct.

[33] Quine, for example, has on occasion worried about the best way of formulating the axioms of the propositional calculus, and whether to prefer economy of vocabulary – by using the Sheffer stroke as the only connective – at the cost of making the formulation of the axioms more complex.

[34] This situation can also be ruled out on more systematic grounds. If '_ is a raven' and '_ is a metal' are disparate properties, then they have no uniquely individuating property in common. But if so, then one could never have evidence for something being a raven or a metal, and ruling out any stronger or weaker claim, because then such evidence would have involved, per impossibile, the observation of just such a uniquely individuating property.

[35] It might be objected that the claim about hyraxes is unconfirmable, hence the hyrax law is not a genuine law. It is clear from the example, however, that the unconfirmability of the hyrax law is merely a consequence of current experimental limitations, and that the law is in principle confirmable.

[36] Note that in none of these three kinds of case have the disjunctive generalisations or laws involved had disjunctive consequents in addition to their disjunctive antecedents. These 'semi-disjunctive' generalisations do seem to have a genuine role to play in the testing and confirmation of scientific theories. It may be the case that fully disjunctive generalisations, such as (3), cannot fill this role. One function which only semi-disjunctive generalisations can perform is that of unifying separate predicates by providing a third predicate whose extension contains them both. This unification can even be inter-theoretical, as is the case of semi-disjunctive 'bridge principles' which, for instance, identify a mental state with a disjunction of physical states. See Fodor (1974), and Seager (1991) for more detailed discussion of the status of these bridge principles, and whether they ought to be counted as genuine laws.

[37] Carroll (1994, p. 47).

[38] See e.g. Armstrong (1983, p. 145) and his discussion of 'genuine' versus 'derived' laws.

[39] Thanks to participants at the 1996 AAHPSSS Conference at the University of Melbourne, and to David Lewis and Gideon Rosen for helpful comments on earlier drafts of this paper.

REFERENCES

Armstrong, D.M. (1983), *What is a Law of Nature?*, Cambridge: Cambridge University Press

Armstrong, D.M. (1978), *A Theory of Universals*, Vol. II, Cambridge: Cambridge University Press

Carroll, J. (1994), *Laws of Nature*, Cambridge: Cambridge University Press

Davies, P. (1996), 'The Day Time Began', *New Scientist* **150**, no. 2027, 30–5

Dretske, F. (1977), 'Laws of Nature', *Philosophy of Science* **44**, 248–68

Fodor, J. (1974), 'Special Sciences (or: The Disunity of Science as a Working Hypothesis)', *Synthèse* **28**, 97–115

Goodman, N. (1983), *Fact, Fiction and Forecast*, Cambridge, Mass.: Harvard University Press

Hempel, C. (1965), *Aspects of Scientific Explanation*, New York: Free Press

Lewis, D. (1983), 'New Work for a Theory of Universals', *Australasian Journal of Philosophy* **61**, 343–77

Mill, J. S. (1846), *A System of Logic*, London: Parker

Molnar, G. (1969), 'Kneale's Argument Revisited', *Philosophical Review* **8**, 79–89

Owens, D. (1989), 'Disjunctive Laws', *Analysis* **49**, 197–202

Ramsey, F. (1978), 'Law and Causality', in *Foundations: Essays in Philosophy, Logic, Mathematics and Economics*, London: Routledge, 128–51

Sanford, D. (1970), 'Disjunctive Predicates', *American Philosophical Quarterly* **7**, 162–70

Schlesinger, G. (1985), *The Intelligibility of Nature*, Aberdeen: Aberdeen University Press

Seager, W. (1991), 'Disjunctive Laws and Supervenience', *Analysis* **51**, 93–8

van Fraassen, B. (1989), *Laws and Symmetry*, Oxford: Oxford University Press

Walters, R. (1967), 'Laws of Science and Lawlike Statements', in *Encyclopedia of Philosophy*, vol. 4, New York: MacMillan, 410–4

JOHN FORGE

LAWS OF NATURE AS RELATIONS BETWEEN QUANTITIES?*

INTRODUCTION

I will assume here that there are such things as laws of nature and quantities in the world. Some of the former certainly seem to be relations between the latter. The gas laws, some of the other laws of thermodynamics, Ohms law, perhaps some dynamical and electrodynamic principles, all appear to state relations between quantities.[1] For example, pressure, volume and temperature are quantities, and the gas laws describe relations between values assigned as measures of these. It is therefore to be expected that such laws will be those things in the world that make (some) numerical laws true, a numerical law being an expression of the form:

1. $v(\mathbf{x}) = f(u(\mathbf{x}), w(\mathbf{y}), \ldots)$

where $v(\mathbf{x})$, etc., are values of variables and f is some function of these values. So it certainly *looks* as if these numerical laws will be true if the quantities in question are indeed related. Maybe the underlying reality is different from the surface appearance – that is a possibility. What seems beyond doubt is that *some* relation underpins numerical laws.

To address these matters we need some account of what it is for something to be a quantity: unless we start out with some idea of what a quantity is, it hardly appears possible to frame any ideas of how quantities might be related. To understand what a relation is we need to know what its terms are. Moreover, being able to form the basis on which to put forward a plausible account of laws of nature would certainly count in favour of a particular theory of quantity – this might even be taken as a requirement or desideratum of any satisfactory theory of quantity. I will consider here no less than *three* theories of quantities and ask whether they can provide a basis for a plausible account of laws of nature. These are: the class nominalist theory that takes quantities to be certain ordered classes[2], D.M. Armstrong's theory in which quantities are taken to be *properties* – substantial or real properties, properties not reducible to classes of objects – and finally a theory that takes quantities to comprise substantial relations of a certain sort, i.e. relations that are more than just classes of ordered pairs. That theory I have called elsewhere 'BP' (Forge, 1995), as it is derived from some ideas about quantities due to John Bigelow and Robert Pargetter. I would stress the word 'derived' here, because BP is by no means the same as the theory they advocate.

H. Sankey (ed.), Causation and Laws of Nature, 111–124.

We shall see that BP needs some further elaboration, if not modification, when called on to ground an account of laws of nature.

Under what conditions would we say that an account of laws of nature as relations between quantities is plausible? In the first place, it should do the job indicated above, namely it should be such as to show how numerical laws can be true. This actually amounts to a *formal* requirement to the effect that values of the variables and the functional relationship between these values can be interpreted in terms of the laws. It will, I hope, quickly become clear what this amounts to. What can be said right now, by way of illustration, is that if the relation between quantities proposed by the account is such that we cannot make proper sense of the functional relationship in candidate numerical laws, then we would judge the account not to have satisfied the formal requirement. Then, in the second place, we would expect to be able to use the account to show how the predictive and explanatory functions we usually associate with laws can be understood, to show how laws can underpin counterfactuals, etc. We might refer to these as *substantive* requirements.

I will be much less concerned here with the substantive requirements. This is partly because they are somewhat contentious. For instance, some philosophers, a small minority admittedly, do not think that science explains, yet among the large majority who think that it does, there are partisans for a number of different theories of explanation. This would evidently lead to several different versions of the substantive requirement that laws explain. Moreover, the formal requirement is prior, in the sense that an account that does not satisfy it will not get past the starting gate. I believe this is the case with the Armstrong account, as I hope to show in Section 2. The class nominalist position is, on the other hand, formally adequate; but as it is a variant of the regularity theory, I believe that it is not substantially adequate. This is not the place for an overall critique of the regularity view.[3] So, the burden of the section which follows is to demonstrate that the class nominalist position is formally adequate for at least one type of numerical law and then to reveal that it is a regularity view of laws. Assuming that disqualifies it, and given that the Armstrong account does not satisfy the formal requirement, it seems that we must pin our hope on BP, or look elsewhere. I would stress that it is not a waste of time to consider the class nominalist and the Armstrong account. The latter has, of late, been very influential and hence deserves attention; the former is certainly much less influential than it used to be, but its formal adequacy is a point in its favour and discussion thereof serves to introduce a number of considerations that are relevant to the two other contenders as well.

1. CLASS NOMINALISM AND THE REGULARITY ACCOUNT

If the class nominalist theory of quantity leads to the regularity account of laws, then many philosophers would think this to be a good reason to give up on it! The regularity account holds that laws are just certain constant conjunctions of events or states of affairs, and in this sense a law is a regularity and nothing more. Various attempts have been made to narrow down the class of all regularities so as to capture those members that seem to most resemble real laws of nature. This was characteristically done by philosophers with a 'linguistic turn', who focussed on the form of law statements, the sort of predicates they contained, their deductive relations with other such statements, and so on. This is (also) not the place to summarise all these twists and turns, for our purpose here will be to see how the class nominalist view of quantities gives rise to the notion that laws of nature are just regular connections between quantities. We need to begin with a few commonplaces about measurement.

When objects or systems that exhibit quantities are measured on some scale, then they can be *linearly* ordered on the basis of the values assigned to them. They might even be linearly ordered literally, for instance they might be lined up on the floor, from largest or biggest on the left to the smallest or least on the right. So, whatever else quantities are, it is evident that they involve *order*. But unless we adopt the operationist standpoint or unless scaling is merely nominal[4], there must be something which holds prior to the assignment of numerical values which determines whether the assignment is correct. To put the matter slightly differently, the linear order of the group of objects we have chosen to consider was already there, we discovered it and 'realised' it on the floor. The operationist thinks that measurement operations define quantities and so he or she would deny that there is any more to the orders associated with groups of objects than the operations by means of which they are manipulated and compared. So the operationist and the realist disagree here, with the former denying that there is anything to discover. But the difficulties with operationism are considerable, well-known and fatal, so we can forget about that interpretation.

The class nominalist takes the order of quantities at face value and asserts that a quantity is simply an ordered class of physical objects. Let q be a quantity, then the class nominalist will take q to be the pair $\langle Q, \geq_q \rangle$, where Q is a class of physical objects and $\geq_q$ is a linear order on Q.[5] The theory is *nominalist* because it refuses to say any more about $\geq_q$, for instance, it refuses to interpret $\geq_q$ as a universal, natural kind, or whatever. $\geq_q$ is said to be some real relation that exists independently of measurers or measurement, but just what sort of relation this is is left unsaid.[6] So if it happens that $\mathbf{a} \geq_p \mathbf{b}$, i.e. if object **a** is located 'higher up' in the order of p than object **b**, then the class nominalist cannot offer any reason, cause or explanation for this occurrence. The order is just what is the case; it is just how things are and nothing more can be said. Indeed, the theory may appear attractive precisely because it does not, unlike the theories to be considered in the sequel, entail any ontological commitment beyond individuals and classes. Whether such commitment is really needed is something we will come back to; what we need now are some (schematic) examples of numerical laws to see how this theory of quantity satisfies the formal requirement.

The simplest numerical law will be a relation between just two terms, such as

2. $p(\mathbf{x}) = f(q(\mathbf{x}))$

where $p(\mathbf{x})$ and $q(\mathbf{x})$ are real values on some scale of measurement, such as mass in grammes, volume in cubic centimetres, etc. $p(\mathbf{x})$ is the value assigned to **x** as a measure of the 'degree' to which it exhibits or possesses quantity p, likewise for $q(\mathbf{x})$. f is a function from $\{q(\mathbf{x})\}$ to $\{p(\mathbf{x})\}$, and this can be understood in the usual set-theoretical sense as a class of ordered pairs. There will (normally) be no difficulty in determining which function f is, as this will be clear from the form of the law. For example if 2 is

3. $p(\mathbf{x}) = cq(\mathbf{x})^{-1}$

then we have a rule for constructing the function. In general, the rule may well be difficult to apply, which is to say that the equation may contain integrals, expansions, etc., and be hard to solve. But that need not concern us here: the point is that the numerical law determines f, so we know what the relation is. It is important to realise that it is necessary to include the *constant* c in 3. Examples of constants in physical laws are the gas constant R in the gas laws, and in other thermodynamical laws, and the gravitational constant g. Constants are needed to make sure that expressions are dimensionally correct, to make sure that the value on the right hand side of the equation is commensurable with the value on the left hand side. If it turned out that the units of q were the units of p to the power of -1 then c would be unnecessary. I'm not aware of any binary numerical law that does not require a constant.

According to the class nominalist, the values p(**x**), q(**x**), etc., mark or designate the 'place' of the object **x** in the orders $\gtrsim_p$ and $\gtrsim_q$ of the quantities p and q respectively. Thus, it seems that what 3 is saying is that these 'places' correspond, in the following sense: Suppose **a** $\gtrsim_p$ **b**, for some objects **a** and **b**, which means that **a** has 'more' p than **b** does. In which case, p(**a**) ≥ p(**b**); the value assigned to **a** as the measure of the amount of p it has will be larger than that assigned to **b**. But 3 tells us that the value assigned as a measure of the amount of q **a** has is 1/p(**a**) times the constant c. Then, whatever c is, it follows that q(**b**) ≥ q(**a**) and hence that **b** $\gtrsim_q$ **a**. What 3 says is that if one object has more p than another, then it has less q. Translating that into the language of orderings, it follows that if **x** $\gtrsim_p$ **y**, then **y** $\gtrsim_q$ **x**, for all **x**, **y**. Suppose 3 is true in the sense that it refers to a law of nature Lpq, which we can represent as $\langle Q, \gtrsim_p, \lesssim_q \rangle$. Is it enough to say that Lpq is the law of nature which is such that this condition holds? Has formal adequacy been established?

Now if **x** $\gtrsim_p$ **y** and **y** $\gtrsim_q$ **x**, then it follows, as above, that p(**x**) ≥ p(**y**), q(**y**) ≥ q(**x**), assuming that these assignments are made correctly. It may appear that this is far too weak, for the only limitation placed on p(**x**) is that it is greater than or equal to p(**y**), etc., whereas in actual fact there will be a value, some particular p(**x**), that is the *correct* p-value for **x**. Just which value is the 'correct' one depends on the scale of measurement and that will embody certain conventional choices. For instance, object **a** might be chosen as the standard and assigned one unit. Hence, for all **x**, if **x** $\gtrsim_p$ **a** and **a** $\gtrsim_p$ **x**, then p(**x**) = 1. When this is not the case, when for instance, **x** $\gtrsim_p$ **a**, but not **a** $\gtrsim_p$ **x**, then it will be necessary to see how many **a**'s can be combined together so that they are equal in p to **x**. Suppose this can be done and that the number is 4, then p(**x**) = 4. This presupposes that p is an additive or *extensive* quantity, like weight or length.

Such quantities are said to be *fundamentally measurable* in that it is possible to set up scales for them in such a way that does not assume that any other scales have already been established. If q is not extensive, it is not fundamentally measurable, and in that case it must be assumed that there is some other quantity which is fundamentally measurable, and hence which is extensive, on which the scale of p is based. What is meant by 'based' here need not worry us. The upshot is that we can guarantee that the correct value is assigned to **x** provided there is a rich enough system of standards. If the extended system of standards, as it is called, is not rich enough, then this is due to limitations of our measurement technology. And that is not a problem exclusive to class nominalism.

In a little more detail, let us assume that there is a rich enough system of standards.

Then for any **x** there will be n objects $\mathbf{a}_1, \mathbf{a}_2, \ldots, \mathbf{a}_n$ such that $\mathbf{a}_1 \geq_p \mathbf{a}_2$, $\mathbf{a}_2 \geq_p \mathbf{a}_1$, etc., and $\mathbf{x} \geq_p \mathbf{a}_1 \oplus_p \mathbf{a}_2 \oplus_p \ldots \oplus_p \mathbf{a}_n$, and conversely, where $\oplus_p$ signifies concatenation. It follows that **x** will have the same 'place' in $\geq_p$ as the n-fold composition of **a** and hence that p(**x**) = np(**a**). Suppose **y** has the same place as an m-fold composition of **a**, then p(**y**) = mp(**a**). So when we look a little more closely at the way scales are established and values assigned we realise that, although all we can infer from 3 is that p(**x**) > p(**y**) if $\mathbf{x} >_p \mathbf{y}$, there will be other relations involving **x** and **y** that fix their location in $\geq_p$, and hence their p-values, precisely. So it does appear that class nominalism gives rise to an account of laws that is formally adequate in that it can provide a suitable basis for interpreting numerical laws.

The class nominalist account of laws is a straightforward extension of the theory of quantity: whereas the former are ordered classes of objects, the latter are 'correspondences' between these orders. We have only considered one form of numerical law, namely 3, but there seems no reason to suppose that others cannot be dealt with. Thus, whatever problems there might be with this proposal, it does not appear to be inadequate in the formal sense. However, it *is* evidently a version of the regularity view. That view of laws, it will be recalled, does not posit any connection between the states of affairs in question over and above constant conjunction. What this amounts to in the case of Lpq is that the state of affairs $\mathbf{a} \geq_p \mathbf{b}$ just *is* accompanied by $\mathbf{b} \geq_q \mathbf{a}$; it is not that $\mathbf{b} \geq_q \mathbf{a}$ *must* follow $\mathbf{a} \geq_p \mathbf{b}$ or that they *must* co-occur, or that one *causes* the other. So one might wonder how this conception of laws of nature could inform scientific explanation, ground prediction, support counterfactuals and so forth. From the substantive point of view it is surely inadequate.

2. QUANTITIES AND PROPERTIES: THE ARMSTRONG ACCOUNT

2 and 3 purport to represent the relationship between the 'degrees' to which objects possess quantities p and q. While the class nominalist reduces 'degrees' to 'places', a more straightforward approach holds that exhibiting a quantity amounts to having a special sort of property, namely a *quantitative property*, and that the degree to which the quantity is exhibited is determined by 'how big' that property is. For instance, it might appear that if **x** has q, then necessarily it has some property Qi: having q is just having Qi.[7] It would then seem natural to associate with q a strictly ordered class of quantitative properties {Qi}. In fact, q might simply be {Qi}, and likewise for p. The states of affairs represented by 2 would then seem to be the relationship between objects having certain pairs of quantitative properties: Qi and Pi such that Pi = FQi, where F signifies the relationship. This is very close to the Armstrong account; Armstrong holds that quantities are special sorts of property and that laws are relations between such properties.[8]

Armstrong is not, however, content to leave the analysis at that level. His reason is that there may be some values of p(**x**) and q(**x**), or some of the Q- and P-properties, that are never in fact instantiated, which is in fact quite likely. So it is quite possible that nothing ever has, had or will have some particular Q, say Qj, and (hence) nothing has ever had or will have Pj. The law Pj = FQj is therefore uninstantiated. Armstrong works within a tradition that proscribes uninstantiated properties, which is a consequence of his identification of properties with universals and because he does not admit uninstantiated universals. He also construes laws as universals – Pi = FQi is a first-order state of affairs and this Armstrong also takes to be a universal (Armstrong, 1983, p. 88). There are no difficulties yet: it will be necessary to assume that {Qi} is not everywhere dense, as there will be some missing properties, and it will be necessary to deny that there is any such law as Pj = FQj, but these are not really problems.

Along with many others, Armstrong believes that laws support counterfactuals, counterfactuals such as "If something had Qj, then it would have Pj = FQj also". But since Pj = FQj does not exist, it cannot support the counterfactual. To accommodate counterfactuals, Armstrong ascends to the level of second-order states of affairs and posits laws of the form **P** = **FQ**, where **P** and **Q** are now second-order or determinable properties (Armstrong, 1983, p. 113). For example, "having mass" is a second-order property, while "having such and such mass" is first-order. The idea, then, is that appeal can be made to the second-order law to support the counterfactual. Also appeal to the higher-order law *unifies* the collection of first-order laws, it binds them together as instances of the same kind. This seems important, for otherwise there would be a whole lot of laws about particular pairs of properties with nothing to connect them to one another as the same kind of law. But we do not think of laws as being just about particular things; laws are general. For all this to work it is not only necessary that the determinate properties Pj and Qj would be instances of **P** and **Q**, but that the relation **F** is 'handed on' to F. But we need now to focus on this relation and ask if it is perspicuous.

Thus far we have taken the expression 'Pi = FQi' to refer to just one 'law', the law pertaining to Pi and Qi. Now think of i as a variable which ranges over all quantitative properties, and hence let 'Pi = FQi' refer to all the first-order states of affairs of objects having Pi's and Qi's (this could be made explicit by adding universal quantifiers but as we are not being formal here I will not do so). We can then think of F as a function, as a class of ordered pairs ⟨Qi,Pi⟩. But which P-property is to be paired with a given Q-property? Let Qk be a particular Q-property, which property is Pk? The answer may seem obvious: Pk = FQk. But what is F? We have said that F is a class of ordered pairs, but which class? There are a great many functions from {Qi} to {Pi} and it is necessary to know which is F. Before, when we were dealing with numerical laws, it was possible to appeal to a rule, to a principle of mathematics that specified which value goes with which, as in 3. Is there any such rule that we can appeal to here?[9]

It is tempting to take quantitative properties to be things like "being 450 grammes in mass". So, referring back to 3, if an object has the property "being 2 units of Q", then we can infer that it has "being 1/2 units of P". The problem with this suggestion is that this reifies scales of measurement. Individual assignments of values are entirely matters of convention, depending on the choice of the standard unit and on the value assigned to that unit. Thus one kilogramme is assigned to the kilo standard, but some other value could have been assigned to it and some other object, some other lump of platinum perhaps, could have been chosen as the standard. Obviously it is a mistake to admit "being 450 grammes in mass", "being 1 pound in mass", etc., as *different* properties. They are clearly one and the same property and these expressions are just different names for the same thing. The property is the one to which it is correct to assign "being 450 grammes in mass", "being 1 pound in mass", etc., where 'correct' is understood in terms of the various conventions and operations associated with the scales in question.

We must therefore acknowledge that these properties cannot be 'inherently numerical'; numbers are *assigned* to them by measurement. This can be done in a non-arbitrary manner provided that {Qi}, etc., are ordered classes, for then the condition is that the *order* of the numbers reproduces the order of the properties – where the class nominalist has ordered classes of objects, the property theorist has ordered classes of properties. But if we accept that properties are not 'numerical' prior to measurement, then we cannot specify an analytical form which shows how to associate P's with Q's. The function f in 3 is specified by the operation "take the inverse of a (rational) number". This operation does *not* apply outside the domain of the numbers; it certainly does not apply to 'non-numerical' properties. Is there any other way to specify F? Is it, for instance, possible to do so in terms of some correspondence between the orderings on {Qi} and {Pi}?

We can say the following: If **x** has Qk and Pk and **y** Ql and Pl, then if Qk $\geq_q$ Ql, then Pl $\geq_p$ Pk, given that 3 is the case. There will be correspondences between the orderings of the two classes of properties and these may suffice to determine the function F, they may select F from all the possible functions from {Qi} to {Pi}. But this way of dealing with the matter seems to presuppose a theory of quantity that is, at least to some extent, *relational*, and as such it may seem at odds with the original Armstrong account. Clearly, the specification depends on the fact that the classes in question are *ordered*, i.e. that their members stand in certain relations to one another. Indeed, it is only the relations that are relevant here – it is quite possible, as we will see, to do without the properties. But we must not anticipate the relational theory of quantity just yet; we must also consider Armstrong's proposal to introduce the second-order law **P** = **FQ**, and see if that works.

There do, however, appear to be objections to Armstrong's proposals for the second-order law. Here there are just two second order universals and the function **F**. But, once again, what is **F**? Recall that, for Armstrong, the second-order state of affairs of **P** standing to **Q** in **F** is supposed to issue in various first-order states of affairs and to support counterfactuals about missing values. Thus F is supposed to be determined by **F** insofar as F is an instance of **F** – or F(Qi) is an instance of **F**(Qi). However, the second-order law cannot be

4. $\mathbf{P} = \mathbf{Q}^{-1}$

because, as has been noted already, the inverse operation applies only to numbers. Moreover, the role of the constant c should be recalled here as well: the terms of the numerical law are scale values that must be rendered commensurable. There is evidently no counterpart to the notion of such a constant at the level of second-order universals because the presence of these constants in numerical laws is due to the fact that the values are values on scales of measurement. Thus, as has been stressed already, the form of a numerical law depends on which scales of measurement are employed.[10]

It was suggested above that it might be possible to come up with a specification of F in terms of correspondences between the orderings in {Qi} and {Pi}, although that suggestion was not worked out. It should be clear that ascent to higher-level universals will *not* help to implement that idea. For one thing, the fact that the Qi, Pi are instances of the **Q** and **P** does not serve to establish the ordering in the classes {Qi} and {Pi}. And neither does **F**, because **F** is a relation between **Q** and **P**, not Qi and Pi.

Maybe, finally, one could say something like this: There exists a second order state of affairs of **P** standing to **Q** in a certain relation **F** which has a number of first-order instantiations. These first-order states of affairs are particular P-properties being connected to particular Q-properties, which in turn has the consequences that anything that has, say Qi, necessarily has Pi, and that the measured values of these satisfy 3. We can say this, we can write it down, but it is hardly informative or helpful. What does the trick? How does the second-order state of affairs issue in first-order laws of nature? So it would appear that however appealing the prospects of this account might be from the standpoint of the substantive issues of explanation, etc., it is not formally adequate and must be rejected for that reason.

3. QUANTITIES AS RELATIONS: THE BP ACCOUNT OF LAWS

Armstrong has a 'properties view' of quantities, in that he takes quantities to be a special sort of property. He acknowledges, of course, that objects that have quantities stand in certain *relations* to one another, but he thinks that this can be understood with reference to properties: for example, **x** will stand to **y** in "has more mass than" if **x** has a 'bigger' mass property than **y**. The class nominalist takes exactly the opposite tack, reducing properties to ordered classes. BP shares this emphasis on relations, but it does not subscribe to any kind of reductive schedule or brief to reduce properties to relations. Moreover, the sorts of relations entertained are not just ordered classes in the sense of set theory. BP takes as its starting point the general ontology of Bigelow and Pargetter (Bigelow and Pargetter, 1990) on which relations are universals. So, as with Armstrong, BP makes a commitment to more than individuals and classes.

BP, in its original version (Forge, 1995), takes quantities to be certain 'structures of universals' in the following sense: quantity $q = \langle Q^*, \{sQ\colon s \in S\}, \oplus_q \rangle$, where $\{sQ\colon s \in S\}$ is a class of *numerical relations* on which S is an index set. So what is a numerical relation? The idea, which is due to Bigelow and Pargetter, is that if **x** and **y** have q that they stand to one another in a relation of the form "has s times as much q as", where s is a real number. Thus what it is for **x** and **y** to have q is that they stand to all other objects that have q, and hence to each other, in some numerical relation sQ, where sQ is "has s times as much q as". If an object **z** does not stand in any sQ, then it does not have q, it lacks that quantity. Thus far, this is a thoroughly *relational* theory; **x**, **y**, etc., do not stand in the sQ because they have some quantitative properties, with **x**'s property being s times 'bigger' than **y**'s.[11] The class of objects having q is the class Q*, which contains Q, the union of the domains and ranges of the sQ, and this makes it explicit that only objects standing in the sQ have q. Notice that the idea of a numerical relation does not presuppose, or reify, scales of measurement. If **x** has twice as much mass as **y**, then **x** could be assigned 900 grammes, or 2 pounds, or . . . , with **y** being assigned 450 gramme, or 1 pound, or . . . [12]$\oplus_q$, as above, is the binary operation which forms composites. So if for some **z**, **x**2Q**z** and **y**1Q**z**, i.e. if **x** has twice as much q as **z** and **y** the same amount of q as **z**, then **x**$\oplus_q$**y**3Q**z** if q is extensive.

One might ask, then, whether BP is a good theory of quantity, whether it does accord well with scientific practice, whether it is better than the property theory and so on. I will assume that BP fares well enough with respect to these questions. However, one can also ask just what is meant by the term 'structure of universals' and whether this in turn is a universal. The entity $q = \langle Q^*, \{sQ: s \in S\}, \oplus_q \rangle$ is a relational structure which comprises a class of objects, a class of relations on that class and a special binary relation. If q is identified with that structure, as it just has been, then q is not itself a universal, for it does not seem that q is the sort of thing that has instances. It is of the first importance not to conflate aspects or elements of the *representation* of physical states of affairs, informed by convention, etc., with the states of affairs themselves. It is certainly convenient to *represent* q as a relational structure, as this collects together all the 'objects' associated with q into one entity that can then be characterised by suitable axioms (cf. Forge, 1995, p. 603, where, I confess, it is said that sQ are transitive whereas they are not). Of course it does not follow that q really is a relational structure. It seems that BP must commit itself to the real existence of the sQ and to (something like) $\oplus_q$, which suffices to define Q*, but to the structure as a whole? This is where BP needs some elaboration.

There is another line one can take here. One could also admit the property "is an sQ numerical relation" where s is any real number – s acts here as a variable on the real numbers – as a *second-order universal* and maintain that all and only Q-numerical relations instantiate this property. Indeed, there will be such a property if the sQ all have something in common. And they do: the sQ are all numerical relations of a certain kind. And what this means, in a theory like Bigelow and Pargetter's that admits higher-order universals, is that the property in question exists. Call it **sQ** – there will be **sP**'s as well. The alternative line is then to equate q, p, with the collection of sQ's, sP's, with the collections being identified in terms of **sQ**, **sP**.[13] However, the 'instantiation problem' has not disappeared. If $q = \{sQ: s \in S\}$, then q is still a set, and sets have members, not instances (the same difficulty confronts Armstrong if he takes $q = \{Qi\}$).

But while the problem of saying just what q is is proving a little difficult, at least we have a necessary and sufficient condition for something *having* q: that is, standing in some (at least one) sQ. Perhaps the 'instantiation problem' might be overcome by the following move: we acknowledge that there are objects, composites, numerical relations holding between them and properties of the numerical relations, but we take quantities to be just representations. Strictly speaking there are no such things in the world as quantities! That would be a somewhat radical and unexpected manoeuvre. Perhaps we should leave the 'instantiation problem' to one side and move on to laws of nature and, for the time being, still think of them as relations between quantities.

According to 3, the value assigned as a measure of q-ness is inversely proportional to the value assigned as a measure of p-ness. Now a value is assigned to an object in virtue of the numerical relation in which that object stands to certain conventionally chosen standards – this must be how things work according to BP. For example, suppose **x** stands in 2Q to the object **u** chosen to represent the unit on that scale, namely that **x** has twice the q of that standard. This means that q(**x**) = 2, where the units are omitted. From 3 it follows that p(**x**) = ½, where the units are again left out and the constant c is set to one, and hence that **u**2P**x**. In general, if **x**sQ**y**, then it follows that **y**sP**x**, which is a remarkably straightforward result.[14] Now let Lqp be a relation that holds such that if **x**sQ**y**, then necessarily **y**sP**x**. Lqp is the obvious candidate for the law of nature that obtains between q and p.

What might Lqp be such that, if **x**sQ**y**, then necessarily **y**sP**x**? If all the relations we admit are universals, and if Lqp is a relation, then it, too, must be a universal. So it must have instances. As a first attempt, then, suppose Lqp = (sQ,s^{-1}P), where s^{-1}P is just the converse of sP, i.e. if **x**sP**y** then **y**s^{-1}P**x**. The brackets are supposed to signify that the pair are related in some yet to be specified way; they do not denote set-membership. The instances of Lqp are then simply the states of affairs **x**sP**y**, **y**s^{-1}P**x**. But if we understand Lqp in this way, then it would only comprise one pair of numerical relations. The terminology here is misleading, for it looks as if sQ is *any* numerical relation, with s being any real number. But numerical relations are determinates, with s being some *particular* real number. So if Lqp is understood in the manner just suggested, it will be necessary to introduce lots of particular laws and the problem of unifying them will have to be faced.

It seems preferable to proceed as follows: Recall the second-order properties **sQ**, **sP**, the properties "is an sQ numerical relation", etc. We also need a relation of the form $\mathbf{s^{-1}P}$, namely the relation "is an s^{-1}P numerical relation". There is no objection (compare the Armstrong account) to introducing or describing properties in this way. For the real and rational numbers there always exists an r such that $r = s^{-1}$ for any s.[15] So **s-1P** is just a different name for **rP**. The law Lqp could then be written as (**sQ**,**rP**) on condition that $r = s^{-1}$, but it is surely easier to express it as (**sQ**,$\mathbf{s^{-1}P}$). All this makes it quite explicit what sort of thing Lqp is now taken to be: it is a relation between two second-order properties of a certain special sort.

The instances of Lqp are thus co-occurrences of particular sQ's and sP's and, since these in turn are universals, an instance of Lqp amounts to there being certain particular states of affairs co-occurring, namely **x**sQ**y** and **y**s^{-1}P**x**. This follows because we only admit instantiated universals, which means that **sQ**, $\mathbf{s^{-1}P}$ only occur as embodied in some state of affairs: the instantiation of higher-order universals implies the existence of some state of affairs involving particular objects. **s**, as it figures in (**sQ**,$\mathbf{s^{-1}P}$) is now a variable that ranges over the real numbers. As **sQ** is a property of all and only q-numerical relations, these will exhaust the instances of that property, and likewise for $\mathbf{s^{-1}P}$. Thus Lqp will capture all the (instantiated) sQ's and s^{-1}P's.

The account now seems to be formally adequate, at least with respect to 3. In one sense it has been easy to establish this, for it was no trouble to show how numerical relations could be used to underpin 3. In fact, the final proposal has rather more in common with Armstrong's account than might have been expected at the beginning. On the other hand, the fact that a universals theory was used as a framework led to the 'instantiation problem', which threw the enterprise a little off course. Then there is the question as to whether the suggestion that laws are relations between *quantities* has been given up. These matters will be taken up in the section which follows.

4. DISCUSSION AND FURTHER ELABORATION

In view of the fact that it seems remarkably easy to demonstrate formal adequacy using numerical relations, there may be a suspicion that this is a consequence of 3. Perhaps 3 is especially simple or atypical or in some way unrepresentative. It is not entirely clear how we should respond to this objection, short of going through all possible numerical laws and showing that it is possible to infer the form of the law.

Indeed, there is no guarantee that this will be possible, as some may be exceedingly complicated. However, it is worth considering two more examples.

So, suppose we have

5. $p(\mathbf{x}) = cq(\mathbf{x})^2$

It is necessary, again, to see which states of affairs of objects, **x,y**, . . . standing in q and p are related. Suppose **x**s**Qu**, where **u** is the unit for q, then the value assigned to **x** as a measure of q-ness on the u-scale is s units. That is because **x** has s times as much q as does **u**. Setting the constant to unity, it follows from 5 that **x** has s^2 units of p-ness ($p(\mathbf{x}) = 1.s^2$). Hence if **x**r**Pu**, where **u** is again the standard for the measurement of q, the $r = s^2$. In general, if **x**s**Qy** and **x**r**Py**, then $r = s^2$.[16] The law of nature in this instance can then be expressed as **(sQ,s2P)**.

Now suppose

6. $p(\mathbf{x}) = c_1\ln.c_2q(\mathbf{x})$

Once again, we need to see which of the rP **x** and **u** stand in, given that **x**s**Qu**. **x** has s units of q; it therefore has e^s times more p-ness than **u**, and hence the relevant P-relation is e^sP. The general case follows as before and hence the law is **(sQ,e^sP)**. These examples do not seem to present any great problems and so we can conclude that the present proposal does seem to be formally adequate.

Is it also substantively adequate? This question cannot be discussed here in any great depth, but it should be said that one of the attractions of any application of a universals theory is that it provides much more substance to work with than is available on an approach such as class nominalism. For instance, let it be supposed that the relation that holds between the second-order properties is (so-called) contingent natural necessity – this was hinted at above. Then provided that the universals theory can say enough about this conception for it to be reasonably perspicuous – there is surely no reason why Armstrong's ideas on this cannot be adopted – then we can take full advantage of it in giving an analysis of explanation. Why is it the case that **x**s^{-1}**Py**? Because it is the case that **x**s**Qy** and in virtue of Lqp the explanandum state of affairs therefore *had* to eventuate. This is only one way the analysis of explanation could be implemented. There are others that may well be better, but at least this shows how we can take advantage of the conception of natural necessity. Supporting counterfactuals furnishes another example.

Support for counterfactuals has traditionally been a touchstone for the adequacy of a proposal about the nature of laws. Let us go back to 3 and suppose that there are no **x,y** such that, for some s, **x**s**Qy** is ever the case – since the s are real numbers, it is overwhelmingly likely that not every numerical relation will have instances. Assuming that 3 is a law of coexistence, then it will not be the case either that **x**s^{-1}**Py**. However, it seems now that we should be able to claim that if only it *were* the case that **x**s**Qy**, then it *would* be the case that **x**s^{-1}**Py**. How is this claim to be supported? By appeal to Lqp. Lqp has some instances, and hence it can be admitted as a *bona fide* universal. These instances are particular pairs of numerical relations, pairs which are such as to underlie the regularity between the observable states of affairs. Granted that these *have to* co-occur – that a particular sQ *necessitates* the instantiation of its partner – then we can claim that if only some uninstantiated sQ were to have been instantiated, so would the corresponding sP. The argument here is essentially that of Armstrong.

CONCLUSION

Finally, by way of conclusion we should address the question as to whether this proposal amounts to the *rejection* of the idea that laws are relations between quantities. In the previous section it became clear that if we used a universals theory as a point of departure and construed quantities, etc., as universals, then it would be necessary to revise or modify BP. BP identified quantities with relational structures and although that seems to amount to a convenient representation, it is by no means obvious that it will really do to suppose that these objects exist 'in the world' in the way in which quantities are normally thought to exist. The difficulty here was dubbed the 'instantiation problem'. In the spirit of BP it is surely that quantities should be identified with the collections of numerical relations: having a quantity is to stand in a numerical relation, so these things are what are important. The difficulty here was finding a way to collect them together, and this was where the second-order properties were convenient. It must then be acknowledged that these properties are not themselves quantities; they are properties of those things which themselves comprise quantities. Hence, strictly, if laws of nature are relations between these properties, then they are not relations between quantities in the sense that the terms of the relations are not quantities.

But this is to be pretty strict. Relations are established between the numerical relations in virtue of the connection between the properties, but we need to be careful how we understand 'in virtue' here. The properties, being second-order, have no life of their own in a theory that only allows instantiated universals. But there is no question here about any 'priority of existence': the relationship between the second-order properties did not somehow come into existence first and *establish* the connections between the numerical relations. There is a sense in which the relation between the properties is logically prior, but that is not the issue. Given all this, then it is, I believe, quite proper to speak of the present proposal as one that sees laws as relations between quantities.

Griffith University

NOTES

* I am grateful to audiences in Melbourne and Pittsburgh for their comments on versions of this paper. Brian Ellis was kind enough to read a draft and, as always, his remarks were perceptive, to the point and most helpful.

1 It is much less obvious that quantum theory contains expressions that seem to designate laws of nature. The way in which that theory represents physical states of affairs is much less 'direct' than most other areas of physics. So if Schrödinger's equation, or other eigenvalue equations, do indeed refer to relations between quantities, then it will be necessary to show this. And that will not be easy, since it presupposes that we have been able to give a satisfactory realist interpretation of the theory. At the other end, so to speak, of the scale, it seems that spacetime theories are not really about relations between quantities at all; rather, they are structural principles that characterise the domain in which objects exhibit quantities. So these are two examples of types of theories that do not fall within the scope of the discussion that follows.

2 A case can, I think, be made for attributing this view to luminaries such as Ellis, Smart, Hempel and

Campbell, though I hasten to add that Ellis for one has given it up. Although not a luminary, I also subscribed to it (see Forge, 1987).

[3] Plenty of people have done so, perhaps the most remorseless is D.M. Armstrong – see the first part of his (1983).

[4] Nominal 'scaling' is simply naming with numerals, as in a football team, and any associated ordering is fortuitous.

[5] This is a first approximation. I have argued that the class nominalist needs to think of quantities as sequences of ordered classes, but the reasons for that proposal need not detain us here – see Forge (1987, pp. 307–8).

[6] As I refused to do in Forge (1987, pp. 300–2).

[7] I should add a note here on terminology. In the previous section, Q, P, etc., denoted classes of objects, while in this section Qi, Pi, etc., denote properties, and in the section that follows sQ, sP, etc., denote relations. This should not cause confusion.

[8] His terminology is different from mine. Thus he would represent the law holding between Pi and Qi variously as N(QiPi), which contains no explicit reference to F, and (we can infer, see pp. 111 and 114) N(Qi, a Pi such that Pi = FQi), which reads "necessarily, being a Qi and being a Pi such that Pi = FQi". For his theory of quantities, see Armstrong (1983; 1988).

[9] Armstrong believes that there is. He believes that the first-order laws can be deduced from the second-order law, Armstrong (1983, p. 114).

[10] Ellis has shown how non-standard scales of measurement can change the form of familiar laws, such as the law of gravitation. See Ellis (1966, pp. 137–8).

[11] We can easily define quantitative properties in terms of the sQ. The best way is to chose one object and, by convention, assign it a quantitative property called 1q. Then if **x** stands to 1q in rQ, it is assigned property rq. Armstrong does things the other way round: if **a** has twice the mass of **b**, he maintains that this is because **a** has the monadic property "being two kilogrammes in mass" while **b** has "being one kilogramme in mass", Armstrong (1988, p. 308).

[12] It is obvious that this works for ratio scales, but what about interval scales? Suppose **x** is assigned a temperature of 10°F and y 5°F; **x** is clearly not twice as hot as **y**. There is, however, an absolute measure of hotness, namely the absolute scale, and the extent to which **x** is hotter than **y** can be expressed on this scale. The issue is thus not so much about comparing values on interval scales as finding some 'absolute measure', and this in turn does not depend on the quantity in question being intensive – density, for example, is intensive and it is clear that objects stand in numerical relations with respect to their densities. If there are any physical quantities that appear to have no such 'absolute measure', and I am not able to think of any, then they will not be accommodated by BP and it will, for instance, be necessary to find some reductive account for them.

[13] What about $\oplus_q, \oplus_p, \ldots$, is it not necessary to include these as parts of q,p, . . . ? Perhaps not. Let $\oplus$ be the universal concatenation relation that forms all possible composites from all individual physical objects. Some of the latter have q, and hence so do their composites. We can then define $\oplus_q$ with reference to $\oplus$ and sQ – it is obvious how to do so – and so it seems that we do not really have to entertain a special concatenation relation for each quantity.

[14] Notice that the states of affairs **xsQu** and **usPx** constitute a necessary condition for the values such as q(**x**) = 2, p(**x**) = 1/2 to be correct. They are also sufficient. Thus, the assignment of values is much more 'direct' according to this scheme of things than it was for the class nominalist.

[15] Except if s = 0. The fact that 1/r always exists except if r = 0 is one of the axioms of a field of numbers.

[16] Just to make sure this works, let's try some values. Suppose q(**x**) = 4.3 units and q(**y**) = 1.2 units, then **x**3.58Q**y** and hence we infer **x**12.81P**y**. p(**x**) = 18.49, p(**y**) = 1.44, 18.49/1.44 = 12.84 ≈ 12.81, which shows that the inference is correct.

REFERENCES

Armstrong, D.M. (1983), *What is a Law of Nature?*, Cambridge: Cambridge University Press

Armstrong, D.M. (1988), 'Are Quantities Relations? A Reply to Bigelow and Pargetter', *Philosophical Studies* **54**, 305–16

Bigelow, J. and R. Pargetter (1990), *Science and Necessity*, Cambridge: Cambridge University Press

Ellis, B. (1966), *Basic Concepts of Measurement*, Cambridge: Cambridge University Press
Forge, J.C. (1986), 'David Armstrong on Functional Laws', *Philosophy of Science* **53**, 584–7
Forge, J.C. (1987), 'On Ellis' Theory of Quantity', in Forge (ed.), *Measurement, Realism and Objectivity*, Dordrecht: Reidel
Forge, J.C. (1995), 'Bigelow and Pargetter on Quantities', *Australasian Journal of Philosophy* **73**, 594–605

CATHY LEGG

REAL LAW IN CHARLES PEIRCE'S "PRAGMATICISM"

(OR: HOW SCHOLASTIC REALISM MET THE SCIENTIFIC METHOD)

"Inquiry must react against experience in order that the ship may be propelled through the ocean of thought . . . "
(*Charles Peirce*[1])

INTRODUCTION

Charles Peirce is arguably one of the most brilliant minds America has produced. His original "pragmatist" approach to philosophy received a contemporary popularisation in the work of James, but recognition of Peirce himself by philosophers of science has been patchy. Yet pragmatism was developed by Peirce precisely from his reflections on scientific method, and the question of what it is about scientific inquiry that has made it so spectacularly successful, and his account has a great deal to offer debates such as those concerning the nature of scientific laws and causation. In this paper I shall introduce some of the main features of Peirce's distinctive brand of pragmatism and argue that it differs significantly from contemporary popularisations of pragmatism, such as that found in the work of Richard Rorty. In particular, I shall argue that with respect to scientific laws, Peirce's view is far from the antirealism which strikes many philosophers of science as an unsatisfying feature of pragmatism. Rather, it presents an original and challenging philosophical approach to scientific law, an approach which falls in the tradition of scholastic realism, as exemplified most notably by Duns Scotus (though, of course, considerably updated by Peirce), and yet does not fall prey to weaknesses of the scholastic realist tradition with respect to scientific law. For instance, Peirce's theory retains the ability to separate semantics from ontology in the investigation of laws of nature which seems required by modern naturalism, and also avoids the postulation of abstract entities whose epistemology can only ever be mysterious.

Because of the broad-brush approach that I am taking to Peirce's ideas, I am advancing on a very wide philosophical front, and will inevitably raise issues that I do not have definitive answers to. However as I explain Peircean pragmatic realism, perhaps the reader will get a glimpse of why this might be deliberate.

H. Sankey (ed.), Causation and Laws of Nature, 125–142.

PART ONE: "NEOPRAGMATISM" VS "PRAGMATICISM"

RORTY

For the purposes of this paper I am going to take Richard Rorty as representing pragmatism's contemporary revival. Before I discuss him, I'll note that Rorty claims James and Dewey as his intellectual forebears as regards pragmatism. In fact he says that Peirce's contribution to pragmatism was, "merely to have given it a name, and to have stimulated James" (Rorty, 1982, p. 161).[2]

Rorty sees three main strands to pragmatism. The first is what he calls "anti-essentialism" applied to traditional objects of philosophical investigation such as truth and knowledge. From the dawn of philosophy until pragmatism entered the scene, Rorty suggests, philosophers busied themselves about trying to discover or define what truth is. What never occurred to them was the possibility that they were looking for something which did not exist:

> Let me illustrate this by James's definition of "the true" as "what is good in the way of belief." This has struck his critics as not to the point, as unphilosophical, as like the suggestion that the essence of aspirin is that it is good for headaches. James' point, however, was that there is nothing deeper to be said: truth is not the sort of thing which has an essence. (Rorty, 1982, p. 162)

If truth has no essence, this would suggest that it is at best an irreducibly family-resemblance concept. The meaning of truth just differs with respect to science, ethics, religious questions, and the attempt to discover what those differing truths have in common is doomed.

Moreover, the concern to remove inquiry from a pedestal of transcendental correspondence to the facts, and to explain it as a human activity which answers to human concerns, also appears in Rorty:

> When the contemplative mind, isolated from the stimuli of the moment, takes large views, its activity is more like deciding what to do than deciding that a representation is accurate. James' dictum about truth says that the vocabulary of practice is ineliminable, that no distinction of kind separates the sciences from the crafts, from moral reflection, or from art. (Rorty, 1982, p. 163)

Thus, the second strand to pragmatism, Rorty claims, is that "there is no epistemological difference between truth about what ought to be and truth about what is, nor any metaphysical difference between facts and values, nor any methodological difference between morality and science" (Rorty, 1982, p. 163). Rorty claims that a non-pragmatic philosophy makes the mistake of believing that rationality consists in being constrained by a rule, in obeying mechanical principles in order to obtain true beliefs. If on the other hand we are to allow room for human judgement and dialogue to play a role in reaching the truth, we need to allow that human faculties such as desire, appetite and will (which are a necessary part of human judgement and dialogue) are inseparable from the truth-seeking process. Another way of putting this is that the notion of truth is 'internally', that is, definitionally, related to the notion of an inquirer.

Rorty's third aspect of pragmatism, however, is the claim that to be a pragmatist is to hold that there are no constraints on inquiry save conversational ones. As it stands this is a very bald claim. Rorty goes on:

> [there are] no wholesale constraints derived from the nature of the objects, or of the mind, or of language, but only those retail constraints provided by the remarks of our fellow-inquirers . . . . The pragmatist tells us that it is useless to hope that objects will constrain us to believe the truth about them, if only they are approached with an unclouded mental eye, or a rigorous method, or a perspicuous language. (Rorty, 1982, p. 165)

What does it mean to model inquiry on conversation in this way? For Rorty, to see inquiry as conversation rather than as mechanically tracking the world's things is to recognise that we exist in a particular time period, that has particular epistemological methods and sets of beliefs, and that the products of our inquiry are *contingent* because of that (Rorty, 1982, p. 166). The term 'contingent' is an important one for Rorty, the recognition of which is for him an important part of understanding the human condition. Rortian contingency with respect to inquiry is a special case of the general philosophical concept of a fact which will not obtain, or a property which will not accompany its possessor, in all possible worlds. He means by it that a group of people faced with exactly the same "external world", but with a different culture and epistemology (Rorty calls this a different "starting point" (Rorty, 1982, p. 166)) would come to a different conclusion about what is true:

> Pragmatists follow Hegel in saying that, "philosophy is its time grasped in thought". Antipragmatists follow Plato in striving for an escape from conversation to something atemporal which lies in the background of all possible conversations. (Rorty, 1982, p. 174)

Rorty sees the embrace of contingency as an escape from philosophers' need to "ground" their epistemological practices – their need to provide a deductive justification for the fact that they "go on" in particular ways, and they to do "first philosophy". He points out that philosophers can "ground" just about anything. In fact at each period in philosophy, one will see philosophers "grounding" the presuppositions of the day, and given that many of these presuppositions have been quite contrary to one another, the activity does not seem to mean much:

> Since the time of Kant, it has become more and more apparent to non-philosophers that a really professional philosopher can supply a philosophical foundation for just about anything. This is one reason why philosophers, in the course of our century, become increasingly isolated from the rest of culture. Our proposals to guarantee this and clarify that have come to strike our fellow-intellectuals as merely comic. (Rorty, 1982, p. 169)

Rorty sums up his own position by negatively defining it against what he calls *Platonism*.[3] It is fair to say that this picture is a crude "criterial" account of truth, meaning that the philosopher sets out to offer necessary and sufficient conditions for a statement's being true purely in terms of a relation said to hold between the statement and a world of utterly external or transcendental things. Rorty dislikes this view for its monism about truth, its claim that truth is a descriptive rather than a normative notion, and its refusal to recognise that all truth-claims are made in a particular context which includes a whole set of notions about what truth should be. Pragmatism therefore, is defined as the rejection of all these claims.

This variety of pragmatism about truth is unsatisfactory as an account of the truth of scientific laws. When Rorty claims to be antiessentialist about truth, he means that different "forms of thought" which possess their own distinctive subject matters and their own integrated sets of terms do not just possess different sets of truths; they possess different notions of what truth is. It is therefore fair to describe Rorty as committed to *pluralism* about truth. Yet there are many different theories or frameworks within science which meet the pluralist's criteria of possessing their own distinctive subject matters and their own integrated sets of terms (such as physics, chemistry and biology[4]). Rortian pluralism seems to suggest that these theories could never converge and produce unified theories. For if they are aiming at different ends (namely $\mathrm{TRUTH}_{\mathrm{physics}}$, $\mathrm{TRUTH}_{\mathrm{chemistry}}$ and $\mathrm{TRUTH}_{\mathrm{biology}}$), how could one make sense of them merging in this way? This does not seem to be borne out by the history of science. To give just one example, it was astronomy's unification with the mechanics of ordinary-sized objects *via* Newton's famous analogy between an apple and the moon which enabled the theory of gravitation to emerge.

As for Rorty's claim of contingency, this seems to be straight-out antirealism. Each age does have its own set of scientific beliefs. One can "naturalistically" explain why certain discoveries were made by certain groups. For example one may argue that geometry was born "because" the human race became agricultural, and needed to measure up plots of land (particularly in Egypt where the Nile flooded each year, and it was necessary to redraw everybody's farming plots after each flood). Astronomy was born "because" people began to navigate by the stars. However, this view is compatible with differing groups of people being led to get epistemically *lucky* in differing ways, but all somehow latching onto differing parts of the one truth. Rorty's claim is stronger than this. The contingency question may perhaps be put – is the truth of any "contingent" theory indexed to the culture that uses that theory, or can we make any sense of it being true or false overall? Where two theories give differing accounts of the one phenomenon, is there a right answer, or are the two views "incommensurable"? To say that the views are commensurable is to commit to realism, while to say that they are not is to commit to antirealism. Rorty is explicit in his (seemingly global) disavowal of correspondence to the facts (and thus commensurability) on the part of our true theories, and it appears that his antirealism could not be more striking:

> For the pragmatist true sentences are not true because they correspond to reality, and so there is no need to worry what sort of reality, if any, a given sentence corresponds to – no need to worry about what "makes" it true. (Rorty, 1982, p. xvi)

In the next section I will argue that this version of pragmatism is quite different from Peirce's idea of it. Even in Peirce's lifetime he found it necessary to distance himself from the uses his term was being put to. He wrote that, "the word begins to be met with occasionally in the literary journals, where it gets abused in the merciless way that words have to expect when they fall into literary clutches", and concluded:

> So, then, the writer, finding his bantling "pragmatism" so promoted, feels that it is time to kiss his child goodbye and relinquish it to its higher destiny; while to serve the precise purpose of expressing the original definition, he begs to announce the birth of the word "pragmaticism", which is ugly enough to be safe from kidnappers. (Peirce, 1960, 5.414)

To be specific, I shall argue that Peirce is not a pluralist about truth, he does not believe that value considerations should inform our decisions about the truth (though he does believe that truth is a normative notion in a special sense I shall outline). As for the claim that truth is contingent (in the sense of indexed to cultures and times) – that is so close to the opposite of what Peirce had to say that it is ironic that both views could come to be called "pragmatism".

PEIRCE'S ACCOUNT OF TRUTH

Peirce is a monist about truth. He claims that this is presupposed by scientific method:

> . . . all the followers of science are animated by a cheerful hope that the processes of investigation, if only pushed far enough, will give one certain solution to each question to which they apply it. One man may investigate the velocity of light by studying the transits of Venus and the aberration of the stars; another by the oppositions of Mars and the eclipses of Jupiter's satellites; a third by the method of Fizeau . . . They may at first obtain different results, but as each perfects his method and his processes, the results are found to move steadily together toward a destined centre. So with all scientific research. Different minds may set out with the most antagonistic views, but the process of investigation carries them by a force outside themselves to one and the same conclusion. (Peirce, 1955, p. 38)

Unlike Rorty, Peirce offers a positive definition of truth – which is the natural limit of this process of settlement of opinion through inquiry:

> The opinion which is fated to be ultimately agreed to by all who investigate, is what we mean by the truth, and the object represented in this opinion is the real. (Peirce, 1955, p. 38)

This idea of truth as "ultimate opinion" has made its presence felt in analytical philosophy lately. But here we can see it in its original context as a scientific idealisation. At times in science it proves useful to describe a process as tending over an indefinite time period towards some limit. For example, when a hot pie is taken out of the oven and introduced to room temperature, it cools at first very quickly and then more and more slowly as the difference between the pie's temperature and room temperature decreases. According to the mathematical equations involved, however, despite its continual approach to room temperature, the pie never quite reaches that temperature, even over an infinite time period. A limit may be defined as the end-point (whether actual or, as in the case of the pie, ideal) of any such indefinitely long and continuous process.

A limit theory of truth, then, defines truth as the end-point of the approach of theories held by inquirers over an indefinitely long period of time to a complete representation of reality. By "end-point" here is meant the theory or set of beliefs that the inquirers would hold at that point. Truth-seeking, then, is a real-world process with the same tendency exemplified by the pie and the room for a feature of one system to approximate over an indefinite period of time ever more closely to a feature of another. Moreover, at least in the case of Peirce's particular limit theory of truth, the notion of truth is a notion internal to science. It is a notion which scientists must understand and employ (*albeit* at a more abstract level than the particular scientific terms which define their specialty) in order to perform scientific work.

A number of objections may arise at this point. Robert Almeder has claimed that Peirce does not actually have "a limit theory of truth" (Almeder, 1980, pp. 49–80). He claims that it is important to distinguish between the notion that truth consists in the indefinite approximation of a theory towards an ultimate true theory, and the notion that truth consists in the indefinite approximation of the probability that that theory is correct towards 1. He claims that Peirce only ever argued for the second interpretation:

> . . . for Peirce, what is approached as a limit is not truth as the product of the final irreversible opinion . . . rather what is approached as a limit is the probability value of 1 that the proposition is true. (Almeder, 1980, p. 60)

In arguing this way, Almeder is concerned to defend Peirce's theory of truth against an objection put forward by Quine that although the notion of approximation to a limit is defined for numbers, to try to apply it to theories is meaningless. Almeder cites Quine:

> Peirce was tempted to define truth outright in terms of scientific method, as the ideal theory which is approached as a limit when the (supposed) canons of scientific method are used unceasingly on continuing experience (5.407). But there is a lot wrong with Peirce's notion . . . there is a faulty use of numerical analogy in speaking of a limit of theories, since the notion of limit depends on that of "nearer than", which is defined for numbers and not for theories. (Quine, 1960, p. 23, cited Almeder, 1980, p. 63)

Almeder suggests that Peirce escapes this charge as his theory of truth is amenable to an entirely non-analogical numerical interpretation.

To Quine's objection that it makes no sense to speak of a theory approximating a limit, Peirce has an answer rather different to Almeder's. He accepts that it is indeed difficult to gain any pragmatic grasp on the notion of the absolute nearness of a theory to the truth. How on earth is such a quantity to be measured? In what units might it be expressed? However, the notion of a theory being *nearer to the truth than one's present theory*, is something scientists work with on a daily basis.[5]

Thus, although it is indeed impossible for anyone to step outside the infinite process of inquiry and make judgements about the satisfactoriness *tout court* of any theory, by using relative judgements the community of inquiry seesaws between replacement of unsatisfactory theories and refinement of (currently) satisfactory theories and inches, generation by generation, ever closer to a complete representation of reality. As long as one is clear that this is all that is meant by "a limit theory of truth", Peirce can justly be said to hold a limit theory of truth.

Having answered Almeder's objection about how one should interpret Peirce's view of truth, a more substantive objection may arise. The objection is that this view is just as antirealist as Rorty's. For Peircean truth is defined in terms of people's beliefs, rather than in terms of any sort of "external world". One might claim, "I can imagine a possible world where something (call it "proposition X") is true but never discovered".

To begin with, we need to note that Peirce's claim is not that proposition X will be accepted in 10 years, 1,000 or even a million, just that if inquiry were to continue forever, it would be stumbled upon. The progress to the end of inquiry is a progress of discovery that is not linear but nevertheless ever-increasing. Inquiry does not reach its limit until discovery cannot grow any further. Therefore, for proposition X to be left out of the picture is the same as saying that it is not *capable* of being discovered.

PEIRCE ON BELIEF, DOUBT AND MEANING

Still, to really attack this objection I need to outline Peirce's account of belief and meaning. Peirce defines truth in terms of settlement of opinion, which he also refers to as the "fixation of belief". But what is belief? Peirce adopted his account of belief in the 1850s from his friend Charles Bain, and subsequently claimed that from this account of belief, pragmatism is "a mere corollary". This account defines belief as a *habit*. This means that a belief is not merely a proposition in somebody's head. It is a genuine disposition to act in accordance with that proposition.

Similarly, *doubt* is a real disturbance in a belief – real in the sense that it is a reluctance to act upon a belief in situations where something is at stake, based on some unpleasant surprise that the acceptance of that belief has led us to receive. Peirce claims that it is extremely important for philosophy to distinguish between genuine doubt and what he calls "paper doubt" – doubt whipped up in the philosophy classroom by reading Descartes. Merely thinking that one doubts something does not mean that one really doubts it.

In this way, Peirce repudiates so-called "first philosophy" – the attempt to provide a deductive grounding for one's beliefs, or one's epistemic practices, which Descartes introduced into the modern era. For such philosophy pretends to start from ignorance of all *a posteriori* matters, yet such an epistemic position is unoccupiable, and will result in the philosopher continuing to hold their initial prejudices but imagining that they have given them some sort of deductive justification:

> We cannot begin with complete doubt. We must begin with all the prejudices which we actually have when we enter upon the study of philosophy. These prejudices are not to be dispelled by a maxim, for they are things which it does not occur to us can be questioned. Hence this initial scepticism will be a mere self-deception . . . and no-one who follows the Cartesian method will ever be satisfied until he has formally recovered all those beliefs which in form he has given up . . . Let us not pretend to doubt in philosophy what we do not doubt in our hearts. (Peirce, 1955, p. 229)

On this rejection of first philosophy, then, Rorty and Peirce agree.

The pragmatic account of *meaning* is contained in Peirce's famous "pragmatic maxim":

> Consider what effects, that might conceivably have practical bearings, we conceive the object of our conception to have. Then, our conception of these effects is our whole conception of the object. (Peirce, 1955, p. 31)

Again, he claims that this view comes straight from scientific practice, referring to it as an "experimentalist's view of assertion". Peirce claimed that thinkers who have received their training in thinking in the scientific laboratory are quite different from those "whose education has been largely a thing learned out of books", in that:

> [of the experimentalist] Whatever assertion you may make to him, he will either understand as meaning that if a given prescription for an experiment ever can be and ever is carried out in act, an experience of a given description will result, or else he will see no sense at all in what you say (Peirce, 1960, 5.412)

Interestingly, meaning here is a modal notion. The meaning of an idea such as "hard" consists in all the possible effects that something's being hard might have. (For example, one will bump one's leg on a hard table, two hard billiard balls will bounce off each other rather than fusing . . . and so on). This might sound verificationist. It is to some degree.[6] However, Peirce's maxim should be understood as a conditional claim whereby *if* we knew all the possible effects of a particular conception, then we *would* know its entire meaning.

In fact, however, we never do know the full meanings of our terms, particularly scientific ones, and the way science progresses is by clarifying and extending the meanings of scientific terms (such as 'force', 'electricity') by exploring the way situations with respect to which the current use of the term is indeterminate behave. This is called scientific experiment. Where this differs from verificationism[7] is that verificationists such as Schlick, Carnap and (in particular) Ayer seemed to believe that we could apply the "principle of verification" to our terms now, and thereby make a ruling once and for all about which terms bore a legitimate and useful meaning. Metaphysical, ethical and religious terms were among those subsequently accused by verificationists of being nonsensical. I contend that this is another form of first philosophy. On the other hand, the view that meanings permit of infinite clarification and development Peirce called "synechism", and it is one of the most fundamental aspects of his thought. Moreover, that such clarification and development is not solely conducted *a priori* is a large determining factor in the greater fallibilism with respect to meaning found in Peirce's self-styled "prope-positivism" (see for instance, Peirce 1960, 5.411–5.434).

Having outlined Peirce's conceptions of belief, doubt and meaning, I will say two things to the objection that Peirce's notion of truth just does not do justice to our intuition that something might be true though no-one in principle could ever find this truth out. First of all, if one accepts Peirce's account of meaning, this possibility turns out to be without any meaning. We have seen that the meaning of a proposition consists in its experiential consequences (including possible ones). However, if "proposition X" had any possible experiential consequences, for any possible observer (not necessarily a human being) it would be knowable. So it must have no experiential consequences at all, and therefore no meaning. Is this really such a difficult thing to rule out? The hypothesis is that a fact might obtain that makes not the slightest difference to anything anywhere in the universe. Is this such a loss to our ontology?

One might argue, if "facts in themselves" (unexperiencable facts) are eliminated by the pragmatist theory of meaning in this way, won't the same thing happen to the concept of the real? How can a Peircean pragmatist give any meaning to this concept? On the contrary, Peirce believed that the real has a continual impact upon our experience. Real things, he thought, are those which have their characteristics independent of what is thought about them, whereas fictions are exactly as we think they are. Therefore, we experience the real every time we confidently expect the world to behave in a certain way, and it gives us a rude surprise. (Science, then, investigates the real by encouraging such rude surprises under controlled conditions where they do us no harm.)

Secondly, to address this charge of antirealism, we need to think about what it means to analyse truth in terms of completely fixed belief. Based on his experience as a working scientist, and deep discussions with working scientists in all areas of science, Peirce believed that it is actually very hard to fix belief (*really* fix belief – not just now but for all time). It is not possible to cheat the real world with a shoddy or incomplete theory. Our beliefs, *via* our actions, are continually meeting the tribunal of experience. So, for instance, if one decides to believe that one's lover is faithful despite evidence to the contrary, further nasty epistemic surprises are likely to crop up that are less capable of being ignored. If the human race decides that God punishes in this life those who work on a Sunday, it is going to have to suppress a tendency in people to notice that this is not borne out by their experience. Of course it is quite possible for totalitarian regimes to set up a set of false beliefs, and force people to adhere to them over a long period of time. Peirce's claim is merely that *in the long run*, these situations can't last. Peirce calls the totalitarian possibility "the method of authority" for fixing belief. Here is his description of its downfall:

> But no institution can undertake to regulate opinions upon every subject. Only the most important ones can be attended to, and on the rest men's minds must be left to the action of natural causes. This imperfection will be no source of weakness so long as men are in such a state of culture that one opinion does not influence another – that is, so long as they cannot put two and two together. But in the most priest-ridden states some individuals will be found who are raised above that condition. These . . . possess a wider sort of social feeling; they see that men in other countries and in other ages have held to very different doctrines from those which they themselves have been brought up to believe; and they cannot help seeing that it is the mere accident of their having been taught as they have, and of their being surrounded with the manners and associations they have, that has caused them to believe as they do and not far differently. Nor can their candour resist the reflection that there is no reason to rate their own views at a higher value than those of other nations and other centuries; thus giving rise to doubts in their minds. (Peirce, 1955, p. 18)

I quoted this passage at such length as it makes an interesting comparison with Rorty. What Peirce is here describing is a recognition that one's views are contingent in the sense described by Rorty. That is, one realises that one believes as one does in some sense *because* of the time and culture in which one was born. Yet for Rorty this means that one should stop searching any further for the truth about what one believes. Contingency shows that there is no justification to be had, and one believes many things because one just can't help it. For Peirce on the other hand, a recognition of contingency is a spur to start *really* inquiring about the beliefs in question. In a sense, then, when it comes to the big questions about which entire cultures differ, in urging us to embrace the contingent views of our environment, Rorty is advocating the "method of authority" for fixing belief.[8]

In short, then, Peircean pragmatism claims that the only way to fix belief for the long term is to accept the existence of the real, and try to track it with one's theories:

> To satisfy our doubts, therefore, it is necessary that a method should be found by which our beliefs may be determined by nothing human, but by some external permanency – by something on which our thinking has no effect . . . Our external permanency would not be external, in our sense, if it was restricted in its influence to one individual. It must be something which affects, or might affect, every man. And though these affections are necessarily as various as are individual conditions, yet the method must be such that the ultimate conclusion of every man shall be the same. Such is the method of science. (Peirce, 1955, p. 18)

Thus, through the difficulty of fixing belief in the long run, the world does transcendentally constrain inquiry towards the representation of real things.

To return to Rorty, he is with Peirce on the need to abandon "first philosophy" – that is, seeking a deductive justification for our beliefs, or a means to know with certainty which of those beliefs are true. Peirce and neo-pragmatism agree that such a feat is not possible and that past attempts to achieve it are largely exercises in rationalising one's contingent presuppositions. Nevertheless, it can be argued that Rorty slides illicitly from "no deductive justification of our beliefs" to "no truth of the matter" about contingent beliefs. This is a slide which perhaps comes naturally to the mind of the analytic philosopher, trained as we are on the whole to respect deduction so highly, and unfamiliar as we are with merely probabilistic reasoning. Nevertheless, Peirce's scientific as well as philosophical training ensured that the slide from abandoning first philosophy to abandoning inquiry was not one which he was willing to make.

Also, Rorty was with Peirce on the need to define truth in some sort of internal sense which ensured that we could make sense of it being a state to which we could attain. Yet he was wrong to think that "internal" and "transcendental" were opposites.[9] Peirce's idea of the limit of inquiry, by defining truth as something inquirers necessarily attain, but no particular set of inquirers attain, brilliantly combines these two:

> . . . reality is independent, not necessarily of thought in general, but only of what you or I or any finite number of men may think about it . . . on the other hand, though the object of the final opinion depends on what that opinion is, yet what that opinion is does not depend on what you or I or any man thinks. (Peirce, 1955, p. 39)

Still, what justifies our belief in such convergence? Isn't it possible that we all might have a lot of true beliefs at some stage, and then make the mistake of adopting a poor scientific method, and subsequently drift into collective epistemic lunacy?

Nothing justifies our belief in convergence in the sense of a deductive demonstration. It is a *hypothesis*. Peirce thought that by adopting the hypothesis of convergence he could make sense of all he knew of the history of science, and also of the real-life experience he had had working in the scientific laboratory, where a sense of progress was palpable. This progress was towards more detailed, more explanatory and more predictive beliefs. Based on his intimate acquaintance with "men of science" since his boyhood, Peirce was able to report that scientists at the coal-face never doubt that their inquiries are converging, and that when they differ, those differences can be eventually settled. According to Peirce, given that doubt is a living phenomenon that arises in situations where one actually uses the beliefs one is doubting, this *datum* should be respected:

> Experience of the [scientific] method has not led us to doubt it, but, on the contrary, scientific investigation has had the most wonderful triumphs in the way of settling opinion. These afford the explanation of my not doubting the method or the hypothesis [i.e. convergence] which it supposes; and not having any doubt, nor believing that anybody else whom I could influence has, it would be the merest babble for me to say any more about it. If there be anybody with a living doubt upon the subject, let him consider it. (Peirce, 1955, p. 16)

(One may read between the lines here and consider how this person might get on querying convergence. If they did manage to show that convergence was not the case, would they then try to convince others of their finding . . . ?)

Peirce thought that the history of science showed that when inquirers think that if the Universe contains any universal facts, they are well hidden, then science stagnates. Whereas when inquirers adopt the optimistic hypothesis of convergence, they start to make real discoveries:

> . . . [a] moral factor of the method of science . . . is the self-confidence of it. In order to appreciate this, it is to be remembered that the entire fabric of science has to be built up out of surmises at truth. All that experiment can do is to tell us when we have surmised wrong. The right surmise is left for us to produce. The ancient world under these circumstances, with the exception of a few men born out of their time, looked upon physics as something about which only vague surmises could be made, and upon which close study

would be thrown away. So, venturing nothing, they naturally could gain nothing. But modern science has never faltered in its confidence that it would ultimately find out the truth concerning any question in which it could apply the check of experiment. (Peirce, 1960, 7.87)

I have now sketched the Peircean account of truth, belief, doubt and meaning, and argued that, although Peirce's take on realism is unique, his view is realist.

PART TWO: SCIENTIFIC LAWS

SCHOLASTIC REALISM

Peirce was quite explicit about the need for a philosophy which is informed by scientific method to embrace realism about universals. In fact he claimed to be a scholastic realist in the mediaeval sense:

The works of Duns Scotus have strongly influenced me. If his logic and metaphysics, not slavishly worshipped, but torn away from its mediaevalism, be adapted to modern culture, under continual wholesome reminders of nominalistic criticisms, I am convinced that it will go far towards supplying the philosophy which is best to harmonise with physical science. But other conceptions have to be drawn from the history of science and from mathematics. (Peirce, 1955, p. 2)

This claim might seem perverse. For it is widely believed that the scientific revolution in the seventeenth century was made possible by a new fallibilist spirit which entered European culture at that time, and that philosophy as practised in the mediaeval monasteries followed to a large degree the "method of authority" for fixing belief outlined earlier (the authorities concerned being the church and the works of Aristotle). Moreover, Duns Scotus of all the mediaevals has a reputation for a proliferation of arid logical distinctions. What could be less scientific? However, Peirce had an enormous respect for Scotus as a logician, and it is a testament to his living by his pragmatist principles that in judging ideas he looked not at their source, or even necessarily their justification by those who put them forward – but the fruitfulness of those ideas for further inquiry.

The term 'realism' originated in the mediaeval period, when intellectual Europe was rocked by a dispute over the ontological status of universals (such as whiteness, or cathood, to give two homely examples). The controversy concerned whether it makes sense to say that whiteness or cathood are "one over many" – that is, genuinely and wholly located in separate places. Conversely, it concerned the question whether cats or white things all have anything in common that is real, rather than being grouped together by human linguistic convention, or by the fact that they all produce the same idea in us. Mediaeval nominalism may therefore be defined as the view that whiteness and cathood are not truly "one over many", while (very crudely) mediaeval realism is the claim that they are.

Scotus had a subtle middle position between full-blooded realism about universals, and nominalism. He claimed that the distinction between a thing and the universals it embodies is not a *real* (or as Scotus also put it, a numerical) distinction, for the two cannot have independent existence. But it is not a pure linguistic convention on our part, either. It is *formal*. By this he meant that there is something in the thing which licenses our predication of the universal, but that something is not itself a wholly separate entity. Scotus referred to such predication-licensing features as "common natures".

Thus, if we consider cats, Scotus thought that cats were composed of matter, and that they embodied the universal or "common nature" of cathood.[10] The common nature of any particular cat is thus not "one over many", for cathood can not be separated ontologically from the multitude of cats in which it finds expression, and if all cats were by some freakish and appalling disaster to be annihilated, cathood would disappear as well. On the other hand, cathood is not "many over many" either, as all cats do share something which really renders them similar to one another in a way that, say, a cat and Socrates are not similar. Scotus therefore ascribes to his common natures a "lesser unity" – somewhere between one and many.[11,12]

Thus, Peircean scholastic realism with respect to scientific law is the claim that all instances of a given scientific law share some real general feature – some feature which we can discover, understand and project into the future (but which is not separable ontologically from the things which embody it). Nominalism is the claim that our laws do not pick out real general features of the world, but summarise the behaviour of all particular things, when placed under certain circumstances. Peirce thought we needed to postulate scholastic realism for explanatory reasons. He thought that only this hypothesis could explain our practice of scientific experimentation. In a lecture he gave at Harvard in 1903 he spells this out very clearly *via* the rhetorical device of a mock experiment.[13] I'll run through this argument in some detail. He begins:

> Suppose we attack the question experimentally. Here is a stone. Now I place that stone where there will be no obstacle between it and the floor, and I will predict with confidence that as soon as I let go my hold upon the stone it will fall to the floor. I will prove that I can make a correct prediction by actual trial if you like. But I can see by your faces that you all think it will be a very silly experiment. Why so? Because you all know very well that I can predict what will happen, and that the fact will verify my prediction. (Peirce, 1960, 5.93)

Peirce has urged his audience to admit that they have almost certain knowledge of a future event (the falling of the stone) based purely on present conditions (that is, the fact that it is being held over the floor by a hand that is about to release it, and the fact that the law of gravity obtains). He now invites them to consider just how remarkable that is:

> But how can I know what is going to happen? You certainly do not think that it is by clairvoyance, as if the future event by its existential reactiveness could affect me directly . . . as an event scarcely past might affect me. You know very well that there is nothing of the sort in this case. Still, it remains true that I do know that that stone will drop, as a fact, as soon as I let go my hold. If I truly know anything, that which I know must be real. It would be quite absurd to say that I could be enabled to know how events are going to be determined over which I can exercise no more control than I shall be able to exercise over this stone after it shall have left my hand, that I can so peer into the future merely on the strength of any acquaintance with any pure fiction. (Peirce, 1960, 5.94)

To further drive home the point that we do know what will happen to the stone, Peirce offers his audience a bet of a hundred to one that the stone will not fall. No-one takes him up on it. What exactly does this show, however? We obviously gained our knowledge of what the stone was about to do from observing stones and other heavy objects falling in the past. Peirce argues that we have two possible hypotheses at this point. There is the nominalist hypothesis, that the falling of those stones was a mere regularity in nature of which nothing more can be said than, it happened. Then there is the realist hypothesis, that the stones fell "for a reason". That is, the stones fell because they all possessed some real general feature which we can discover and then use to explain and predict future fallings:

With overwhelming uniformity, in our past experience . . . stones left free to fall have fallen. Thereupon two hypotheses only are open to us. Either

1. the uniformity with which those stones have fallen has been due to mere chance and affords no ground whatever, not the slightest for any expectation that the next stone that shall be let go will fall; or

2. the uniformity with which stones have fallen has been due to some active general principle, in which case it would be a strange coincidence that it should cease to act at the moment my prediction was based on it . . .

Of course every sane man will adopt the latter hypothesis. if he could doubt it in the case of the stone – which he can't – and I may as well drop the stone once for all – I told you so! – if anybody doubts this still, a thousand other such inductive predictions are getting verified every day, and he will have to suppose every one of them to be merely fortuitous in order reasonably to escape the conclusion that **general principles are really operative in nature**. That is the doctrine of scholastic realism. (Peirce, 1960, 5.101)

The nominalist will probably wish to have a few things to say at this point. He may protest that this is a very poor refutation of nominalism, as Peirce's experiment is perfectly compatible with the nominalist hypothesis. So what if the stone falls – nominalism never said it wouldn't! Nominalists are not committed to lawlessness, their claim is merely that laws are convenient summaries of patterns amongst real particulars, rather than invocations of real general entities in Nature. Surely one needs to be able to create an experimental result *incompatible* with a given hypothesis in order to be able to refute it?

Peter Skagestad has discussed this objection against Peirce's stone experiment, arguing against Manley Thompson who has claimed that the argument is confused (Skagestad, 1980). Skagestad discusses ways in which one *might* argue that Peirce's public use of the stone is not a real experiment. Firstly, the result of the experiment (that the stone will fall) is known beforehand. Scientists experiment because they genuinely don't know what will happen in a given situation, and wish to find out. Secondly, the result is compatible with both the nominalist and the realist hypotheses, and therefore can hardly be said to discriminate between them. What, then, is going on here?

Skagestad points out that Peirce is basing his argument for realism not on the fall of the stone alone but on the conjunction of the stone's fall and our confident belief that this will occur. Thus, Skagestad calls Peirce's strategy:

. . . a genuine thought-experiment on a phony physical experiment, and the very phoniness of the physical experiment is part of the experimental outcome of the thought-experiment. (Skagestad, 1980, p. 536)

He claims that although Peirce's argument does not prove realism in a deductive sense, it is intended to provide inductive support for it:

The nonfulfilment of an inductive prediction [such as that a particular stone will fall] does not overthrow realism, but the nonfulfilment of *all* inductive predictions would overthrow it; hence . . . each fulfilment of an inductive prediction counts as evidence in its favour. (Skagestad, 1980, p. 536)

Nevertheless, Skagestad claims that the hypothesis of realism is not a hypothesis within science. This is because it is not a hypothesis that any *particular* real law holds, but an overarching hypothesis that there are real laws *of some kind*. Therefore, the only way the hypothesis can be refuted is by the inductive refutation of all laws of nature. But such a refutation would leave us with no empirical science at all. He therefore suggests that Peirce's argument was confused in its formulation, as it suggested that one single event (the dropping of his stone) could prove realism, whereas realism's proper proof is *via* a vast induction over all law-determining scientific experiments ever performed (Skagestad, 1980, p. 537).

Considered as an induction in this way, Peirce's argument seems terribly weak. If he were arguing inductively from a single stone's falling to a law of gravitational attraction (which has an infinite number of applications) the argument would be weak enough. But it seems that Peirce is arguing from a single stone's falling to the existence of laws of nature *tout court*. What could be less conclusive?

I disagree with this interpretation of Peirce's stone-experiment. Skagestad is right to point out that Peirce's argument against the nominalist is not *deductively* successful. For this to be the case, the result of his experiment would have to be inconsistent with the nominalist hypothesis, and we have seen that this is not so. However, are deduction and induction the only alternatives here? For Peirce they most decidedly are not. Peirce thought that a complete account of logic (and "logic" for him meant the methodology of inquiry) needed to add a third term, which he called *abduction*. Abduction is similar to what nowadays is often called "argument to the best explanation".[14]

It seems to me that what Peirce is offering us in his stone-experiment is a demonstration that the realist hypothesis is an abduction[15] which, despite its merely hypothetical status, provides an *explanation* of our scientific practice in some deep sense which nominalism cannot supply. Skagestad is right, I think, to note that it is not the falling of the stone *per se*, but the combination of our belief that it will fall, and the stone's remarkable conformance to that belief which his argument really rests on. *How do we explain this ability to predict the future?* The nominalist hypothesis that particular things have been noticed by us to fall into certain regular patterns, which we call laws, says nothing about what those particular things will do in the future, as Hume famously noted. The nominalist hypothesis explains and justifies retrodiction but not prediction. The realist hypothesis, on the other hand, by postulating something real (something which Peirce calls "active general principles") whose very identity stretches into the future, does explain our confident expectation that stones and all other massive objects will fall to the floor whenever unsupported. Therefore, it is the better hypothesis.

Unlike induction, argument to the best explanation can draw on single cases, and yet be quite strong. According to the legend, it was the falling of a single apple from its tree that prompted Newton to look at it, think of the moon and suddenly conceive a certain very strong hypothesis which explained the moon's behaviour. It was not the case that Newton lay under the tree waiting for more apples to fall so that he could make sure his hypothesis was a good one. It is not, however, right to use the term 'proof' for an abductive argument (unlike a deductive or strong inductive argument). For the task of abduction is to produce hypotheses rather than proofs. In this regard, it is worth noting that although Skagestad uses the term 'proof' for Peirce's argument, Peirce, although he begins by saying he will attack the problem of scholastic realism "experimentally", does not use the term himself.[16]

I shall also qualify Skagestad's claim that for Peirce realism is not a hypothesis within science. It *is* "within science" in the sense that his acceptance of the hypothesis shows exactly the same logical form as scientists' acceptance of any hypothesis (for instance, the acceptance of Newton's second law by the scientific community, when that occurred). What is the logical form of the acceptance of hypotheses for Peirce? Inquiry recognises something as needing an explanation. Abductive inference generates hypotheses which would explain the phenomenon if true. The inquirer then tries to discriminate experimentally amongst different hypotheses which explain the phenomenon equally well. *Via* his stone-experiment Peirce seems to be suggesting that realism is an overarching abduction in science, yet to be challenged by any other hypothesis which does the job of explaining our predictive success with respect to the world.

Nevertheless, there is an important sense in which realism is a meta-explanation with respect to science, in the sense that if we are not realists about general entities, we have no business framing hypotheses with future consequences and setting out to test them *via* experiment. And without this, science could not get off the ground. Thus scholastic realism, though it is a hypothesis in terms of its logical structure, is a hypothesis that makes all other hypotheses possible. We saw that this was also the case with the hypothesis of convergence. Again the issue emerges – is this not a blatantly circular analysis? Again the answer comes back – this is not an analysis! This is a scientific hypothesis. Many of the observations which justify our support of classical mechanics are couched in the very theoretical terms which that theory makes possible (for example, mass, force, velocity and momentum). However, for philosophers to say, "What a circular theory!" would be impertinent, as the theory *worked*. It provided testable future consequences which, for a long time, were found to be verified. (It still does work in most everyday situations).

Peircean pragmatism sidesteps the demands of the deductivist philosopher by making justification future-directed. If one's view corresponds to the facts, it will be adopted and used eventually. When it comes to hypotheses, Peirce advocated a characteristically surprising yet apt usage of the Biblical maxim, "By their fruits shall ye know them". It is difficult to conceive at first just what a different approach to philosophy is being presented here. It is not possible to send one's ideas into the future by justifying them oneself, Peirce thought. This is a hangover of Cartesianism. The only thing that can ensure that one's ideas endure, is to submit them to something much greater and more permanent than oneself, first by clarifying and testing one's ideas against the Real as much as possible, and then by trusting to future inquiry, by converging on the Real, to converge on one's ideas. Much of this paper has been concerned with the question of whether Peirce was a realist. I submit that such epistemic surrender is realism in its purest form.

It might be protested at this point that "future-directed justification" is no justification at all, that Peirce cannot offer any conclusive reason why we should accept *his* theory. All he can give us is whatever explanatory force it has, and his sincere belief that it will end up at the end of inquiry. Indeed – Peirce is quite consistent here! Reading about Peirce's life, it is difficult to believe that a philosopher who suffered such conspicuous professional ruin, such poverty, such discouragement of his ideas from his enemies and misunderstanding by his friends, could nevertheless have kept on developing those ideas with ceaseless energy until he died. Why did he do this? Peirce worked at the American Coast and Geodetic Survey for about twenty years, but in 1890 a report which he prepared on measurement, which contained insights on scientific methodology which he had gleaned from the study of measurement over all those years, was rejected as being far too long and wilfully philosophical. Faced with this humiliating event Peirce resigned. Apparently when the Coast Survey moved offices many years later they found Peirce's report in a paper parcel in one of the cupboards, and written on it was the inscription, "save for history."[17] We will see . . .[18]

Australian National University and University of Sydney

NOTES

[1] Charles Peirce (1960, 8.118).
[2] A remark somewhat chilling in its apparent insouciant ignorance.
[3] This summary is presented in, 'Introduction: Pragmatism and Philosophy', (Rorty, 1982, pp. xiii-xvii).
[4] Or even the many theories that exist within sciences such as physics, such as general relativity, classical and quantum mechanics.
[5] Peirce makes this point (in terms of knowledge rather than truth, but it seems clear that the point is the same) in his application for a grant from the Carnegie Institute in 1902, in a section entitled 'On the Economics of Research':

In the economics of research the "laws" are mere general tendencies to which exceptions are frequent. The laws being so indefinite, at best, there is little advantage in very accurate definitions of such terms as 'amount of knowledge'. It is, however, possible to attach a definite conception to one increment of knowledge being greater than another. To work this out will be the first business of the memoir. (Peirce, MS L75)

[6] Peirce actually became less verificationist over his lifetime. The original pragmatic maxim dates from the 1870s. Here is what he had to say about it in 1902:

The doctrine appears to assume that the end of man is action – a stoical axiom which, to the present writer at the age of sixty, does not recommend itself so forcibly as it did at thirty. If it be admitted, on the contrary, that action wants an end, and that that end must be something of a general description, then the spirit of the maxim itself, which is that we must look to the upshot of our concepts in order rightly to comprehend them, would direct us towards something different from our practical facts, namely to general ideas, as the true interpreters of our thought. (Collected Papers, 5.3)

So here Peirce keeps the notion of belief as habit (and thus as intrinsically general), but instead of a habit of behaviour, belief becomes a habit (or generalising tendency) of thought.
[7] For further discussion of the difference between Peirce's pragmatism and verificationism, see Almeder (1980, p. 20).
[8] This is ironic for such a professed liberal humanist.
[9] As evidenced, for example, in the way he sets up his third definition of pragmatism (Rorty, 1982, pp. 165–6).
[10] He also thought that each cat possessed its own particular individuating principle or haecceity (to which cat owners will no doubt testify), but this feature of his metaphysics is not relevant to the present discussion.
[11] For this explication of Scotistic realism, and its links to Peircean realism I am indebted to John Boler's landmark book on the subject (Boler, 1963).
[12] Papers in this volume seem to indicate some turning in Australian analytic metaphysics from a previous hyper-Humeanism, in the sense of Humeanism according to which there are no necessary connections between distinct existences, back towards mediaeval essentialist views which do countenance such connections. Yet the only option being canvassed seems to be a relatively crude Aquinean ontology of pure matter and essential properties (or 'forms'). Given that the medievals thought so carefully about these issues, it might be worth examining other alternatives such as Scotus' as well. For example, it might be worth think-

ing about Scotus' subtle middle-ground between "distinct existence" and full-blown identity with respect to different instances of the same property – that is, his famous "formal distinction". Another possibility is to consider Scotus' explicit inclusion of haecceities in his ontological mix, and the merits of such a theory with respect to the individuation of entities in scientific contexts.

[13] The usefulness of Peirce's "Harvard experiment" for current discussions of scientific realism has been noted by Robert Almeder (Almeder, 1992, pp. 147–51). He claims that, in this regard, "It is noteworthy that Peirce's argument has received so little attention" (Almeder, p. 150). However Almeder construes Peirce's argument somewhat more broadly than Peirce himself presented it: as an argument not just for real generals, but for the reality of the "external world".

[14] With the following qualification: abduction is argument to the best explanation of something surprising. This is important in spelling out the role abduction plays in inquiry – it steps in when a theory runs up against some evidential obstacle, and generates new hypotheses which explain the obstacle, and are then tested to see if they hold water. See for instance (Peirce, 1955, p. 151), where Peirce sets out the logical form of abductive inference as follows:

> The surprising fact, C, is observed.
> But if A were true, C would be a matter of course.
> Hence, there is reason to suspect that A is true.

This account of abduction might seem to conflict with Peirce's rhetorical use of the stone in the "Harvard experiment". For surely the fact that stones fall when dropped is one of the more unsurprising facts about life on Earth, and it seems that Peirce is in a sense drawing on this extreme obviousness to make his rhetorical point that we all know what will happen to the stone, and thus that we are committed to realism about natural law whether we know it or not. Yet Peirce's apparently simple rhetorical gesture can be seen to work on a yet deeper level, if we consider that what he is drawing attention to by waving the stone above the ground so suggestively is the remarkable fact that we are so unsurprised when objects that are unsupported fall to the ground, given that there are many other perfectly intelligible future possibilities with respect to what those objects might do that we do not even consider. Given nominalism, this fact about us is entirely surprising, and thus Peirce's real rhetorical aim is to uncover this surprise, to draw attention to it, and to suggest realism as the only explanation for it.

[15] Douglas Anderson has argued similarly (Anderson, 1995, p. 91), describing Peirce's response to global scepticism about scientific method (which method Peirce views as inseparable from the hypothesis of realism currently under discussion) as a "transcendental abduction": a phrase of some power.

[16] Thanks are due to Peter Skagestad for generously pointing this out to me in private correspondence.

[17] This anecdote is related by Joseph Dauben in his (1995, footnote 38).

[18] Acknowledgements: In writing this paper I have benefited greatly from discussions on the email list "Peirce-L" (peirce-l@ttacs6.ttu.edu), particularly with Peder Christiansen, Arien Malec, Joseph Ransdell, Peter Skagestad, and (the late and sadly missed) Tom Anderson. Thanks are due also to Howard Sankey, Huw Price and an anonymous referee for helpful criticisms.

REFERENCES

Adams, Marilyn (1987), *William Ockham*, South Bend, Indiana: University of Notre Dame Press

Almeder, Robert (1980), *The Philosophy of Charles S. Peirce: A Critical Introduction*, Oxford: Blackwell

Anderson, Douglas (1995), *Strands of System: The Philosophy of Charles Peirce*, Indiana: Purdue University Press

Boler, John (1963), *Charles Peirce and Scholastic Realism: A Study of Peirce's Relation to John Duns Scotus*, Seattle: University of Washington Press.

Carre, Meyrick H. (1946), *Realists and Nominalists*, Oxford: Clarendon Press

Dauben, Joseph (1995), 'Peirce and History of Science', *Peirce and Contemporary Thought*, ed. Ketner, K., New York: Fordham University

Duns Scotus, John (1987), *Philosophical Writings*, trans. Allan Wolter, Indianapolis: Hackett Publishing Company

Haack, Susan (1992), "'Extreme Scholastic Realism': Its Relevance to Philosophy of Science Today", *Transactions of the Charles S. Peirce Society*, **XXVIII**, No. 1, 19–50

Hausman, Carl (1993), *Charles S. Peirce's Evolutionary Philosophy*, Cambridge: Cambridge University Press
Hookway, Christopher (1993), 'Belief, Confidence and the Method of Science', *Transactions of the Charles S. Peirce Society*, **XXIX**, No. 1, 1–32
Josephson, Jon and Susan (1994), *Abductive Inference*, Cambridge: Cambridge University Press
Peirce, Charles S. (1955), *Philosophical Writings of Peirce*, ed. Justus Buchler, New York: Dover Publications
Peirce, Charles S. (1960), *Collected Papers*, vols 1, 2, 5, 6, Cambridge, Mass.: Harvard University Press
Peirce, Charles S. (1992), *Reasoning and the Logic of Things: Cambridge Conferences Lectures, 1898*, ed. Ketner, K., Cambridge, Mass.: Harvard University Press
Quine, W.V.O. (1960), *Word and Object*, Cambridge, Mass: MIT Press
Rorty, Richard (1982), *Consequences of Pragmatism: Essays 1972–80*, Minneapolis: University of Minnesota Press
Skagestad, Peter (1980), 'Pragmatic Realism: The Peircean Argument Reexamined', *Review of Metaphysics*, **XXXIII**, No. 3, 527–40
Skagestad, Peter (1983), 'Peirce and Pearson: Pragmatism vs. Instrumentalism', *Language, Logic and Method*, ed. Cohen, R.S. & Wartofsky, M.W., Dordrecht: Reidel, 263–82
Skagestad, Peter (1987), 'Peirce's Conception of Truth: A Framework for Naturalistic Epistemology?', *Naturalistic Epistemology*, ed. Shimony, A. & Nails, D., Dordrecht: Reidel, 73–90

DAVID LEWIS

FINKISH DISPOSITIONS

I. THE CONDITIONAL ANALYSIS REFUTED

The analysis stated

All of us used to think, and many of us still think, that statements about how a thing is disposed to respond to stimuli can be analysed straightforwardly in terms of counterfactual conditionals. A fragile thing is one that would break if struck; an irascible man is one who would become angry if provoked; and so on. In general, we can state the *simple conditional analysis* thus:

> Something x is disposed at time t to give response r to stimulus s iff, if x were to undergo stimulus s at time t, x would give response r.

Simple indeed – but false. The simple conditional analysis was decisively refuted by C.B. Martin. The refutation has long been a matter of folklore – I myself learned of it from Ian Hunt in 1971 – but now it has belatedly appeared in print.[1]

How a disposition can be finkish

Dispositions come and go, and we can cause them to come and go. Glass-blowers learn to anneal a newly made joint so as to make it less fragile. Annoyances can make a man irascible; peace and quiet can soothe him again.

Anything can cause anything; so stimulus s itself might chance to be the very thing that would cause the disposition to give response r to stimulus s to go away. If it went away quickly enough, it would not be manifested. In this way it could be false that if x were to undergo s, x would give response r. And yet, so long as s does not come along, x retains its disposition. Such a disposition, which would straight away vanish if put to the test, is called *finkish*. A finkishly fragile thing is fragile, sure enough, so long as it is not struck. But if it were struck, it would straight away cease to be fragile, and it would not break.

Any finkish disposition is a counter-example to the simple conditional analysis. The thing is disposed to give response r to stimulus s; it is not true that if it were to undergo s, it would give response r. The analysandum is true, the alleged analysans is false.

H. Sankey (ed.), Causation and Laws of Nature, 143–155.

How a lack of a disposition can be finkish

Suppose instead that we have something that is not yet disposed to give *r* in response to *s*. It might gain that disposition; and *s* itself might be the very thing that would cause it to gain that disposition. If the disposition were gained quickly enough, while *s* was still present, it would at once be manifested. So the counterfactual analysans is true: if the thing were to undergo *s*, it would give response *r*. And yet, so long as *s* does not come along, the dispositional analysandum is false: the thing has not yet gained the disposition to give response *r* to *s*. This time, it is the lack of the disposition that is finkish, but again we have a counter-example to the simple conditional analysis.

Dispositions with finkish partners

Dispositions, as Martin has often emphasized, can come in pairs: *x* is disposed to respond to the presence of *y*, and *y* is disposed to respond to the presence of *x*, by a response *r* given jointly by *x* and *y* together.[2] In a nice case, where the simple conditional analysis works, we can express this by a counterfactual: if *x* and *y* were to come into one another's presence, they would jointly give response *r*.

(Or, more generally: if *x* and *y* were to enter into such and such relationship . . . . But let us stick to the case where the relationship is a matter of proximity.)

For example, I and a certain disc are so disposed that if I and it came together, it would cause in me a sensation of yellow. We could say that it is disposed to influence me; or that I am disposed to respond to it. Or both. Or we could say that the two-part system consisting of me and the disc is disposed to respond to the coming together of its parts. In the nice case, where the simple conditional analysis works, it does not matter which we say.

But in a finkish case, perhaps the coming together of me and the disc would alter my dispositions, or the disc's dispositions, or both, so that if I and it came together, there would be no sensation of yellow. The disposition of the two-part system to respond to the coming together of its parts is finkish in just the way we have already considered.

Nothing new yet. But suppose we want to speak not only about the dispositions of the two-part system but also about the dispositions of the two parts, of me and of the disc. It might be that the coming together would alter my dispositions, but would have no effect on the disc's dispositions. Then my disposition to respond to the disc would be finkish, but the disc's disposition to influence me would not be.

Yet if the disc's disposition is not finkish (that is, if it is not itself a counterexample to the simple conditional analysis) why would it not be manifested? Because it is a disposition to influence me-as-I-*would*-be-if-I-had-not-lost-my-own-finkish-disposition; and that is not how I would be if I and the disc came together. Because of the finkishness of my disposition, the unfinkish disposition of the disc can have no occasion to be manifested.

Saul Kripke has imagined a special shade of yellow, 'killer yellow', which, thanks to some quirk of our neural wiring, would instantly kill anyone who set eyes on it.[3] If what I have just said is right, then, whatever else may fairly be said against a dispositional theory of colours, the case of killer yellow does not suffice as a refutation.

Resisting the refutation: a dilemma about timing?

Philosophers being what they are, not everyone will find Martin's refutation of the simple conditional analysis immediately convincing.

One line of resistance begins with a dilemma about timing. A thing might have a finkish disposition to give response *r* to stimulus *s*. Since the disposition is finkish *s* would cause it to go away. But would it go away instantly?

If no, there would be a little time after the advent of *s* and before the disposition goes away. During this little time, before the disposition goes away, we would have *s* and we would still have the disposition. Then would we not have *r* after all? Then is not the conditional analysans true despite the finkishness of the disposition?

If yes, on the other hand, the case seems to involve a kind of instantaneous causation that is contrary to the normal ways of the world. The resister may protest with some justice that the case is fantastic, that we are not entitled to firm linguistic intuitions about such far-fetched cases, and accordingly that the case is not a convincing refutation.

We might reply by proposing a case in which the disposition would be gone by the time *s* arrived, but not by means of instantaneous causation. Rather, the finkishly disposed thing would somehow see *s* coming. Some precursor of *s* would cause both *s* and the loss of the disposition.

But then the resister can insist that the counterfactual analysans, if properly interpreted, is true after all. If we counterfactually suppose that *s* happens at time *t*, and we hold fixed the actual course of events before *t*, our supposition does not include *s*'s precursor. Then neither does it include any side-effects of *s*'s precursor, such as the loss of the disposition. Under the supposition of *s* without *s*'s precursor, *r* would have followed. It is a familiar point that backtracking counterfactual reasoning, which runs from a counterfactually supposed event to the causal antecedents it would have to have had, is sometimes out of place. The resister need only insist that the counterfactuals whereby we analyse dispositions must not be backtrackers.[4]

Our best hope for an uncontroversial case of a finkish disposition (though I myself also accept the controversial cases that work by instantaneous causation) will be to return to the first horn of the resister's dilemma. That means that *s* would arrive at least a short time before the disposition went away. Does it really follow that we would have *r*? Not necessarily. Sometimes it takes some time for a disposition to do its work. When stimulus *s* arrives and the disposition is present, some process begins. (It might be a process of accumulation: of charge, of neurotransmitter, of tiny cracks, of vexation, etc.) When the process reaches completion, then that is, or that causes, response *r*. But if the disposition went away part-way through, the process would be aborted. In such a case, the disposition to produce *r* can be finkish, without any need either for instantaneous causation or for backtracking. (However, the disposition to begin the process is not finkish.) So the resister's dilemma about timing is answered, and the refutation of the simple conditional analysis is unscathed.

Martin's principal example in his (1994) is an 'electro-fink': a machine connected to a wire that makes the wire instantly become live if touched by a conductor; or, if operating 'on a reverse cycle', makes the wire instantly cease to be live if touched by a conductor. It is instructive to see how to amend this example so that it withstands the resister's misgivings. (a) We remove Martin's stipulation that the electro-fink reacts instantaneously. Quickly is good enough. Then the electro-fink on a reverse cycle need not be anything more remarkable than a (sensitive and fast-acting) circuit breaker. (b) We respecify the effect to which the wire is finkishly disposed not as any flow of electrical current but as flow of a certain wattage for a certain duration – as it might be, enough for electrocution. This is a process that can be aborted by breaking the circuit part-way through.

Resisting the refutation: a compound disposition?

A different line of resistance suggests that if something is finkishly disposed to give response *r* to stimulus *s*, what it really has is a compound disposition. It has a state that at least resembles a disposition to give response *r* to *s*. Our resister, since he accepts the simple conditional analysis, will think it inaccurate to call this state a disposition. (I shall signal his terminological scruples with inverted commas.) At any rate, this first 'disposition' is embedded in a second disposition. The thing is disposed to lose the first 'disposition' in response to *s*.

Now the resister is struck by the difference between the first 'disposition' all by itself and the first 'disposition' when it is embedded in the second. He implores us not to be over-impressed by such similarities as there are, and instead to heed the difference the second disposition makes to the overall dispositional character of the thing. When we say that the thing is disposed to give *r* in response to *s*, he thinks we are misled by thinking of the first 'disposition' in abstraction from the second.

Well, that may be so, or it may not, in the sort of case the resister has in mind. (I myself think it is not so.) Be that as it may, there is a different sort of case. It may be that the thing would lose the first 'disposition' in response to *s*, but *not* because of any second disposition of that thing; rather because of something wholly extrinsic.

A sorcerer takes a liking to a fragile glass, one that is a perfect intrinsic duplicate of all the other fragile glasses off the same production line. He does nothing at all to change the dispositional character of his glass. He only watches and waits, resolved that if ever his glass is struck, then, quick as a flash, he will cast a spell that changes the glass, renders it no longer fragile, and thereby aborts the process of breaking. So his finkishly fragile glass would not break if struck – but no thanks to any protective disposition of the glass itself. Thanks, instead, to a disposition of the sorcerer.

I have replied to the resister by wielding an assumption that dispositions are an intrinsic matter. (Except perhaps insofar as they depend on the laws of nature. I myself would wish to insist on that exception, but this is a controversial matter that need not be considered now.) That is: if two things (actual or merely possible) are exact intrinsic duplicates (and if they are subject to the same laws of nature) then they are disposed alike. I have used this premise twice over. Suppose the sorcerer's protected glass and another, unprotected, glass off the same production line are intrinsic duplicates (and both subject to the actual laws of nature). Then they are disposed alike. Certainly the unprotected glass is disposed to break if struck; therefore, so is the sorcerer's glass. Certainly the unprotected glass is not disposed to lose its fragility if struck; therefore, neither is the sorcerer's glass.

I do not deny that the simple conditional analysis enjoys some plausibility. But so does the principle that dispositions are an intrinsic matter. The case of the sorcerer sets up a tug-of-war between conflicting attractions, and to me it seems clear that the simple conditional analysis has the weaker pull.

At least in such cases, Martin's refutation succeeds. I myself think it succeeds in other cases as well. But to refute an analysis, one counterexample is all we need.

Whither?

Once we scrap the simple conditional analysis, what should we say about dispositions? Martin's own response is radical: a theory of irreducible dispositionality. Properties are Janus-faced: each of them has, inseparably, a qualitative (or 'categorical') and a dispositional aspect. Since dispositionality is irreducible, it is not to be explained in terms of the causal and nomological roles of properties, but rather *vice versa*.[5]

Those who are disappointed with the usual menu of theories of lawhood and causation might do well to try out this new approach. But those of us whose inclinations are more Fabian than revolutionary, and who still back one or another of the usual approaches to lawhood and causation, may well suspect that Martin has over-reacted. If what we want is not a new theory of everything, but only a new analysis of dispositions that gets right what the simple conditional analysis got wrong, the thing to try first is a not-quite-so-simple conditional analysis. Rather than starting with irreducible dispositionality, as Martin does, we shall start with fairly widely shared ideas about properties, causation, lawhood, and counterfactuals; and on this foundation we shall hope to build a reformed conditional analysis of dispositions.

II. A REFORMED CONDITIONAL ANALYSIS

Causal bases

Suppose that a certain glass is (non-finkishly) fragile; and it is struck; and so it breaks. The breaking presumably was caused; and caused jointly by the striking and by some property *B* of the glass. We call this property *B*, a property which would join with striking to cause breaking, a *causal basis* for the fragility of the glass.

Three comments. (a) Different fragile things may have different causal bases for their fragility. (b) Strictly speaking, it is the having of the property that does the causing: a particular event, or perhaps a state of affairs. To speak of the property itself as a cause is elliptical. (c) What causes what depends on the laws of nature. If lawhood is a contingent matter, as many but not all of us think it is, then it is also a contingent matter which properties can and which cannot serve as causal bases for fragility.

Prior, Pargetter, and Jackson have argued convincingly for the thesis that all dispositions must have causal bases.[6] Let us assume this. Or at any rate, let us agree to set aside baseless dispositions, if such there be. Our goal, for now, is a reformed conditional analysis of based dispositions – including finkish ones.

(Prior *et al.* argue from a simple conditional analysis of dispositions. But that flaw in their argument is not a serious one. Though wrong as an analysis, the simple conditional analysis remains true as a rough and ready generalization: fragile things that are struck do for the most part break, and those that are unstruck would for the most part break if they were struck. So, despite the possibility of finkish fragility, still for the most part we must posit causes for the breakings that fragile things do or would undergo.)

A finkish disposition is a disposition with a finkish base. The finkishly fragile glass has a property *B* that would join with striking to cause breaking; and yet the glass would not break if struck. Because if the glass were struck, straight away it would lose the property *B*. And it would lose *B* soon enough to abort the process of breaking.

Then is it true to say, as I did, that *B* "would join with striking to cause breaking"? Yes and no. What I meant, when I said that, was that if the glass were struck and retained *B*, then *B* together with the striking would cause breaking. That much is true. And yet it is also true that if the glass were struck it would not retain *B*. Thus the possibility of finkishness rests on a logical peculiarity of counterfactuals: their 'variable strictness'.[7] It can happen that two counterfactuals

If it were that *p*, it would be that not-*q*
If it were that *p* and *q*, it would be that *r*

are true together, and that the truth of the second is not merely vacuous truth. Because the first counterfactual is true, the supposition that *p* and *q* is more far-fetched, more 'remote from actuality', than the supposition just that *p*. But we are not forbidden to entertain a supposition merely because it is comparatively far-fetched. Variable strictness means that some entertainable suppositions are more far-fetched than others.

The finkish lack of a disposition works in a parallel way. The glass has no causal basis for fragility, therefore it is not fragile. Yet it would break if struck. Because, if it were struck, it would straight away gain some property *B* that would serve as a causal basis for fragility. And *B* would arrive in time (though maybe only just in time) to join with the striking to cause the glass to break.

(But will not the striking be over and done with by the time *B* arrives? Not necessarily. And even if it is, *B* could join with after-effects of the striking to cause the breaking. Then the striking would still be a cause of the breaking via a causal chain passing through the after-effects.)

Once we appreciate that finkishness pertains, in the first instance, to particular causal bases and to lacks of particular causal bases, we are in a position to describe a variety of finkishness that has so far escaped our notice. Suppose that B_1 and B_2 are two alternative causal bases for fragility. As it actually is, the glass has B_1 and lacks B_2. But if it were struck, it would undergo a swap: straight away, it would lose the property B_1 and gain the property B_2. It finkishly has one basis for fragility and it finkishly lacks another. Yet it is not finkishly fragile, at least not in the sense of being a counter-example to the simple conditional analysis. It is fragile thanks to the basis B_1. If struck, it would be fragile thanks instead to the substitute basis B_2. If struck, therefore, it would break. But its breaking if struck would not be a manifestation of the fragility it has when not struck, because if it were struck it would come to be fragile in a different way.

We need to add something to our characterization of finkish fragility, so as to distinguish it from the different situation just considered. As follows: the finkishly fragile glass has a property *B* that would join with striking to cause breaking, yet the glass would not break if struck. Because if it were struck, it would lose *B*, *and it would not gain any substitute basis for fragility*.

Toward an analysis: beginning

Once we have accepted the thesis that all dispositions must have causal bases, it is an easy step to conjoin the converse thesis and to say, for instance, that something is fragile if and only if it has some causal basis for fragility. That biconditional, generalised and spelt out, shall be our reformed analysis of dispositions. In saying what it means for a property to be a causal basis for fragility, or whatever, we shall need a counterfactual conditional. But the conditional part of our reformed analysis will come at the end. Before that, we need a beginning and a middle.

The beginning of any analysis is an analysandum. Ours shall be as follows:

Something x is disposed at time t to give response r to stimulus s iff . . . .

The noteworthy thing about our analysandum is what it is *not*. Our plan is to answer one question without getting entangled in another. The question we want to answer is "What is it to *have* such-and-such disposition (as it might be, the disposition to break if struck)?" The question we want to leave unsettled is "What *is* a disposition?"

Once we accept that a disposition must have a causal basis, we might choose to say, as Armstrong has done, that the disposition *is* its causal basis. That choice has the advantage of delivering a straightforward account of the role of dispositions in causal explanation: the fragility of the glass, along with the striking, are the causes that jointly cause the breaking. On the other hand, that choice has the drawback that what we would offhand think was *one* disposition, fragility, turns out to be different properties in different possible cases – and, very likely, in different actual cases.[8]

Or we might instead choose to say, as Prior and her allies have done, that the disposition is the second-order property of having some suitable causal basis or other.[9] That way, fragility is indeed a single property common to all fragile things, actual or merely possible. However, the drawback of this choice is that if fragility is the second-order property, it is far from clear how it plays a role in causal explanation. When the struck glass breaks, do we want to say that the breaking is caused *both* by the second-order property which is the fragility *and* by whatever first-order property is the causal basis for the fragility in that particular case? It's not a case of overdetermination, after all! But neither should we want to say, as Prior *et al.* do, that fragility is causally impotent.

If forced to choose, I would side with Prior against Armstrong; and I would dodge the overdetermination-or-impotence issue by appeal to some fancy and contentious metaphysics. (Thus. Let us speak of the *relata* of the causal relation as 'events', whether or not that is altogether appropriate as a matter of ordinary language. Sometimes an event, in this sense, is a having of a certain property by a certain thing.[10] Now we can say that just one event joins with the striking to cause the breaking, so there is no overdetermination. This one event is a having of the causal basis. But also, perhaps in a different sense, this same event is a having of the second-order property. Two different properties are had in the same single event. So the second-order property is not impotent.) This may work, but it is complicated and contentious and best avoided for as long as possible. Our choice of an analysandum is meant to allow us to remain neutral in the disagreement between Armstrong and Prior. When a glass is fragile, it has two properties. It has some first-order property which is a causal basis for fragility; it also has the second-order property of having some causal basis for fragility or other. We need not say which of these two properties of the glass is its fragility.[11]

If we remain neutral in the disagreement between Armstrong and Prior, not only do we refuse to say which properties are dispositional; equally, we refuse to say which properties are *non*-dispositional, or 'categorical'. So we would be unwise to speak, as many do, of 'categorical bases'. Because if we then saw fit to go Armstrong's way, and to identify the disposition itself with its causal basis (in a particular case), we would end up claiming to identify dispositional with non-dispositional properties, and claiming that dispositions are their own categorical bases! Rather than risk such confusion, we do better to eschew the alleged distinction between dispositional and 'categorical' properties altogether.

Our chosen analysandum has another advantage: generality. Suppose instead that we had taken some particular example of a dispositional concept: the concept of a poison, say, or the concept of fragility or the concept of a lethal virus. A dispositional concept is the concept of being disposed to give such-and-such response to such-and-such stimulus. So the first problem we face in analysing any particular dispositional concept, before we can turn to the more general questions that our particular example was meant to illustrate, is the problem of specifying the stimulus and the response correctly.

We might offhand define a poison as a substance that is disposed to cause death if ingested. But that is rough: the specifications both of the response and of the stimulus stand in need of various corrections. To take just one of the latter corrections: we should really say "if ingested without its antidote". Yet the need for this correction to the analysis of 'poison' teaches no lesson about the analysis of dispositionality in general.

(Some, for instance Johnston, might doubt the need for the correction.[12] They say that a disposition may be masked by something that prevents the response even when both the stimulus and the causal basis are present; in this way, we get failures of the conditional analysis even when the causal basis is not finkish. One who is prepared to speak of masking might stay with the simple definition of a poison as a substance disposed to cause death if ingested, but might say as well that the disposition of poisons to kill is masked by antidotes. Perhaps we have no substantive issue here, but only a difference between styles of book-keeping. But if so, I think the masker's style is less advantageous than it may seem. For even if we say that the poison has the disposition spelt out in the simple definition, and we say as well that this disposition is masked by antidotes, do we not still want to say that the poison has the further disposition spelt out in the complicated corrected definition?)

Or, to take fragility: we have said so far, and we shall go on saying, when greater precision is not required, that being fragile means being disposed to break if struck. But what of this story (due, near enough, to Daniel Nolan)? When a styrofoam dish is struck, it makes a distinctive sound. When the Hater of Styrofoam hears this sound, he comes and tears the dish apart by brute force. So, when the Hater is within earshot, styrofoam dishes are disposed to end up broken if struck. However there is a certain direct and standard process whereby fragile things most often (actually, nowadays, and hereabouts) break when struck, and the styrofoam dishes in the story are not at all disposed to undergo that process.[13] Are they fragile? To say so would be at best a misleading truth, and at worst an outright falsehood; and I have no idea which. However, my purpose in raising this question was *not* to answer it, but rather to insist that it is merely the question of which response-specification is built into the particular dispositional concept of fragility. Once again, it affords no lesson about dispositionality in general.

To show this, I turn to a case that goes differently. A certain virus is disposed to cause those who become infected with it to end up dead before their time, but *not* to undergo the direct and standard process whereby lethal viruses mostly kill their victims. For this virus does not itself interfere with any of the processes that constitute life. Rather, it interferes with the victim's defences against *other* pathogens – whereupon those other pathogens, like the Hater of Styrofoam, do the dirty work. Do we call this a lethal virus? Of course we do. After all, my story of the virus is not just another philosophical fantasy! It is the true story of HIV, slightly simplified.

We should not think, therefore, that dispositional concepts generally have built-in response-specifications requiring a direct and standard process. The concept of fragility does. (Though whether it is built in as a matter of truth-conditions or as a matter of implicature remains unclear.) The concept of a lethal virus does not.

Toward an analysis: middle

We begin our analysans with a restricted existential quantifier over properties:

. . . iff, for some suitable property *B* that *x* has at *t* . . . .

'Suitable', of course, is a mere place-holder. We want to restrict the quantification to properties that can serve as causal bases for a disposition.

We need to require that *B* is a property (a having of) which can cause something. But we shall provide for this later, in the conditional part of the analysis: we shall say counterfactually what *B* would cause. So it is unnecessary to add a requirement of causal potency at this point as well.

Some would deny that negative properties, such as the absence of force or fear or food, can do any causing. Should we then impose a restriction that properties suitable as causal bases for dispositions must be entirely positive (whatever that means)? No. For everyone agrees that negative properties make some sort of difference to what happens, and the difference they make is causal. Martin puts the point thus: "Absences and voids are causally *relevant* but not causally *operative*".[14] I myself would draw no such distinction between 'causation' and 'causal relevance'. But if others can make good on this supposed distinction, let them by all means help themselves to it. Anyhow, call it what you will, what matters is that we not omit the causal difference-making of negative properties from the causal roles of bases for dispositions. Therefore we want no restriction to positive properties.

What we do need to require is that *B* is an intrinsic property of *x*. Earlier, we considered and accepted a principle that dispositions are an intrinsic matter. If causal bases could be extrinsic then it could happen, contrary to that principle, that two intrinsic duplicates (subject to the same laws of nature) were differently disposed, because of some difference in their extrinsic causal bases.

We illustrated the principle that dispositions are an intrinsic matter by the case of the sorcerer and his protected glass. But to illustrate the principle further, and to placate those who will not be convinced by fantastic examples, I offer the case of Willie. Willie is a dangerous man to mess with. Why so? Willie is a weakling and a pacifist. But Willie has a big brother – a very big brother – who is neither a weakling nor a pacifist. Willie has the extrinsic property of being protected by such a brother; and it is Willie's having this extrinsic property that would cause anyone who messed about with Willie to come to grief. If we allowed extrinsic properties to serve as causal bases of dispositions, we would have to say that Willie's *own* disposition makes him a dangerous man to mess about with. But we very much do not want to say that. We want to say instead that the disposition that protects Willie is a disposition of Willie's brother. And the reason why is that the disposition's causal basis is an intrinsic property of Willie's brother.

If we insist that dispositions must have intrinsic causal bases, we run a risk of surprises. It just might turn out, for example, that electrons are not after all disposed to repel one another. Because it just might turn out that negative charge, the causal basis of the repulsion, was an extrinsic property involving the state of the surrounding aether. How bad would that be? Not so bad, I think, that we ought to buy immunity from such surprises at the cost of saying the wrong thing about dangerous Willie.

Toward an analysis: end

Now at last we reach the conditional part of our reformed conditional analysis, the counterfactual which says that property B is a causal basis for x's disposition to give response r to stimulus s. We shall proceed by successive approximations; asterisks will mark attempts due for subsequent rejection.

Even if B is finkish and would go away in response to s, the counterfactual supposition we want to consider is that s arrives and B nevertheless remains. How long? Long enough to finish the job of causing r, however long that job may take.

> * . . . for some time t' after t, if x were to undergo stimulus s at time t and retain property B until t', x would give response r.

The quantificational prefix and the antecedent are now in final form, but the consequent still will not do.

For all that the analysans in its present form tells us, x might finkishly lack fragility: it might be that x would break if struck, but no thanks to any disposition that x already had when unstruck. Yet our quantified counterfactual might come out true. B might be some property entirely unconnected with the breaking: x's colour, say. Or B might be connected in the wrong way with the breaking: logically instead of causally. For instance, B might be the property of either being unstruck or breaking (provided we understand the first disjunct as well as the second in a way that makes it intrinsic). To exclude such inappropriate choices of B, we amend the consequent:

> * . . . s and x's having of B would jointly cause x to give response r.

(In case we have chosen to circumvent the alleged impotence of the second-order property in the way considered earlier, we had better say that 'x's having of B' here is to be understood in the sense in which an event is a having of the causal basis, not the different sense in which that same event is a having of the second-order property.)

There is one more problem. (Martin pointed it out to me. At least, I think this is the problem he had in mind.) It involves what we might call a finkish partial lack of a causal basis. The glass has property *B* but it lacks property *B′*. *B* and *B′* together would constitute a causal basis for breaking if struck; that is, striking and having *B* and having *B′* would together cause breaking. *B* alone is not a causal basis: striking and having *B* would not suffice to cause breaking. But the lack of *B′* is a finkish lack. If the glass were struck, straight away it would gain *B′*; and in addition it would retain *B*; and so it would break. And *B*, together with the striking, would be a cause of the breaking. Not, indeed, the complete cause; but a part of the cause is still a cause, so our analysans in its present form is satisfied. And yet because of the lack of *B′* it seems false that the unstruck glass is fragile. In short, the problem of finkish lacks has reappeared within our conditional analysis of what it is to be a causal basis.

The solution is to make one final amendment to the consequent of our counterfactual. We have the notion of a complete cause of an effect. (Mill called it the 'whole cause'. I use a different term to mark that we need not be committed to Mill's own analysis.) We can introduce a restriction of that notion: a cause complete in so far as havings of properties intrinsic to *x* are concerned, though perhaps omitting some events extrinsic to *x*. For short, 'an *x-complete* cause'. In the example just considered, the striking plus *x*'s having of *B* would indeed be a cause of the breaking, but not an *x*-complete cause. So our amended consequent is:

> . . . *s* and *x*'s having of *B* would jointly be an *x*-complete cause of *x*'s giving response *r*.

Putting all the bits together, our reformed conditional analysis runs as follows:

> Something *x* is disposed at time *t* to give response *r* to stimulus *s* iff, for some intrinsic property *B* that *x* has at *t*, for some time *t′* after *t*, if *x* were to undergo stimulus *s* at time *t* and retain property *B* until *t′*, *s* and *x*'s having of *B* would jointly be an *x*-complete cause of *x*'s giving response *r*.

An unlovely mouthful! But I think there's reason to hope that it will do the job.

Being oppositely disposed

A surprising, but unobjectionable, consequence of our reformed conditional analysis is that the same thing, at the same time, may be disposed in two opposite ways: as it might be, to break if struck and also not to break if struck. Of course, one of the two opposite dispositions will have to be finkish. Further, it will have to be the kind of finkish disposition that involves a compound disposition rather than an extrinsic intervention. That may not be the best kind for convincing the resister, but I myself still think it is one possible kind of finkish disposition.

The finkishly fragile glass has intrinsic properties *B* and *B**. *B* is an *x*-complete causal basis for breaking if struck; *B** is an *x*-complete causal basis for losing *B* if struck, and also for not breaking if struck. Thanks to *B*, the glass is finkishly disposed to break if struck. Yet thanks to *B** it also is non-finkishly disposed not to break if struck.

An unsatisfactory reformulation

Given that dispositions must have causal bases, and given that causal bases must be intrinsic, we might hope to stay closer to the simple conditional analysis. How about this, for instance?

> The glass is fragile iff, if it were struck and its intrinsic character were unchanged, it would break.

Martin has warned us that it will not help just to insert a '*ceteris paribus*' into the simple conditional analysis, because when the time comes to say explicitly what is to be held fixed, we shall want to say that it is the dispositional character of the glass that is to be held fixed – and if we say that, our conditional analysis of dispositions becomes circular.[15]

But that was not what we said – rather we said that the intrinsic character was to be held fixed. So Martin's warning does not apply. (Or not unless intrinsic character must somehow be analysed in terms of dispositions, which seems unlikely.)

Holding fixed the intrinsic character means holding fixed all the intrinsic causal bases (and all the lacks thereof) which underlie the dispositions (and lacks of dispositions) of the glass. That would solve the problem of finkishness.

But the solution doesn't work, because holding fixed the intrinsic character of the glass means holding fixed altogether too much. If the glass were struck and its intrinsic character were unchanged, it would indeed retain the intrinsic causal basis of its fragility. But also it would be not at all deformed, not at all compressed, not at all afflicted with vibrations or shock waves, etc. So it would *not* break.

What it would do is astonish a sufficiently knowledgeable observer. We can agree that the glass does have a disposition to astonish such an observer – an extremely finkish disposition, with the entire intrinsic character of the glass as its causal basis. That is not the only disposition the glass has for responding to being struck; and not the most noteworthy disposition. Yet it is this disposition, and not any opposite disposition, that our present proposal deigns to notice.[16]

Princeton University

NOTES

[1] Martin (1994). See also Shope (1978); Johnston (1992).

[2] See, e.g., Martin (1996, pp. 62ff).

[3] The example occurs in unpublished lectures. I am obliged to note that I am not reporting the whole of what Kripke said in those lectures. For further discussion of killer yellow, see my (1997).

[4] See my (1979).

[5] See Martin (1996, pp. 62ff), (1993) and elsewhere.

[6] Prior, Pargetter and Jackson (1982, pp. 251–253). Earlier discussions of dispositions and their causal bases include Quine (1960, pp. 222–226); Armstrong (1968, pp. 85–88), (1973, pp. 11–16); Mackie (1973, pp. 129–148), (1977, pp. 361–370).

[7] Stalnaker (1968); Lewis (1973).

[8] Armstrong, (1973, pp. 14–16).

[9] Prior, Pargetter and Jackson (1982 pp. 253–256); Prior (1985, pp. 82–95).

[10] For details, see Lewis (1986).

[11] Mumford (1996), gives a reply to Martin which agrees to a considerable extent with mine, but which is built upon an answer to the very question that I have taken care to bypass, namely, the question what dispositions are.

[12] Johnston (1992, p. 233).

[13] Cf. A. D. Smith (1977, p. 444)

[14] Martin (1996, p. 64).

[15] Martin (1994, pp. 5–6).

[16] Thanks are due to C.B. Martin, Allen Hazen, Daniel Nolan, Barry Taylor, and others; and to the Boyce Gibson Memorial Library and Ormond College.

REFERENCES

Armstrong, D.M, (1968), *A Materialist Theory of the Mind*, London: Routledge & Kegan Paul

Armstrong, D.M. (1973), *Belief, Truth and Knowledge*, Cambridge: Cambridge University Press

Johnston, Mark (1992), 'How to Speak of the Colors', *Philosophical Studies* **68**, 221–263

Lewis, David (1973), *Counterfactuals*, Oxford: Blackwell

Lewis, David (1979), 'Counterfactual Dependence and Time's Arrow', *Noûs* **13**, 455–476

Lewis, David (1986), 'Events' in David Lewis, *Philosophical Papers*, Volume II, Oxford: Oxford University Press

Lewis, David (1997) 'Naming the Colours', *Australian Journal of Philosophy* **75**, 325–342

Mackie, J.L. (1973), *Truth, Probability, and Paradox*, Oxford: Oxford University Press

Mackie, J.L. (1977) 'Dispositions, Grounds, and Causes', *Synthese* **34**, 361–370

Martin, C.B. (1993), 'Power for Realists' in J. Bacon, K. Campbell, and L. Reinhardt, eds., *Ontology, Causality and Mind*, Cambridge: Cambridge University Press

Martin, C.B. (1994), 'Dispositions and Conditionals', *The Philosophical Quarterly* **44**, 1–8

Martin, C.B. (1996), 'How It Is: Entities, Absences and Voids', *Australasian Journal of Philosophy* **74**, 57–64

Mumford, Stephen (1996), 'Conditionals, Functional Essences and Martin on Dispositions', *The Philosophical Quarterly* **46**, 86–92

Prior, Elizabeth W., Robert Pargetter, and Frank Jackson (1982), 'Three Theses About Dispositions', *American Philosophical Quarterly* **19**, 251–257

Prior, Elizabeth (1985), *Dispositions*, Aberdeen: Aberdeen University Press

Quine, W.V. (1960), *Word and Object*, Cambridge, Massachusetts: MIT Press

Shope, Robert K. (1978), 'The Conditional Fallacy in Contemporary Philosophy', *Journal of Philosophy* **75**, 397–413

Stalnaker, Robert (1968), 'A Theory of Conditionals' in N. Rescher, ed., *Studies in Logical Theory*, Oxford: Blackwell

Smith, A.D. (1977), 'Dispositional Properties', *Mind* **86**, 439–445

BARRY TAYLOR

COMMENTS ON DAVID LEWIS: 'FINKISH DISPOSITIONS'

Lewis's paper addresses the topic of dispositions, and Martin's views on them (see Martin, 1994), with his usual pellucid elegance. He contends that Martin has succeeded in refuting the 'simple conditional analysis' of dispositions. I agree. It is noteworthy how Lewis's exposition of Martin's case makes a small but significant improvement on the original. For Martin's counterexamples all involve some trickery with time – a divine agent with foreknowledge, an electro-fink capable of wreaking its effects instantaneously – which may seem to leave a ray of hope for the friend of the simple conditional analysis. Lewis's adaptations use no such tricks, and he makes it clear that hopes of saving the simple conditional analysis by claiming foul play with time are illusory.

The causal bases of dispositions loom large in Lewis's paper, both in the negative part in his exposition of finkish counterexamples, and in the positive part, where they figure crucially in the complex conditional analysis proposed to supplant the discredited simple version. It is worth remarking, however, that the negative point can be made without appeal to causal bases. Take Lewis's Sorcerer, for example. When his favourite glass is about to be struck, Lewis tells us that "quick as a flash, he will cast a spell that changes the glass, renders it no longer fragile, and thereby aborts the process of breaking". That is, Lewis's Sorcerer aborts the process by altering the causal basis of fragility in the glass. But he could abort it equally well by interfering, perhaps in different ways on separate occasions, with the causal chain linking the base with the breaking. (If it is objected that there is no such chain, the causation being immediate, let the example be changed; as Lewis says, to refute an analysis, one counterexample is all we need.) So the special status of the causal basis plays no crucial role in the Sorcerer counterexample, which could accordingly be adapted to refute the simple conditional analysis even of dispositions which have no causal base.

But are there any baseless dispositions? Lewis is inclined to think not, citing a 'convincing' argument by Prior, Pargetter and Jackson (1982) to that effect. At the same time, he prefers to adopt a conciliatory tone on the issue, inviting us to agree to "set aside baseless dispositions, if such there be", and aim at a reformed conditional analysis restricted to dispositions which *are* causally based. It is doubtful, however, whether Lewis can really afford the luxury of conciliation over this point. For we will be left with no analysis at all of baseless dispositions, if such there be. Yet, since counterexamples with a similar structure are what give rise to difficulties in analysing baseless and based dispositions alike, it seems plausible that an adequate solution should

H. Sankey (ed.), Causation and Laws of Nature, 157–159.

equally apply to both. It would be much nicer if Prior, Pargetter, and Jackson could be relied upon to have shown that baseless dispositions are a myth.

But, so far as I can see, their argument to this end is not available, since it relies upon the now-discredited simple analysis of dispositions. The argument comes in two parts, the first establishing that dispositions in deterministic worlds are causally based, the second generalising the result to indeterministic ones. In the first part, we are invited to suppose that our actual world is deterministic, and to consider an actual glass *A* and ask what happens in worlds closest to ourswhen *A* is knocked. Since, it is argued, these closest worlds must resemble our own sufficiently enough to be every bit as deterministic as it is, in such worlds it will be either determined that *A* breaks, or that *A* does not break. "In the latter case", they argue, "clearly *A* is not fragile" (Prior, Pargetter and Jackson, 1982, p. 251). But here they rely on the discredited principle. Thanks to Martin, we now know that there is a way in which *A* might be fragile and yet not break when struck in closest worlds, even deterministic ones; for it might be finkishly fragile.

Perhaps, however, I am, like others before me, making unnecessarily heavy weather of this issue. Dummett has suggested (e.g. Dummett, 1976, sec. 3) that part of the point of adopting dispositional discourse – of speaking categorically of bravery or fragility, rather than conditionally of what would happen if danger were faced or if striking occurred – is precisely to record a belief in an underlying state of things, causally responsible for the way they behave under relevant contingencies. Thus to say "Jones is brave" is to say that there is a mechanism in Jones in virtue of which he would behave bravely if confronted with danger, and accordingly is to say more than does the mere conditional "if Jones were to face danger he would behave bravely". Dummett draws various consequences from this point which are irrelevant for present purposes, and of course his discussion takes no account of the complications finkish dispositions entail. But we can still appeal to the central idea that dispositional discourse essentially involves the *hypothesis* of an underlying causal basis – a causal base, we can now say, having the characteristics required in Lewis's analysans; and hence that the idea of a baseless disposition is incoherent.

A final comment, still about causal bases. Lewis requires his causal bases to be intrinsic, citing weakling pacifist Willie to make out his case. His notion of intrinsicality, as explained in other places (see Lewis, 1986, pp. 61–9), depends on a primitive notion of natural properties.[1] I have had a shot at criticizing that notion elsewhere (Taylor, 1993), and will not repeat the arguments here. Suffice it to say that I would advocate an account which would make the naturalness, and hence the intrinsicality, of properties very much a matter of degree. The case of pacifist Willie would accordingly lead me to require that the properties figuring in a causal basis not be at the very bottom of the intrinsicality scale, but my relative notion of intrinsicality would allow me to be more flexible than Lewis about what may count as a causal basis. In view of the counterintuitive things Lewis admits he may have to say about the dispositions of electrons to repel one another if negative charge turns out not to meet his strict standards for intrinsicality, this seems to be one up for the relative notion.

University of Melbourne

NOTES

[1] Added in proof: even as these comments were being prepared, Lewis was at work in collaboration with Rae Langton on an alternative characterization of the intrinsic: see Lewis and Langton (1998). But as Section 3 of that paper makes clear, the new account is as reliant as the old on a distinction between natural properties and the rest.

REFERENCES

Martin, C.B. (1994), 'Dispositions and Conditionals', *The Philosophical Quarterly* **44**, 1–8

Prior, E.W., Pargetter, R.J. and Jackson, F.C. (1982), 'Three Theses about Dispositions', *American Philosophical Quarterly* **19**, 251–7

Dummett, M. (1976), 'What is a Theory of Meaning? II', in Gareth Evans and John McDowell (eds), *Truth and Meaning: Essays in Semantics*, Oxford: Oxford University Press, 67–137

Lewis, D.K. (1986), *On the Plurality of Worlds*, Oxford: Blackwell

Lewis, D.K. and Langton, R. (1998), "Defining 'Intrinsic'", *Philosophy and Phenomenological Research* **58**, 333–45

Taylor, B. (1993), 'On Natural Properties in Metaphysics', *Mind* **102**, 81–100

NOTES

[illegible]

REFERENCES

[illegible]

J. J. C. SMART

LAWS AND COSMOLOGY

THE PROBLEM

The main purpose of this paper is to seek a reconciliation between two apparently conflicting views of mine. I have argued (for example, Smart, 1963) for realism about theoretical entities, for example electrons, protons, photons, possibly space-time points, perhaps the 'Ψ'-wave of Schrödinger's equation and so on. Quine has also plausibly argued that we should believe in mathematical entities, since in physics we quantify over them no less than over electrons and protons. I except cases in which in physics the existential quantifications are part of merely pretence discourse. Perhaps in spherical astronomy talk of the celestial sphere should be treated in this way. Alternatively the celestial sphere could be thought of realistically as a sphere whose centre is that of the earth while the stars and planets are thought of (or correlated with) points or small areas on the sphere's surface in their lines of sight. I am not concerned in this paper to delimit fact from fiction in scientific discourse. The settlement of borderline disputes can wait for another occasion. My main argument for scientific realism is the cosmic coincidence argument. Would it not be a cosmic coincidence if the world were merely *as if* there were electrons, protons, etc?[1]

However I have defended a form of regularity theory of laws of nature (Smart, 1993). Many of my Australian colleagues reject regularity theories. They think that the regularities would be cosmic coincidences unless the laws were natural necessities. Consider the proposition that all electrons have the same charge. On the regularity view would not this look like a colossal cosmic coincidence? Should I not therefore use the cosmic coincidence argument in order to infer that law statements express natural necessities? I have tried to argue (Smart, 1985) that there are two sorts of cosmic coincidences, the good ones and the bad ones, and that the law regularities are examples of the good ones. It is only the bad (or ad hoc) ones that I reject. The good ones contribute to the simplicity of total theory or are potentially explicable by other good ones or else by an ultimate theory that is simple and comprehensive. The simplicity and cosmic comprehensiveness is enough to save us from attributing cosmic coincidence. Or so I argued.[2]

In this paper I am not concerned to defend realism about theoretical entities, though I do defend it, but to give further sense to the notion that even on a regularity view of laws of nature these need not be seen quite as cosmic coincidences.

H. Sankey (ed.), Causation and Laws of Nature, 161–169.

NATURAL NECESSITIES

So should I use the argument from cosmic coincidence to argue for natural necessities, perhaps espousing D.M. Armstrong's view of laws of nature as expressing relations of necessitation between universals? Do these relations explain the regularities, thus saving the regularities from the charge of cosmic coincidence? I have a temptation to say that the explanation has the philosophical vice of being too much, to use a phrase of John Wisdom's, 'made to measure' (Wisdom, 1953, p. 50). There is a sense in which a regularity theorist, provided that he or she believes in the appropriate universals, can say that laws express a relation between universals. Consider the case of $E = mc^2$. Trivially this has to be interpreted as follows: E is the energy (or change of energy) of a body in joules, m is its mass (or change of mass) in kilograms and c is the velocity of light in metres per second. E, m and c are all pure real numbers. Together with the surrounding chit chat this equation indirectly states a relation between energy and mass. (I'll leave out velocity because ideally we ought to have units of space and time such that $c = 1$.) Consider another example, Maxwell's equations. Here the equations involve pure mathematical entities, not only real numbers but vectors and operations such as 'div', 'curl' and differentiation. But the story is the same: these statements about relationships between purely mathematical entities together with the surrounding chit chat state a relation between universals. Even a regularity theorist could agree to this. I have argued (Smart, 1993) that perhaps one should believe in certain determinable properties which seem to me to be needed and happily postulated by theoretical physicists. Thus 'has a spin of' or 'has a charge of' comes out as referring to spin or charge. It can be adapted to Quinean canonical notation by introducing the predicate '... has ... of ...', where '... has ... of ...' is a triadic predicate relating a physical entity to a property and to a mathematical entity, such as a real number or a vector or tensor.[3] (More accurately one would need to make it tetradic, bringing in reference to a standard, say of length, or a convention related to angular momentum, as in the case of spin.) These properties are scientific posits. They are not second grade entities, as are the properties which are introduced by the dubious notion of meaning, and so belief in them should be immune to Quine's distrust of properties. According to this approach the universals in which I believe are determinable properties, not determinate ones. So in a sense I can accept that in the above mentioned trivial sense fundamental laws of nature express relations between universals. The 'fundamental' is important. I see no need to postulate a universal of rabbithood or perhaps even a universal of nucleic acidhood. Also what universals we believe in will be subject to changes in fundamental physics. So my acceptance of universals is a bit fuzzy. But given such an acceptance, a regularity theorist who believed in certain determinable properties in fundamental physics could say, rather trivially, that fundamental laws express relations between universals. Armstrong's theory of laws as relations between universals must be more meaty than this.

Armstrong has the problem, pointed out by van Fraassen (1989, pp. 103–9), of how the fact that 'N(F,G)', where N signifies a relation between properties F and G, implies '$(x)\,(Fx \supset Gx)$'. Actually there is a horrible use-mention mistake in this notation, since I have used 'F' and 'G' in the second sentence as predicate letters, not as names of properties. It could be fixed up by understanding 'Fx' etc. as 'x has F', etc. I suppose that people who believe that there are properties of rabbithood and mammalhood would agree that '$(x)\, x$ has rabbithood $\supset x$ has mammalhood'. However in the absence of the considerations about the pure mathematical entities that I have canvassed above the extra ontology seems unnecessary. There still remains the question of how we are to understand Armstrong's necessity predicate 'N'. If it is postulated simply to explain the occurrence of a regularity is it not, as I hinted earlier, too much 'made to measure'? Does not the use of the necessity predicate look like a piece of magic? Indeed Armstrong came to suggest himself that "the reference to nomic necessitation did little explanatory work" (Armstrong, 1993, p. 172). He came to prefer to rely on the concept of *causation*, which he thinks we grasp through direct experience of pressure on the body and the operation of the will, and that we grasp the notion of nomic necessitation by analogy with the experience of instances of causation of these sorts. I have my doubts as to whether the notion of causation plays a big part in theoretical physics and cosmology, and as to whether if it does at all it ought to do so. I would also have Humean doubts as to whether we can have immediate experience of objective necessity. Hume himself believed that we do have an immediate experience of necessity as the propensity of the mind to pass naturally from idea of a cause to idea of an effect. This would be an example of a psychological regularity, the necessity we feel about it being explained in the same way, and so on up as far as the mind is able to go. Of course such a subjective account of our idea of necessity is no good for Armstrong's purpose, and nor is it for mine. Armstrong needs a stronger notion, and, as I have said, he thinks that he can get it from immediate experience of acts of will and of pressure on the body which, if I understand him, tells him that the cause *makes* the event happen. In this paper I shall not attempt to argue against Armstrong's theory but shall try to explore a different tack.

A MINIMAL NOTION OF NECESSITY

It would be absurd to claim that in ordinary discourse we do not intelligibly use words such as 'necessarily', 'possibly', and more colloquially 'must be' and 'could be'. I follow Quine in treating our ordinary uses of these words as highly contextual (Quine, 1963). The assertibility conditions of sentences containing these words depend in general on background assumptions contextually agreed or shared by conversational protagonists. If I say "necessarily p", e.g. "David Armstrong must be nearly here by now", I am implying that p (e.g. "David Armstrong is nearly here by now") is assertible if p follows from contextually agreed background assumptions. (When the set of background assumptions is null we have logical necessity.) We can interpret 'follows' in the sense of first order logic. (If someone thinks that there are analyticities which supplement the logic we can simply invite him or her to throw them in among the background assumptions.) I am inclined to think that such modal sentences have assertibility conditions rather than truth conditions. This will apply to counterfactuals also. (Compare J.L. Mackie's view of counterfactuals as disguised arguments (Mackie, 1962).) Nevertheless the assertibility conditions can be converted to truth conditions by going metalinguistic. Of course to avoid trivialization the background assumptions have to be chosen with tact. This is a complicated problem which was investigated by Reichenbach (1954). For my part I am content to say "*solvitur ambulando*".[4]

In my 1993 paper I suggested that this minimal Quinean notion of necessity gives a sense of 'necessarily' when a universal regularity is explained by some more general theory which constitutes the background assumptions. It may also, as I suggested, explain a legitimate but very minimal feeling of necessity for laws of nature which are purely experimental, with no explanation of their truth or approximate truth by means of a more general law or deeper theory. This is when it is plausible that there is such a theory, developed in the future up ahead of us in Minkowski space, or perhaps in whatever Platonic heaven theories as abstract structures live. (More of this later.) Consider Ohm's law before the days of solid state physics. Still it surely was plausible that one day its approximate truth would be explained. (It was explicitly approximate when expressed as $E=IR(1+a_1R+a_2R^2)$ with empirical constants a_1 and a_2.)

I must stress that I am not making the silly modal fallacy that '$N(p \supset q)$' implies '$(p \supset Nq)$'. The Quinean notion of necessity is not that of modal logic. It is merely that of the 'must' as in the example of "David Armstrong must be here by now". It is highly contextual and in the present case implicitly refers to some deeper theory. Of course in the end we have to stop at some ultimate theory, known or supposed. Here we need simply the notion not of necessity but of simplicity and symmetry. A cosmological TOE would not strike us as describing a cosmic coincidence.

All right, let us agree that there is this sense in which a regularity theorist could say that laws of nature are necessary. Nevertheless this does not at once satisfy the objection from the cosmic coincidence argument. Would not the deep or more general laws themselves express cosmic coincidences?[5] The minimal sense of 'necessary' may seem to have got us nowhere. However I think that it has got us a *little* way. To see how we may go a bit further, let me bring in some considerations about simplicity.

SIMPLICITY AND ASPIRATIONS TOWARDS A THEORY OF EVERYTHING

Simplicity is still an unclear notion. In particular cases we can give particular examples, e.g. that a circle is simpler than an ellipse, because a circle needs three points to determine it and an ellipse five. A polynomial with four different parameters is simpler than one with five. Einstein's gravitational theory is simpler than Newton's because it explains the identity of inertial and gravitational mass. I do not know of a generally accepted definition of simplicity. The need for it is a weakness in my position. However I shall avail myself of it in an intuitive way and hope that it can be made more respectable. I shall include comprehensiveness as going along with simplicity. Obviously scientists could come up with simpler theories if they ignored phenomena to be explained. Thus simplicity goes along with integration. Identifying genes with stretches of DNA has brought chemistry and neo-Mendelian genetics together in a way which as we all know has increased comprehensiveness of explanation and also immense technological advances.

The ultimate integration will be that of physics with cosmology. This has already gone a fair way. It is true that cosmology is still a hugely controversial subject. However cosmological hypotheses are not always untestable. Hermann Bondi used to like to explain how much can be learned by reflecting on the simple observation that the sky is dark at night. The nineteenth century astronomer Olbers considered an infinite static and isotropic distribution of stars, and if these be considered as point particles the sky should be infinitely bright, and in any case very bright. Now we can think in terms of galaxies not stars, and the fact that the sky is not bright at night shows something wrong with assumptions made by Olbers. Another example of potentiality of cosmological theory for empirical test is that of the discovery of the 60cm cosmic background radiation which is generally taken to have falsified the steady state theory and to have supported the hypothesis of the big bang. Cosmological theories are now, as compared with even only fifty years ago, subject to a great variety of observational tests. They are still quite speculative as compared with normal physics. Perhaps we should say that cosmology is on the speculative end of physics, and maybe metaphysics is on the speculative end of cosmology. (Perhaps this is unfair, as metaphysics does involve a lot of not too speculative philosophical analyses of concepts. But nevertheless speculative in that philosophers still disagree, e.g. on such questions as that of whether there is a flow of time.)

There seems to be a continual integration of cosmology and physics. It would be hard to state whether an expert in general relativity was a physicist or a cosmologist. This has implications for a possible solution to the question of whether laws of nature are cosmic coincidences. This is in two ways. One is to do with aspirations of cosmologists and theoretical physicists to a 'theory of everything'. ('TOE' for short. Note that by a theory of everything I do not mean merely an integration of theories of the four fundamental forces, but one which is a synthesis with cosmology.) The other has to do with the uniqueness of the universe. I here ignore the 'many universe' hypothesis put forward by Brandon Carter to defuse what he called the 'anthropic principle' in cosmology. Believers in the many universe hypothesis can substitute 'our universe' for 'the universe' in the phrase "the uniqueness of the universe". John Bigelow, Brian Ellis and Caroline Lierse have put forward the notion of the universe as one of a kind, if only a one-membered kind. Though I am suspicious of these authors' use of the notions of natural kinds and of essences,[6] I think that there is something important to be learned from their idea.

THEORIES OF EVERYTHING AND THE UNIQUENESS OF THE UNIVERSE

As in the history of Ohm's law, we can attribute necessity conjecturally by referring to theories which do not at the time exist but which we plausibly think will occur in the future. Plausibly we can look forward to a TOE which will unify the four forces of nature, and in particular will unify the theory of the first three forces, which fits in with quantum mechanics, with the geometrically based theory of gravitation. It is plausible that such a theory will be simple and symmetrical and that its laws will hang together well and will not have the appearance of cosmic coincidences. Here I am relying on a metaphysical postulate of simplicity. But we need this postulate already in order to defend realism about theoretical entities against the likes of van Fraassen who holds that all that is needed is that a physical theory be 'empirically adequate' but not also *true*. If it is objected that this metaphysical postulate cannot be validated,[7] the same could be said about the ordinary inductive principle that the future will be like the past. Still there is some evidence that the universe is ultimately simple from the success of science when it has proceeded on the assumption that it is. In the eighteenth century this success was no doubt motivated partly by theological assumptions. However Jane E. Ruby (1986) has argued that the notion of 'law of nature' did not originate only in Kepler's and Descartes' idea of divine governance. She traces it back, with impressive scholarship, to secular sources, such as a naturalized Roman 'lex', the medieval association with 'rule' or 'law' in mathematics, and the Stoic conception of the law of nature. The theological motivation can perhaps be taken as a ladder which on account of the empirical evidence of success can now be thrown away. In expecting simplicity I am talking about fundamental physics. In much of biology, for example, we should *not* expect simplicity. Evolution of organisms is a very opportunistic business, and structures that evolve not only can be complex, as they have to be, but are often slightly reminiscent of Heath Robinson inventions. Consider the extraordinary history of the evolution of the human ear. Nevertheless this expectation of complexity is derived from the theory of evolution, and so there is an overall increase of simplicity in the totality of science.

A theory of everything would be simple in a metaphysical sense. This is the sense in which Einstein's gravitational theory is simple. The TOE will almost certainly not be simple in the sense of 'easy to understand.' It will almost certainly demand fearsome mathematical sophistication. I am reminded of my father (in the middle 1930's probably) asking Eddington how many people would understand a book that Eddington was writing. (I suspect that it was Eddington's *Relativity Theory of Protons and Electrons* but it might have been an early shot at *Fundamental Theory*.) Eddington said 'six.' *A fortiori* what hope for a mere philosopher?

Now the TOE might have a sort of internal necessity in that if any of it were incorrect it would predict a universe qualitatively different from what it is. Philosophers naturally criticized Eddington for what they took to be *a priorism*. However E.T. Whittaker claimed that this is a mistake (Whittaker, 1958, p. 186) and that Eddington "regarded himself as at liberty to borrow anything from *qualitative* physics – he did in fact assume the identity of mass and energy, the theory of the energy tensor . . . the exclusion principle and other propositions of the most advanced physical theory." But out of these qualitative facts Eddington hoped to get pure numbers, those of fundamental constants. Whittaker gives the analogy of how Archimedes deduced the value of π from the non-quantitative axioms of Euclid. I'm not sure of the analogy because it seems to me that the congruence axioms make Euclidean geometry quantitative. Unfortunately Eddington got the wrong numbers.

The incorporation of a TOE into cosmology would exhibit a very believable simplicity. The ordinary idea of inductive reasoning is somewhat Heracleitean, the idea of the universe evolving by passage of time. First an A is a B, then another A is a B, then yet another, and so on and on. Is it not a miracle that each A is a B, one after the other? Whereas on the Parmenidean conception that I advocate, that the passage of time is an illusion, we are left with a simple space-time pattern. The pattern would be particularly nice if the universe were symmetrical in time, between big bang and big crunch, as Huw Price is inclined to take seriously (Price, 1996, Chapter 4). There would be a loss of simplicity if there were exceptions to the pattern, unless everything were an exception so that there was complete randomness. Complete randomness is as simple as complete lawfulness. However it is obvious that the universe is not completely random (unless we are most improbably in the middle of an era of accidental lawlikeness). So if we look for simplicity of space-time patterns rather than Heracleitean evolution the simplicity seems less coincidental. Perhaps not much less, but may be more so if it were seen in the light of a TOE. It helps to think of laws as referring to space-time patterns.

The TOE might be unstable. That is if slight alterations were made to it there would be prediction of a totally chaotic universe. In which case it would be necessary relative to the background assumption that the universe is not chaotic. However this would not be an explanatory necessity. The non-chaotic state of the universe would not explain the TOE any more than (as misusers of the anthropic principle in cosmology have sometimes thought, in a back to front way) the fact that we are here explains the fine tuning of the constants of physics. Still we need not say that the TOE was *necessary*. All we need say is that its simplicity and comprehensiveness prevented it from asserting an implausible cosmic coincidence. At least if we can believe that simplicity makes for probability. Am I to be accused of an unacknowledged theism here? (If I interpret 'necessity' as relative to a TOE of course the TOE cannot be necessary in this sense. I do not mind whether we say it is just simple and comprehensive or we say that it is necessary in the sense of being simple and comprehensive!)

It is here that one might look for help from the Bigelow, Ellis, Lierse (1992) idea that the universe is one of a kind, even though I am doubtful that their notion of kind or essence helps. One could say that it is essential to an electron that it has spin $\pm \frac{1}{2}$. But it is not essential to an electron that it occurred in one of J.J. Thomson's cathode ray tubes. There is a sense in which electrons constitute a natural kind, in that the word 'electron' occurs in fundamental laws of nature. Here we elucidate 'kind' and 'essence' in terms of 'law', not the other way round. What I get from Bigelow, Ellis and Lierse is the attractive idea of thinking of laws as describing patterns in the one and only space-time universe, not of laws as describing 'just one thing after another'. The former way they do not seem so much like cosmic coincidences as they do when we look at them the other way.

One final worry. In making necessity relative to a TOE I may be thinking, as I do not want to do, of a possible world other than the actual world. Suppose that an asteroid hits the earth and Hawking, Penrose and indeed all the human race are exterminated. Then there is not a terrestrial TOE up ahead of us in Minkowski space. Perhaps there is one elsewhere in the universe. Or perhaps in the Platonic heaven where numbers and sequences of them live there is a sequence of 0 and 1 that could be interpreted as a TOE. We could in theory decode it by reference to its proper part which contains familiar concepts and ascend from thence by Ramseyfication. A similar method would have to be used with reference to an extra-terrestrial TOE. We could get in touch by signalling patterns of dots corresponding to $5 + 7 = 12$, etc. and from then on build up mathematical expressions by similar methods, e.g. sending '$e^{i\pi} = -1$' in the hope that the extra-terrestrials would catch on. Then we could send patterns relating to Maxwell's equations and other salient laws, and numbers corresponding to pure constants of nature.[8] Though in fact we wouldn't do it, on my supposition, because we wouldn't be there after the collision with the asteroid.[9] But my story may give sense to the notion of an extra-terrestrial or a Platonic TOE.[10]

Australian National University

NOTES

[1] Bas van Fraassen's chapter 'Gentle Polemics' in his (1980) was in part a witty and amicably satirical reply to my (1963). My (1985) was in part a reaction to this, to which van Fraassen responded on pp. 147–8 of his (1989). He remarks that for my argument simplicity has to be global. In the present paper I partially agree with this. Because of the contextual nature of explanation I think that it is best to state the argument in terms of the notion of cosmic coincidence rather than in the fashionable terms of argument to the best explanation.

[2] A referee has remarked that we could use argument to the best explanation by saying that the best explanation for all electrons having the same charge would be that they all instantiate the same universal. However (a) this would seem to require the postulation of determinate (as well as determinable) properties and (b) more importantly it could be thought a cosmic coincidence if all electrons instantiated the same determinate universal, unless the assertion that all electrons have the same charge would be true by definition and so trivial. The question would recur as that of why all particles of a certain sort were electrons.

[3] This is a modification, by bringing in properties, of a method due to Carnap and Quine, of exhibiting equations in physics as referring to pure mathematical entities. See Quine (1960, pp. 244–5).

[4] In view of the lack of a decision procedure for first order logic perhaps 'solvitur ambulando' is in theory the best we can do.

[5] So also with David Lewis's sophisticated Ramsey type theory, which is a regularity theory of a sort (Lewis, 1983).

[6] This metaphysical sense of 'natural kind' differs from the sense of 'natural kind' in Quine (1969), which is largely to do with the way that evolutionary pressures shape the hard wiring of our perceptual spaces. Quine concludes that it is the mark of a mature science that it no longer needs such a notion of 'kind'.

[7] Perhaps induction can be vindicated. See Clendinnen (1982).

[8] Compare Freudenthal (1960).

[9] Even if there is no asteroid disaster and our descendants get in conversation with intelligent beings a thousand light years away, communication would be hard as they would have to wait two thousand years for a reply to a question.

[10] By a Platonic TOE I do not mean a theory about Platonic entities, but a theory which is itself a Platonic entity.

REFERENCES

Armstrong, D.M. (1983), *What is a Law of Nature?*, Cambridge: Cambridge University Press

Armstrong, D.M. (1993), 'Reply to Smart', in *Ontology, Causality and Mind: Essays in Honour of D.M. Armstrong*, ed. John Bacon, Keith Campbell and Lloyd Reinhardt, Cambridge: Cambridge University Press, 169–74

Bigelow, John, Ellis, Brian and Lierse, Caroline (1992), 'The World as One of a Kind: Natural Necessity and Laws of Nature', *British Journal for the Philosophy of Science* **43**, 371–388

Clendinnen, F.J. (1982), 'Rational Expection and Simplicity', in *What? Where? When? Why?*, ed. Robert McLaughlin, Dordrecht: Reidel, 1–25

Freudenthal, H. (1960), *Lincos: design of a language for cosmic intercourse*, Part I Amsterdam: North-Holland

Lewis, David K. (1983), 'New Work for a Theory of Universals', *Australasian Journal of Philosophy* **61**, 343–77

Mackie, J.L. (1962), 'Counterfactuals and Causal Laws', in *Analytical Philosophy*, R.J. Butler (ed.), Oxford: Blackwell, 68–80

Price, Huw (1996), *Time's Arrow and Archimedes' Point*, New York: Oxford University Press

Quine, W.V. (1960), *Word and Object*, Cambridge Massachusetts: MIT Press

Quine, W.V. (1963), 'Necessary Truth' in Quine (1976), 48–56

Quine, W.V. (1969), 'Natural Kinds', in Quine, *Ontological Relativity and Other Essays*, New York: Columbia University Press

Quine, W.V. (1976), *The Ways of Paradox and Other Essays*, 2nd ed., Cambridge Massachusetts: Harvard University Press

Reichenbach, H. (1964), *Nomological Statements and Admissible Operations*, Amsterdam: North Holland
Ruby, Jane E. (1986), 'The Origins of Scientific "law",' *Journal of the History of Ideas* **47**, 341–59
Smart, J.J.C. (1963), *Philosophy and Scientific Realism*, London: Routledge and Kegan Paul
Smart, J.J.C, (1985), 'Laws of Nature and Cosmic Coincidences', *Philosophical Quarterly* **35**, 272–80
Smart, J.J.C. (1993), 'Laws of Nature as a Species of Regularities', in *Ontology, Causality and Mind: Essays in Honour of D.M. Armstrong*, ed. John Bacon, Keith Campbell and Lloyd Reinhardt, Cambridge: Cambridge University Press, 152–74
Van Fraassen, Bas C. (1980), *The Scientific Image*, Oxford: Clarendon Press
Van Fraassen, Bas C. (1989), *Laws and Symmetry*, Oxford: Clarendon Press
Whittaker, E.T. (1958), *From Euclid to Eddington*, New York: Dover
Wisdom, John (1953), *Philosophy and Psycho-Analysis*, Oxford: Basil Blackwell

D. M. ARMSTRONG

COMMENT ON SMART

Smart says that the main argument for scientific realism is the 'cosmic coincidence argument'. It would be a cosmic coincidence if the world lacked a unified structure but instead is merely *as if* there were the scientific entities (electrons, protons and so on) which, if they had existed, would have articulated that structure. This stand, of course, poses a problem for his own frequently expressed view that laws of nature are nothing more ontologically than cosmic regularities or, as he himself puts it, cosmic coincidences.

But can one get beyond cosmic coincidence in the case of laws? Not very far, Smart thinks. In particular, he rejects the view, argued for by me in the past, that laws are relations of (contingent) necessitation holding between universals. Smart is now prepared to countenance universals, provided that they are not identified with meanings, but are instead introduced on *a posteriori*, on scientific, grounds. In particular, for the fundamental functional laws, he is prepared to postulate *determinable* universals corresponding to the predicates such as 'mass', 'energy' and so forth to be found in the statement of such laws. He rejects such universals as rabbithood, nucleic acidhood, and the other *kinds* (as would I) and also, presumably, such universals as *determinate* masses and energies (as I do not). But I have in recent years become more sympathetic to postulating the corresponding determinable universals. They are needed for functional laws, and the fundamental laws, as Smart indicates, seem to be functional.

But Smart thinks, as many others in the Humean tradition of reflection on the nature of laws of nature think, that the attempt to link the universals involved in laws of nature in some 'direct' manner that entails but is not entailed by a mere universal quantification, does not succeed. I myself am conscious that the attempt may prove an explanatory Arnhem, an 'explanation too far'. Smart is aware that I now largely drop talk of necessitation, and think of a connection of universals on the analogy with singular causation. I am convinced that Hume quite failed to disprove the hypothesis that there is an impression of (singular) causation from which our idea (concept) of singular causality is derived, derived in the best empiricist fashion. I think we do have such impressions (perceptions), in particular in our awareness of forces acting on our own body and, as many have held, in our awareness of the operations of our own will. Hume's own arguments on this score turn out, when actually examined as they almost never are, to be pitifully weak. (See Armstrong, 1997, Ch. 14.6.)

Given this beginning, which Smart of course contests, we can think of, say, arsenic causing death (I do not really think we have universals here, but it will do as an exam-

H. Sankey (ed.), Causation and Laws of Nature, 171–172.

ple) as a causal relation between universals having the following pattern: the ingesting of arsenic is a causer of death to the ingestee. This is to be taken as a relation of universals, and it seems to entail (with the weakening qualification that no other relevant factors interfere with the operation of the arsenic), without being entailed by, the corresponding universal quantification. (See again Armstrong, 1997, Ch. 15.2.)

Smart's alternative seems to be about the best that a 'cosmic coincidence' view of laws can do. It involves a draft on the progress of natural science, not too implausible a draft in my opinion. Suppose that there is found a 'theory of everything', a fundamental law or small set of physical laws that, whether deterministic or probabilistic, together perhaps with boundary conditions, suffice to explain the behaviour of everything there is in every respect. This may be seen as the same pattern, in some sense a simple pattern, repeated everywhere in spacetime. Will this not be as much unification as can be found?

My bridge, that may or may not turn out to be a bridge too far, would then be that this pattern is itself the *very same* causal structure in each of its instances. Being a token of a certain sort of structure, one having certain properties and relations, determines or probabilifies further developments or further aspects of that structure simply in virtue of these properties and relations. The pattern of cosmic coincidences is then no coincidence, because it is explained by the 'causal' relations holding between the universals involved. That seems to me to be a vision worth fighting for.

Of course, if these 'relations between universals' are contingent only, the fundamental law remains unexplained. But that really may be a bridge too far!

University of Sydney

REFERENCES

Armstrong, D.M. (1997), *A World of States of Affairs*, Cambridge: Cambridge University Press

IV.

CAUSATION AND THEORIES OF CAUSATION

D. M. ARMSTRONG

THE OPEN DOOR: COUNTERFACTUAL VERSUS SINGULARIST THEORIES OF CAUSATION

I. INTRODUCTION

All contemporary analytic philosophers owe an enormous debt to David Lewis, not least myself. But in this paper my object will be to criticize his Counterfactual theory of causation, and to argue instead for a Singularist theory.[1]

We are concerned, as Lewis is concerned in his papers 'Causation' (1973, reprinted 1986) and 'Postscripts to "Causation"' (1986) with "causation in particular cases", this event of taking cyanide causing a particular death as opposed to the causal generalization that taking cyanide causes death. Suppose that a particular taking of cyanide, *c*, is followed by the death of that individual, *e*. Then Lewis defines *causal dependence* between these two events as the conjunction of two counterfactuals. In his symbolism these are $O(c) \rightarrow O(e)$ and $\neg O(c) \rightarrow \neg O(e)$, where O( ) is the proposition that the event actually occurs. The first counterfactual is, according to him, trivially true, because $O(c)$ and $O(e)$ are both true. I wonder whether this first counterfactual really is true. But we can avoid this issue in the simple way adopted by Ned Hall (unpublished) by substituting for the first counterfactual just the occurrence of *c* and of *e*.

Causation for Lewis is the ancestral of causal dependence: it is a chain of causal dependences. But non-adjacent links in the chain need not, he says, stand in the relation of causal dependence. An *e* that is not causally dependent on a certain *c* may still have that *c* as a cause, provided only that a chain of causal dependencies links the two. It needs to be added further that the counterfactuals involved must not be "backtracking" conditionals, typically cases where *e* is actually *before* *c* and the speaker is suggesting that for *c* not to have occurred, the past would have had to be different.

Our concern is with the second counterfactual: if *c* had not been, then *e* had never existed, with this proposition stipulated not to be a backtracker. This counterfactual, Lewis asserts, is true if and only if there exists a possible world where *c* fails to occur and *e* fails to occur, a world that is closer in overall similarity to the actual world than any world in which *c* fails to occur but *e* does occur.

In a first draft of this paper, I assumed that talk of possible worlds was to be taken in a realistic way here, because Lewis, after all, does have a realist theory of possible worlds. He has told me, however, that he holds that the relevant counterfactual truths about our actual world are true solely in virtue of features of this world. (Judging by

H. Sankey (ed.), Causation and Laws of Nature, 175–185. *©1999 Kluwer Academic Publishers. Printed in Great Britain.*

a desultory asking around, this information will come as news to a number of philosophers.) As I but not he would put it, Lewis holds that the *truthmakers* for these counterfactuals are to be found solely in this world. *In his theory of causation* the possible worlds enter as mere calculational devices. He gives as an example the way that we might say with truth that a person is a Montague rather than a Capulet, without being committed to the view that these families are actual. The fictional families are used as no more than a sort of calculational device.

This does make it difficult to grasp the ontology of his theory of causation. Some instructions are given for ranking different worlds in terms of overall resemblance. Keeping the laws of nature as constant as possible, and minimizing the size of the miracles required to instantiate the antecedents of the counterfactuals, are the most important ways of maximizing such resemblance. But it is hard to become more precise than this at a general level.

It might be thought that the Lewisian thesis of Humean supervenience, the thesis that causes and laws supervene upon the particular, local, matters of fact, will give all the ontology that is needed. But this is to run together Lewis' theory of causation and Lewis' theory of law. His theory of law is neo-Humean, the Mill–Ramsey–Lewis view that laws are the most systematic and coherent sub-set of the regularities that the world contains. It would seem possible, however, to hold a Lewisian Counterfactual theory of causality, together with a *non–Humean* theory of laws. For instance, one might hold that laws are contingent connections holding directly between universals. (If the connections were necessary, that might be difficult to fit together with counterfactual causality.) As a result, the ontology that lies behind the Counterfactual theory remains somewhat obscure.

2. NEURON DIAGRAMS AND SINGULAR CAUSATION

Lewis discusses at length many types of situation that constitute *prima facie* objections to a Counterfactual analysis and I will shortly discuss three of these. But first a general remark. In the course of these discussions Lewis introduces his justly celebrated "neuron diagrams". These involve imaginary systems of neurons which are hooked up according to certain often quite complex causal patterns of firings and inhibitions of firing. They are then used to illustrate such situations as epiphenomenal events, preemptive causation, causal redundancy and so on.

When we look at these diagrams, we can immediately see that they are possible causal patterns, in most cases empirically possible cases (you could construct such a circuit). Counterfactual theories of causation (and, it may be added, Regularity theories and Probability-raising theories of causation) struggle with these diagrams. Wittgenstein spoke of an open door that we had only to see and go through to escape philosophical confusion (see Malcolm, 1958, p. 44).[2] The solution that I recommend to the problems posed by the neuron diagrams is very simple. Where there is an arrow in a diagram showing that one neuron brings it about that another neuron fires, take it that here there is a genuine two-term relation of singular causation holding between cause and effect. Where there is no such arrow, deny that there is any such relation. This is the open door.

3. NEGATIVE CAUSES AND EFFECTS

I turn aside now to consider, all too briefly, the important topic of negative causation. We say things like "lack of water caused his death" and surely may say them with truth. Here we seem to have a negative cause, and perhaps a negative effect. Lewis calls this "causation by omission" (1986, pp. 189–93). Again, consider those special backward arrows in the neuron diagrams which represent the firing of one neuron as *preventing* the firing of another neuron. Here we seem to have a negative effect.

Yet once one has accepted a Singularist theory of causation one seems almost forced to deny both negative causes and negative effects. They are very unimpressive candidates for terms in a direct singular causal relation. In the case of negative effects one will presumably not want to make them epiphenomenal, so these effects will be negative causes as well.

For Lewis, though, "causation by omission" is causation. There are, after all, Lewisian counterfactuals naturally associated with such cases. In the lack of water case there is "if he had not lacked water, he would have survived". The ready availability of such counterfactuals may, indeed, be thought to be a strength of the Counterfactual theory.

As Lewis recognizes, however, there are also difficulties. If omissions are causes alongside positive causes, then they ought to be events, and, I would add, events that have both causes and effects. (Rather puzzlingly, Lewis says that he does not need them as effects, though he will not bar them.) But then he finds himself in a dilemma which he develops thus. If negative causes are events, they will regularly be highly disjunctive – there will be, for instance, many, many, ways to omit precautions against a particular kind of accident – and disjunctive events look wrong as causes. If, on the other hand, they are not events, then the form of the counterfactual has to be changed. The antecedent cannot be: if event *c* (the omitted event) had *not* occurred . . . , but rather it is: if some event of kind K (the omitted kind) *had* occurred . . . This new sort of counterfactual raises certain problems for Lewis which I will not here rehearse. Nor is he satisfied with an attempt to compromise between these two positions.

We Singularists can be heartened by this. I think that we should argue that here the ordinary *language* of causality conceals the true ontological situation. Causation by omission is, for Singularists, causation by a certain courtesy only. True counterfactuals are certainly associated with such "causation", indeed may be said to constitute it, but omissions and so forth are not part of the real driving force in nature. Every causal situation develops as it does as a result of the presence of positive factors alone.

The lack of water case perhaps sounds, off-hand, like a true cause, with the lack literally doing the causing. Consider, however, some parallel cases. Think of all the persons in the past who died because of the unavailability of contemporary medical treatment. Is it really plausible to think, as the Counterfactualist presumably has to think, that all these unavailabilities were contributing causes of their death? Or that our daily freedom from being hit by large meteorites or attacked by efficient flame-throwers are circumstances that contribute to keeping us alive? In all these cases it seems that true counterfactuals of the right sort are available. But if one tried to draw the equivalent of a neuron diagram in such cases, to show what was keeping a person alive, one would surely wish to include only the positive factors that were operating to produce certain positive effects.

Phil Dowe (this volume) has recently given a very beautiful treatment first of preventings and second of omissions. With his permission, I give a brief resumé. His case of preventing is that where a billiard ball, C, hits another billiard ball, A, which is thereby prevented from hitting ball B. C prevented the collision AB. Here, he says, C → A is genuine causation. But there is also a counterfactual: if interaction CA had not occurred, then A would have caused the interaction AB. This true counterfactual, whose truthmakers are to be found in this world,[3] makes CA a case of prevention. Omissions involve a nested counterfactual claim. Suppose C *fails* to hit A, and as result A goes on and does hit B. Here the genuine causal fact is that A collides with B. But there is also a counterfactual: if C had hit A, then this would have prevented A's colliding with B. "Going into this other, though nearby, world", that is, nesting our counterfactuals, the prevention that did not happen may then be analysed in just the same way as the original prevention. Notice that these counterfactuals already presuppose, and so cannot analyse, singular causation. For prevention: with CA removed, A would have *caused* B to act differently from the way it actually did act. With omissions, we have a counterfactual about the prevention that would have occurred if C had acted causally on A.

We may call these cases *simple* prevention and omission. Assuming that Dowe has analysed all such cases correctly, we can go on to the more complex prevention of preventions, prevention of omissions, omissions of prevention and so on indefinitely.

4. PROBABILISTIC CAUSATION

I pass on to consider Lewis' discussion of "chancy causation" in B. of the 1986 Postscript, the sort of thing that we seem to get in quantum phenomena. The following case gives him difficulty:

> . . . *c* occurs, *e* has some chance *x* of occurring, and as it happens *e* does occur; if *c* had not occurred, *e* would still have had some chance *y* of occurring, but only a very slight chance since *y* would have been very much less than *x*. We cannot quite say that without the cause, the effect would not have occurred; but we can say that without the cause, the effect would have been very much less probable than it actually was . . . I think we should say that *e* depends causally on *c*, and that *c* is a cause of *e*. (1986, p. 176)

The obvious thing to say instead, it seems to me, is that *e* may or may not causally depend on *c*, and that *c* may or may not be the cause of *e*. Suppose, to take the case that seems to have the best chance of empirical instantiation, that besides *c* a potential cause c_1 is also present and it is the latter that gives the smaller chance *y* of *e* occurring. It seems to be a perfectly objective question, when *e* occurs, whether it is *c* or c_1 that is the cause, although it is more likely to have been *c*. It might even be possible to settle the matter empirically. If the counterfactual theory cannot deliver this answer, that seems to be a weakness of that theory.

As he makes clear on p. 180, Lewis would reject this reply. My style of answer is considered there. He points out that in the sort of case just considered two different counterfactuals would compete in cases where c and c_1 were both present followed by the occurrence of e. In one case, where c is the cause, we have if not c, then not e, but it is false that if not c_1, then not e. But with c_1 the cause, the truth-values of these two conditionals are reversed. He says:

> But I reject the presupposition that there are two different ways the world could be, giving us one definite counterfactual or the other. That presupposition is a metaphysical burden quite out of proportion to its intuitive appeal; what is more its intuitive appeal can be explained away.
>
> The presupposition is that there is some hidden feature which may or may not be present in our actual world, and which if present would make true the [relevant] counterfactual.

Lewis goes on at some length discussing candidates for the "hidden feature". But there is an obvious candidate, and not a mysterious one either if we are not blinded by the rhetoric of Hume and the Humeans. It is simply causation in the single case. Either c does this causing, or c_1 does. And some of us believe that we *observe* that this relation holds in particular favourable cases. If so, singular causation is not always hidden. (My favourite candidates for unhidden causality are perception of pressure on our own body and the operation of our own will.)

Notice that we can very well have "probabilistic causation of a genuine chance event" (Lewis, 1986, p. 183) on this Singularist view, although "probabilistic causation" is a rather unhappy and potentially misleading phrase. What we have rather, I suggest, is the probability, less than strictly 1, of causing occurring. This causing is the very same causing that we have with deterministic causing. Event c had a genuine chance of causing event e. Perhaps it did. The causing will then be ordinary singular causation. Event c_1 had a lesser chance of causing e. Perhaps c_1 was lucky – you can't lose them all, as the folk say – and c_1 was the event that caused e. Ordinary singular causing again. An open door, I think. It is a measure of the power of a Singularist theory that it sweeps aside problems about chancy causation so easily.

I find that a little mnemonic is useful: not probabilistic causation but the probability of causing. In cases where an event c has a less than strictly 1 chance of causing e and does cause e then we can say that c "fires". If c fails to fire, there is no causation at all, though there was a chance of it.[4]

5. LATE PREEMPTION

In what Lewis calls "a variety of late preemption" a causal chain goes to completion, and that very completion aborts another causal chain that would have brought about substantially the same result if the first chain had failed to go to completion. You shoot the villain through the heart and he drops dead. But your mate, Crack Shot Willie, was also there and if he had perceived that the villain was not dead, which perception would have caused him to shoot, he would have done the job. The trouble for the Counterfactual theory is that the death does not depend counterfactually on your shooting because if you had bungled the job, Willie would not have bungled it. Yet obviously it was you alone that caused the death.

Lewis considers, but rejects, the idea that the problem is to be solved by saying that the actual effect is "fragile", so that the death Willie would have inflicted is a *different* effect. Instead, he agrees that the situation for him is rather serious. He tries to solve his problem in the following way. He starts by appealing to an intuition (1986, p. 205):

> Intuitively, whether the process going on in a region is causal depends only on the intrinsic character of the process itself, and on the relevant laws. The surroundings, and even other events in the region, are irrelevant. Maybe the laws of nature are relevant without being intrinsic to the region (if some sort of regularity theory of lawhood is true) but nothing else is.

Lewis does say immediately that intuitions of what is intrinsic are not to be trusted very far. I tend to agree. But as already indicated, the *problems* of non-intrinsic theories of causation, in particular those thrown up by neuron-diagram thinking, causal-flow thinking, force us towards an intrinsic view of causation.

The general lines of Lewis' solution can now be indicated, using our particular case. Many people deliberately kill other people by shooting them. In the normal case of such causation, there is no back-up mechanism so that if, contrary to fact, there had been a failure, somebody else would have done the job. In such a case, Lewis suggests, we should allow the paradigms, these normal cases, to rule. Though there is not causal dependence in the back-up cases, we have what Lewis calls quasi-dependence. We include the back-up cases by a sort of courtesy, and call it quasi-dependence. Causation involves dependence *or* at least quasi-dependence.

(We may note that there is a true counterfactual involved: in the absence of the non-paradigmatic back-up system, there would have been true causal dependence. Note also that the coming to be of the actual effect *causes* the abortion of the back-up process. Is there any circularity for Counterfactualism here?)

Lewis is clear that this piece of ingenuity is a modification and a complication of the original analysis. Nor is he very happy with it. But he thinks it is the best that he can do.

6. THE PREVENTION OF PREVENTION

But Lewis has a still higher hurdle to jump. Ned Hall (unpublished) has worked out a whole series of ingenious cases where a process goes to completion, but its potential aborter is itself neutralized. He calls this "double prevention", but I think that phrase misleadingly suggests two preventers of the original process, so that prevention is over-determined. Hall's cases are really meta-prevention. Like double negation his "double prevention" brings us back to an original. So I prefer to speak of prevention of prevention.

His first case involves Suzy, a World War Three bomber pilot who blows up her designated target. Billy, the fighter pilot, escorts her and manages to shoot down Lucifer, the enemy fighter pilot, who would otherwise have shot down Suzy. Billy prevents Lucifer's intended prevention.

This case creates some trouble, but the trouble is nothing to further cases which complicate this, cases involving Hillary, a second escort for Suzy, and various complications for Lucifer, including the essential role his alarm clock plays in getting him into the air. (For one case, given the Counterfactual theory, this clock becomes part of the cause of the *bombing*, Hall argues!) I will not recapitulate his argument. He considers (1) appeal to intermediate events in the actual causal chain, pointing out that there are none that are relevant. He further considers (2) appeal to the "fragility" of the effect, but notes that in his cases the exact nature of the effect – the demolition of Suzy's target – is quite unaffected by all the by-play involving Billy, Lucifer and Hillary. He considers finally (3) appeal to the Lewis amendment to his own Counterfactual analysis just discussed in my previous section, an appeal which he calls the "piggy-back" strategy.

In examining the piggy-back strategy Hall assumes that Billy, who shoots down Lucifer, is part of the intrinsic process which culminates in Suzy destroying her target. It would be difficult for Lewis to deny this, because he accepts prevention as causation. With this given, Hall is able to mount devastating objections. But one of the important morals that he extracts from his argument is that:

> The causal structure of a process is determined solely by the laws of nature, together with the intrinsic character of that process.

Just fine, the Singularist will say.[5] But this principle should lead to the conclusion, at least if you abandon causation by omission, that the causal structure of the intrinsic process that is Suzy's bombing run, and which culminates in the destruction of her target, is determined by that run alone. Billy, Lucifer, Hillary and so on play no part in its causal structure. Without Billy the run would have been very different; he makes the final success of the run possible. But it is Suzy's bombing, and this alone, that *causes* the target to be destroyed.

7. SINGULAR CAUSATION AND COUNTERFACTUALS

I pass now from criticism to construction. Causation may not be analysable by means of counterfactuals, but it is clear that there is an important link between causality and counterfactuals. Let us begin the long task of developing a comprehensive theory of singular causation (not attempted here!) by trying to spell out these links. My suggestion is that if we take a singular causal sequence simply as a singular causal sequence, abstracting, as I think we can abstract, from the further question whether the sequence falls under some law, then we get an entailment along the following lines:

> *c* causes *e* entails that:
>
> if *c* had not existed, then, unless there was in the situation a c_1 that would still have caused *e* in the absence of *c*, and unless *e* would have come to exist without any cause, then *e* would not have existed.

A simplifying form of words is:

> *c* causes *e* entails that:
>
> if not-*c*, then, in the absence of a back-up cause, or *e* coming to exist uncaused, then not-*e*.

It may well be, of course, both that a back-up cause is absent, and that, likely enough, uncaused events never occur. If so, then the stronger if not-*c*, then not-*e* will be *true*. Given that situation it seems fair enough to say that the truth that *c* causes *e supports* the proposition that if not-*c*, then not-*e*. But without the qualifying clauses, this support is not entailment. I think, incidentally, that this already shows that the prospects of a counterfactual *analysis* are dim. For the qualifying clauses make essential use of the notion of causality.

8. THE NATURE OF SINGULAR CAUSATION

But what is this causation that so many analytic philosophers find so mysterious? In what I think is a very important paper 'Probabilistic Causation and the Pre-emption Problem' (1996) Peter Menzies suggests that such causation is a "theoretical entity". Look at the difference a word or two makes! Lewis stigmatized singular causation as a "hidden feature", a phrase well calculated to arouse our Humean reflexes. He also says that it is "mysterious" (1986, p. 182). But the phrase 'theoretical entity' puts a new complexion on matters! And then Menzies goes on to suggest that we should avail ourselves here of the treatment of theoretical entities to be found in the work of . . . David Lewis.

Lewis' best known application of his theory of theoretical entities is to our concepts of mental states. His idea is to collect the platitudes of folk psychology that concern the causal relations of mental states, sensory stimuli and behavioural responses. Call these platitudes when confined to a particular sort of mental state, pain say, the causal role of pain. Pain is a theoretical entity, that entity that plays the causal role of pain. Here the notion of cause is treated as an undefined primitive. But, says Menzies, why not apply the same treatment to causation itself? Collect the platitudes of singular causation in the same sort of way. Causation is that relation between singular events for which the causal platitudes hold.

In previous work Menzies had emphasized not so much counterfactuals as the raising of probabilities: a cause raises the probability of its effect. It is known, however, that there are cases which pose great difficulty for theories of this type. But we know that a cause does *typically* raise the probability of its effect. Menzies therefore offers us the following Lewisian-style definition of the causal relation: it is *the intrinsic relation that typically holds between two distinct events when one increases the chance of the other event* (p. 101). The word 'intrinsic' here is meant to capture the Singularist insight. Menzies says, furthermore, that the notion of increase of chance is to be understood in the counterfactual manner found in Lewis' theory. So the counterfactuals that *typically* hold in the case where one event causes another, but which may fail in cases of late pre-emption and other sorts of unusual case, can be part of the causal platitude.

All very fine, I think. For myself I would favour taking Menzies' "causal platitude" as fixing the reference of the singular causal relation rather than analyzing the concept. Menzies may agree. But I have three points to make. First, I think that we can correct the epistemology involved, not in major fashion but interestingly. Second, I think that Menzies' platitude stands in need of some expansion. Third, and more speculatively, I have hopes that we can get behind the Menzies definition to a deeper, more ontologically revealing, account of the nature of singular causation.

First the epistemology. I begin by noting that Lewis' causal role account of the mind makes the notion of the mental a theoretical one. I have great sympathy with this account, though I might elsewhere niggle about details. But Lewis would not deny that this theoretical entity is also one to which we have some observational access in the first-person case. We have introspective access to some of our own mental goings on. We do not have introspective access to the intrinsic nature of these goings on, certainly not if this intrinsic nature is as Lewis and I believe it is, physical in nature and going on inside our heads. But we do have introspective access to our own mental goings on, perhaps *as* goings on that typically play a certain causal role.

This should embolden us to wonder whether we do not have observational access to singular causation, at least in certain favourable cases. Hume denied this, holding in effect that there is no impression of singular causation from which our idea (concept) of this relation is derived. Critical cases for him are our awareness of forces acting on our own body and our awareness of the sucessful operation of our own will. Hume does consider these cases, but the textual evidence shows that he did this *after* he had developed his own theory of causation – by which time he had a vested interest to protect – and in fact his arguments are not at all strong. I have examined these arguments elsewhere and cannot repeat that examination here, and have further argued for direct (not, of course, infallible) awareness of singular causality in the pressure and the will cases. (See my *A World of States of Affairs*, Ch. 14, Sec. 6, and see also the very valuable discussion in Evan Fales, 1990, Ch. 1.)[6]

Second, it seems to me that Menzies has left out two things from his collection of platitudes which "fix the reference" for this relation. The first is what we might call the same cause → same effect platitude. I mean by this that it is regularly true, even allowing that there are many apparent exceptions in ordinary experience, that from the same *sort* of cause the same sort of event, or at least roughly the same probability distribution of events, will follow. We might call it the Regularity platitude, and can see it as the truth that the Regularity theory of causation grasps but exaggerates.

Perhaps, however, addition of this platitude is implicit in Menzies' section 5, where he follows up his Lewisian-style definition of causation by calling attention to:

> . . . the powerful intuition that causal relations relate events in virtue of specific features of the events themselves, in particular the properties they exemplify. (p. 105)

But the intuition as stated is not enough. We want the idea that not only does causation depend on the properties of the thing doing the causing, but also that from the same properties the same effects tend to flow.

A final platitude that we might include can be linked with the Agency or Manipulative theory of causality. Causes characteristically are to their effects as *means* or empirically possible *means* are to their *ends*. We see, then, that Regularity, Counterfactual, Probability-raising and Agency theories, though all rejected, each do contribute their platitude to fixing the reference of the causal relation.

Third, we come to a deeper, if more controversial suggestion, which I owe to Adrian Heathcote (Heathcote and Armstrong, 1991), that a *deep* account of singular causation can be given as *the instantiation of a law*.

Laws link properties. In the simplest case, the instantiation of a certain property ensures, or in the probabilistic case merely probabilifies, that either that same particular or some related particular instantiates a certain property. That, surely, would *explain* the raising of the probability of the effect that is so regularly involved. (It would also explain the counterfactuals that we associate with causation, and why causes are means to ends.) A law (an entity in the world) provides the ground in nature for an inference. The cause is like the antecedent. If you know the singular cause and the law (which may be probabilistic only and/or may involve *ceteris paribus* conditions of operation) then, in general though not invariably, you can infer a raised probability of the presence of the effect.

I think of this *identity* thesis as an *empirical* identity claim. It will not, however, be like the identification of that which plays the mental causal role with brain processes. The latter is a *contingent* identity claim, because it does not tell us the purported *essence* of the mental. The brain processes are just those entities which happen to play the mental role. But my claim is that instantiation of law gives us the essence (or perhaps only part of the essence – explanation to come shortly) of singular causation. A better model than mind/brain identity will therefore be the identity of heat with molecular motion or water with H_2O molecules. It is a "Kripkean" necessity.

I now indicate very briefly a consequence of combining the identity of singular causes with instantiations of a law (or laws) with the view that laws are relations of universals. Suppose that laws link properties; suppose that the ultimate properties, the properties that divide the world along its ultimate joints, are universals; suppose that the ultimate laws, at least, are linkages of universals. Such a linkage of universals will itself be a universal. As a further result, each instantiation of a fundamental law will be the instantiation of a universal. But, famously, each instantiation of a universal is complete in itself, so the law will be present *completely* in each instantiation. So where singular causation is the instantiation of such a law it will be a completely intrinsic relation. In that case, instead of saying as Hall does:

> The causal structure of a process is determined solely by the laws of nature, together with the intrinsic character of that process

we can say instead:

> The causal structure of a process is determined solely by the intrinsic character of that process.

This result was unsought, but I think it is a very welcome consequence of my theory of laws.[7] By contrast, any Hume-inspired theory of laws makes the law-like nature of an instantiation of the law an *extrinsic* property of the instantiation.

But I end with a problem that my view of singular causation faces. I said two paragraphs ago in an aside that instantiation of a law may be no more than *part* of the essence of singular causation. My reason was this. Not all laws are causal laws, or at least are not obviously so. So it seems that all I am entitled to say is that singular causation is instantiation of *a certain sort of* law. My identification is therefore incomplete. And it would be a sad anti-climax to say that singular causation is instantiation of a causal law!

One very satisfactory solution of this problem from my point of view would be that all the *fundamental* laws are, or can fairly be represented as being, causal laws. It is an attractive hypothesis, one perhaps to be found in the work of Nancy Cartwright (1989), that the whole work of the world is done by causal action and interaction, that the world is a vast causal net.[8] It would be a natural corollary of this thesis that the fundamental laws are all causal.

University of Sydney

NOTES

[1] Michael McDermott (1995) has proposed an amended version of Lewis' theory of causation, with the particular object of overcoming difficulties about redundant causation. Lack of space debars me from considering McDermott's views. I should like to thank Keith Campbell, Nancy Cartwright, John Clendinnen, Phil Dowe, David Lewis, Michael McDermott, Peter Menzies, George Molnar and two anonymous referees for comment and suggestions.

[2] The full quotation runs: "A person caught in a philosophical confusion is like a man in a room who wants to get out but doesn't know how. He tries the window but it is too high. He tries the chimney but it is too narrow. And if he would only turn around, he would see that the door had been open all the time!" I am indebted to Nicholas Smith for tracing the reference. Like all of us, Wittgenstein failed to go through a few open doors himself.

[3] This is my interpolation. Dowe tells me that he does not necessarily agree.

[4] I may well have picked up this term 'fires' from the writings of Nancy Cartwright, who tells me that she sometimes uses it in the same way. In her (1989, p.108) she says that she is "trying to model a very special concept of probabilistic causality, a concept according to which the cause either contributes its entire influence or it does not contribute at all." That is my idea also, although it does not seem so special to me.

[5] An anonymous referee has pointed out that Hall's attractive formulation may be a little too simple. There are many causal processes in which each new stage is determined not only by the earlier state of the process but also by factors external to the process. Consider a planet moving under the influence of a variable gravitational field.

[6] Menzies has since joined me here, although he bases himself on the work of the Louvain psychologist A. Michotte (1963, first published 1946) who holds, on the basis of experimental evidence, that vision yields direct perception of causality. (See Menzies, this volume.) I do not reject Michotte's results, but I think that the causality of our own will, and still more the perception of pressure on our own body, are nearer to being "knock-down" cases.

[7] In the case of a merely probabilistic law I contend that the law, as opposed to the antecedent condition of the law, is instantiated only in the positive cases, only where the law 'fires'.

[8] See also the view of C.B. Martin that the world is a vast power-net (Martin, personal communication). See again, but without that striking phrase, Martin (1993).

REFERENCES

Armstrong, D.M. (1997), *A World of States of Affairs*, Cambridge: Cambridge University Press

Cartwright, Nancy (1989), *Nature's Capacities and their Measurement*, Oxford: Clarendon Press

Dowe, Phil (this volume), 'Good Connections: Causation and Causal Processes', 247–63

Fales, Evan (1990), *Causation and Universals*, London: Routledge

Hall, Ned (unpublished), 'New Problems for an Analysis of Causation'

Heathcote, Adrian and Armstrong, D.M. (1991), 'Causes and Laws', *Noûs* **25**, 63–73

Lewis, David (1973), 'Causation', *Journal of Philosophy* **70**, 556–67, Reprinted in *Philosophical Papers*, Vol. II, by David Lewis, New York: Oxford University Press, 159–72

Lewis, David (1986), 'Postscripts to "Causation"', in *Philosophical Papers*, Vol. II, by David Lewis, New York: Oxford University Press, 172–213

McDermott, Michael (1995),'Redundant Causation', *British Journal for the Philosophy of Science* **46**, 523–44

Malcolm, Norman (1958), *Ludwig Wittgenstein: A Memoir*, New York: Oxford University Press

Martin, C.B. (1993), 'Power for Realists', in *Ontology, Causality and Mind*, ed. John Bacon, Keith Campbell and Lloyd Reinhardt, Cambridge: Cambridge University Press

Menzies, Peter (1996), 'Probabilistic Causation and the Pre-emption Problem', *Mind* **105**, 85–117

Menzies, Peter (this volume), 'Intrinsic versus Extrinsic Conceptions of Causation', 313–29

Michotte, A. (1963 [f.p. 1946]), *The Perception of Causality*, trans. T.R. and Elaine Miles, London: Methuen

F. JOHN CLENDINNEN

CAUSAL DEPENDENCE AND LAWS

In this paper I want to explore how laws, or statements which have some of the characteristics which we associate with laws, can be employed to capture the notion of causal dependence, and also how this latter concept can accommodate a process account of causation. David Lewis (1986[1973]) proposed that we take causal dependence as the basic notion, with causal connection being defined in terms of it. However Lewis equates causal dependence with counterfactual dependence, and there are objections from an empiricist perspective to the semantics of counterfactual conditionals. An account of causal dependence which invokes the regularities that we discover in nature, and does not presuppose counterfactuals, is proposed in Clendinnen (1992).[1] This makes it possible to retain the advantages of a dependence account of causation while avoiding the metaphysical commitments of Lewis's theory. This paper explores the conceptual grounding of such a theory and how it relates to Wesley Salmon's account of causal process. In particular it considers those processes which include irreducibly probabilistic steps.

Let us start by considering the nature of an adequate theory of causation; this should clarify the task, and help to relate what I have to say to other contributions in this volume, indicating similarities and differences of approach.

I. PRELIMINARIES

I take the most basic concepts employed in organizing and describing empirical information to have essentially the same character as the more specific concepts of science. These latter are modified and revised as theory develops. However an influential philosophical perspective sees basic concepts, such as causation and the structure of space-time, as grasped a priori. The thought is that since these concepts are in part constitutive of empirical knowledge, they must be logically prior to it. This view was dramatically challenged by developments in physics in the first decades of this century. The basic concepts which came most naturally, and had seemed beyond challenge, could not be employed to develop a coherent theory that accommodated empirical findings, while a modified conception of space-time which went against our intuitions, did, and resulted in an immense extension of predictive power. It seems that basic concepts, like others employed by science, have no rational guarantee, or immunity from revision in the light of empirical findings.

Doubtless because of the evolutionary adaptation of our species, humans start with ideas that come naturally and are largely adequate. Nevertheless they need to be made precise, and sometimes modified, to ensure a powerful, coherent theory. Above all,

H. Sankey (ed.), Causation and Laws of Nature, 187–213.

our understanding of these concepts must make sense of the way we employ empirical information in applying them. These considerations indicate the nature of a theory of causation, and in particular the role that our intuitions should play. It is largely via our linguistic intuitions that we grasp our naïve concepts, and it is appropriate to take some account of these so that we do not lose sight of what we wanted of each concept. Nevertheless these concepts are not sacrosanct, and we may need to go against intuitions to develop a concept in the best way. A final reliance on intuition would make sense if we took them to be glimpses of something determined by reason, perhaps like Plato's forms. This view can only lead us astray; it ignores the possibility of our taking decisions, in the light of the empirical evidence available to us, about what concepts can best be employed in representing the objective world.

A theory of causation needs to specify a precise concept which captures, as far as possible, the most fruitful aspects of the way we unreflectively talk about causal connections. It is appropriate to characterize this procedure as an 'explication', in the sense spelled out by Carnap in the introduction to his (1950). An explication is a proposal, and so will be more or less valuable, rather than true or false. We may hope that an explication will identify the character of the objective entities or relationships covered by a concept. However there is a limit to how far we can go in this direction. Such an account, if non-tautological, must refer to more basic entities; and the regress of ever more basic kinds of entities must end. What we can hope for is a systematic account of how evidence bears on the claims couched in a given concept, even if it is not ontologically reducible. Thus the minimum requirement for an explication of a concept will be the specification of univocal standards for its application, that is of the assertability conditions for claims employing the concept. Our first goal, in seeking an explication of causation, should be the formulation of a systematic account of the assertability conditions for causal claims, and initially we should not assume that it will be possible to go beyond this.

There is a further feature of explications which must be accepted if we are to avoid setting impossible goals. Is an explication which only determines the use of a term in situations which occur in the actual world adequate? Certainly we should try to allow for unforeseen eventualities; but to seek criteria which would settle applicability in all logically possible situations may very well be counterproductive. Science does not seek such definitions; and as a consequence is sometimes obliged to re-evaluate concepts in the face of discoveries of a kind not anticipated. However this is a more economical policy than trying to spell out rules for the applicability of all concepts which would cover all possible situations. Given that this is accepted practice with the more specific concepts of science, I conclude that an explication of causation is not vitiated simply because it is only viable provided the world has certain very general characteristics which we are confident that it has. Certainly the more contingencies covered the better, but there are other important desiderata which cannot be ignored. Let us first seek an explication which will be satisfactory providing the world has the general character which we believe it has.

1.1 The Role of Judgements About Causal Connections

To evaluate a proposed explication of causation we must consider what role this concept plays in science. According to some, it is very little. They see the concept as inevitably reflecting human concerns and interests. This is true of some causal notions. For instance selecting one out of a number of relevant factors as *the* cause of an event inevitably reflects the interest or the context of the speaker. On the other hand, the distinction between causal connections and non-causal correlations is an objective issue which is of serious concern to science. A reliable sign is distinguished from a cause, even though in both cases there is a sequence in time which is subsumed under a regularity. The distinction is important because it identifies those locations in the network of events where intervention may be efficacious. Our interest in directing, or at least influencing, natural change is obviously served by this distinction. However science is concerned to trace causal connections where there is a non-human intervention in a system (where we perhaps know nothing of the origin of the disturbing event). In particular science is interested in the chain of causation by which the human species came into being. Scientists remain concerned with and confident about the distinction between correlations which are due to one item causing another and those which are not (both perhaps having a common cause). Cases where the distinction cannot be drawn are due to lack of information; and typically it is possible to point to the kind of information which if available would settle the issue. However the rationale of this distinction is not transparent. The trouble is that our causal idiom is so pervasive that we too readily slip into using it in explaining our distinctions. We say "This is only a symptom, for it did not bring E about", and so give one causal judgement to ground another. A satisfactory explication should identify in non-causal idiom the way criteria can settle these questions.

Granted the goal, how do the findings of science bear on an explication of causal relationships? The obvious point is that we must now abandon the notion, once taken to be guaranteed by reason, of universal deterministic causation. We must either allow for probabilistic causation or deny that all events are caused. On the other hand there are vast achievements of science which largely sustain the viability of the naïve notion of causation, and which can underpin an explication of a precise concept. Firstly there is evidence about the extent that regularities, if not always deterministic then at least stochastic, pervade nature. In a universe that was radically less regular we would not only find fewer causes, but the very concept of causation in its present form might have little place. Another impressive general feature of nature that has emerged, and which it is easy to overlook just because it is so ubiquitous, is the way regularities can be systematized by deriving some from others. It may be a logical truth that any body of knowledge can be systematized by taking as an axiom any statement that is not logically derivable from one otherwise accepted, but what logic can by no means guarantee is that the axiomatic system will be much simpler than that with which we start. Nor could it lead us to expect that the set of axioms sufficient at one time would remain sufficient as new regularities were discovered. It is a fact of great significance for science that such a high level of systematization of regularities by derivation is possible.

2. CAUSAL PROCESSES AND LAWS

Wesley Salmon's (1971) was an important step towards a more satisfactory account of statistical explanation. He, together with R.C. Jeffrey and J.G. Greeno, who published papers in the same collection, abandoned the view that an explanation had the logical form of an argument, and held that an undetermined event was explained by exhibiting all the factors which were statistically relevant to the occurrence. Salmon held a frequency interpretation of probability, and allowed that probability statements are laws of nature, deterministic laws being special cases of statistical laws in which the frequency is unity (1971, pp. 78–9). We will return to this work in dealing with probabilistic causation. At this stage we must note that by his (1984) Salmon's perspective has changed somewhat. Causal connections now play a crucial role in explanation, and causal connections are taken to be causal processes (p. 261). Laws of nature still play an essential role, but he insists that mere subsumption under laws, "mere fitting of events into regular patterns" does little towards explanation. What is needed is an account of the causal connections involved, provided that our conception of causation allows for probabilistic causes (pp. 121–4). A plausible view of a process is of a chain of lawfully linked events. However Salmon has rejected this view in favour of the 'at-at' account of process. Part of his motivation is to avoid the paradoxes that arise from an ontology of distinct events. He outlines Bertrand Russell's proposal to solve Zeno's arrow paradox of motion: 'to move from point A to point B is simply to be at the appropriate point of space at the appropriate time – no more no less. The resulting theory is therefore known as "the 'at-at' theory of motion"'(1984, pp. 152–3). Believing that this theory avoided Zeno's paradoxes, Salmon adopted it for all processes. However he also sees this notion of process playing another important role, namely avoiding Hume's strictures against employing concepts such as 'power' in our understanding of nature (p. 147).

Salmon expresses some unease about the account of process he adopts in his (1984). Following Reichenbach he recognized the need to distinguish pseudo-processes, such as moving shadows which exhibit the same kind of spatio-temporal continuity, from genuine causal processes. The distinction was to be made by the mark test: only genuine processes can transmit a modification imposed on the process. However he found it necessary to employ counterfactual conditionals in his specification of the test; many genuine processes do not actually transmit any modification, but would have if marked. This notion of a causal process as one which has the ability to transmit a mark seems to involve one of those powers which Hume had proscribed, and which Salmon was very concerned to avoid. It is in response to this problem that he introduced the 'at-at' theory of processes (p. 147). This theory makes no mention of the states of a process *changing*. Salmon considers the question: "What do we mean when we speak of *transmission*? How does the process *make* the mark appear elsewhere within it?"[2] He answers: "There is, I believe, an astonishingly simple answer. The transmission of a mark from point A in a causal process to point B in the same process *is* the fact that it appears at each point between A and B *without further interactions*" (p. 148). There is no need to use words which carry the notion of power, nevertheless he saw no way of avoiding the counterfactual conditionals, and they involve more than a whiff of powers. He could do no more than say that the difficulty involved in their use was not insurmountable. "Science has a direct way of dealing with the kind of counterfactual assertions we require, namely, the experimental approach" (p. 149). He acknowledges that he has no analysis of these locutions which shows how they may be confirmed by experiment, but advances strong plausibility considerations for believing that any satisfactory analysis would have this character (p. 150).

It is clear that Salmon was not entirely happy about employing counterfactuals; and this is confirmed in his more recent work. This is written primarily in response to a penetrating but sympathetic critique of Salmon (1984) by Phil Dowe (1992), who proposed a *conserved quantity* theory of causation that preserves the central place of process and interaction, but identifies each in terms of quantities which are conserved. Salmon in his (1994) endorses this theory, with only very slight modifications. As his title makes clear, Salmon sees it as having the virtues he had sought in his (1984) while avoiding counterfactuals.

2.1 The Conserved Quantity Theory

This theory is of the general kind that is argued for above: it offers an explication that builds on scientific knowledge, and according to it claims about causal relationships will be settled by scientific findings. Should we simply accept it without further ado? It is an interesting and promising theory, however a good deal more needs to be done to establish that all the connections which we confidently call causal are covered by it. If and when this is done we will have a theory about the nature of causal connections which can be accepted as true. Nevertheless it will not have captured the concept of causation with which we now operate.

When we confidently claim the existence of a causal connection we rarely if ever ground our claim on any quantity having been conserved. It is not simply that knowledge of the conservation does not precede beliefs about the cause. This would not settle anything; for we often jump to conclusions in causal as well as other claims. What is crucial is that when we appraise a causal claim what we seek is different information, not what quantities are conserved. (For instance, to take an example Salmon used to illustrate his account (1994, pp. 303–4), if anyone suggested that it might be the impact of air particles rather than that of a baseball which caused a window to break, we would not seek to check on the momentum of the fragments of glass to settle the matter.) Holding that the proposed theory does not capture our extant concept does not deny that once accepted it may very likely modify our concept of causation, just as our theory of atomic structure has modified our concepts of the various elements. What I stress is that our extant concept of causation is not like some naïve concepts which are so vague that they are of only historical interest once replaced by a clear, theory-based concept. As I have argued above, we have the ability to draw important distinctions between causal and noncausal correlations which we have every reason to believe are objective. Consequently I hold that, as well as further examining the conserved quantity theory, we should seek to make explicit the criteria by which we presently make causal distinctions. This is the aim of this paper.

I see some difficulties with Salmon's claim that causal processes are cut off from the context in which they occur. He tells us that while strictly speaking a process does not enter into any interaction, in some situations we ignore very small interactions and treat what happens as a single, composite process (1994, pp. 308–9). This is no more than a convenient idealization; a bullet moving through the air is actually a sequence of processes, each free of interactions, interspersed with interactions when the bullet collides with air particles (p. 309). Salmon leaves no room for processes which are continually influenced by some external factors. Particles which are subject to forces due to a variable field exemplify this, as do processes like waves consisting of the progression of some modification in a medium. Examples such as these show that generalizations, such as those describing the action of fields and the influence of various properties of the medium, are more important in identifying a process than Salmon allows.

We may be led to ignore the role of laws if we take the movement of a particle as the paradigm of a process, and interactions with other particles as the paradigm of modifications of a process. However even in these cases it is clear that the characterization of what is happening involves appeal to generalizations: that processes of a certain kind will persist in the absence of interactions, and that interactions of certain kinds will occur in certain contexts. The kinds of processes that this view ignores exhibit the role of generalizations very clearly. To identify a process as of a certain kind we must take account of the way changing circumstances may result in its modification. This suggests a view of processes as law governed, continuous sequences of states: the view rejected by Salmon in favour of the at-at theory. Indeed in his (1994) he holds that appeals to laws in a theory of causation are both unnecessary and undesirable. He allows that the conserved quantity theory invokes generalizations, and any such generalization could be false. We, of course, draw on only those which to the best of our knowledge are true, for instance we may invoke the claim that the charge on an electron is constant. He then goes on:

> If the statement about the electron charge is true, then there is a true generalization about the charge on an electron. However, *it makes no difference whether or not that true generalization is lawful; only its truth is at stake*. The problem of laws is the problem of distinguishing true lawlike generalizations from other true generalizations. That is a problem we do *not* have to face. (1994, p. 310)

These remarks follow, and amplify, his claim that any attempt to account for causation in terms of laws so as to avoid appeal to counterfactuals (such as the proposed nomic dependence theory) is very much like jumping out of one frying pan into another. He sees appeals to laws, or lawful regularities, as just as unsatisfactory as appeal to counterfactuals. I think that we must see his unease about invoking laws as one with his concern to conform to 'Humean strictures' against powers and necessities in nature.

2.2 *Kinds of Generalizations*

No doubt a theory which avoided an appeal to laws would have the advantage of avoiding controversy. There are various contested theories about their nature, and it is sometimes argued that science does not employ anything that are laws in any serious sense. We must, however, recognize that there are significantly different kinds of statements that have the form of a generalization. Sometimes a statement of this form is believed just because of a belief that a certain set exhausts the extension of the antecedent and the character of each element of this set has been observed. Such statements have no predictive content and so can play only a very limited role. General statements of the kind which typically concern science have either been established by inductive projection from an empirically established finite pattern, or are derived from wider generalizations which are established in this way. Let us call those generalizations which can only be accepted via an inductive projection at some stage *projected generalizations*. Their predictive content marks them off from other statements of the same logical form, and hence science's concern with them.

I propose to invoke the notion of projected generalizations in the explication of causation and causal processes, and to postpone the discussion of the relationship of this notion to that of lawfulness until later. At that time we will also consider an issue that is perhaps related, namely what can be said about the truth conditions of projected generalizations; so far they have been identified only by assertability conditions. At this stage there is one issue which should be considered. Salmon holds that the conserved quantity theory does not require laws. Can he also claim that it does not require projected generalizations? He allows that an appeal to generalizations is involved in identifying a process as of the conserved quantity kind. The mere fact that the magnitude of some quantitative property of a process remains unchanged over an interval certainly does not establish that the process over that interval is causal. Nor does the fact that the magnitude of the decrease of a quantity of one process equals the gain in the same quantity by another process establish that there has been an interaction (even given spatio-temporal contiguity). These facts will only provide evidence of interaction if the quantity in question is a conserved quantity, and that this is so is a generalization that goes far beyond the observed facts that evidence it. Thus the conserved quantity theory invokes projected generalizations. The theory may be formulated without referring directly to projected generalizations, but some of the notions that are employed in that formulation do involve this concept. Salmon is concerned to deny that laws are involved, but it is clear that projected generalizations are. I will argue that this weaker notion, together with the ordering of generalizations with respect to which are more basic, is sufficient for a nomic dependence theory of causation. So this theory is built only on notions that are uncontroversial and quite central to science.

2.3 The Empirical Basis of Causal Claims

A theory according to which causal claims are grounded in projected generalizations makes the empirical nature of the concept evident. The impossibility of identifying the meaning of a sentence with the means of verifying it has been long recognized. However the defects of this thesis should not blind us to the relationship that does hold between the way we grasp meaning and verification. Understanding an empirical sentence should enable us to recognize, or discover via collateral information, what would count as evidence for or against its truth. Sometimes a sentence we use does not fully meet this standard; perhaps we have only a vague idea of what would confirm it. In this case its meaning needs to be clarified. However if there is not even a hint as to what would confirm or disconfirm a sentence in any possible context then that sentence has no meaning for us.

All this is so of claims about causal relationships. We understand these claims and are normally able to recognize what is evidence for or against them. However there is no generally accepted, unified specification of what constitutes this ability. This is what an explication of this concept should primarily seek to provide. It is not necessary that we be able to identify a set of evidence which is necessary and sufficient for the truth of any causal claim, but for clarity of the concept we need a systematic way of identifying what is evidence. As Hume pointed out, one event being followed by another might suggest causation but is quite compatible with this not being so. Even if we consider the sequence as a necessary condition,[3] we must consider what additional evidence would settle the matter.

David Armstrong, who urges a singularist concept of causation, holds that we perceive causal relations; in this way he claims to show how we know about them. However perceiving, like knowing, is a success concept. Z not being the case is sufficient for the falsity of the claim that Joe perceived that Z. So our believing Joe when he tells us that he is in a state that he takes to be perceiving that Z does not settle the matter (unless it is claimed that perception is of subjective states). If we would add what else we need to justify the claim that Joe perceived that Z, we must specify the assertability conditions for 'Z' (rather than anything about Joe). So to say that we know, or have reasonable beliefs, about causal connections by perceiving them, does not get us very far.

That one event x caused another, y, involves something more than that x and y occurred in a certain spatio-temporal relationship. It is not at all clear that this something more can be identified in experience contemporary with the utterance of the assertion. The claim that we can perceive causal connections is plausible just because we allow that perception is fallible and that additional information, as well as present experience, is finally relevant to the truth of causal claims. That the additional information comes from our beliefs about regularities is consistent with much of what has been written about causation, and it makes clear the empirical character of causal claims. However the way regularities need to be related to a sequence for the latter to be causal is not as obvious as has sometimes been thought.

3. CAUSAL DEPENDENCE

The key feature of a dependence account of causation is that it is two-staged. According to the definitions of both kinds of dependence that David Lewis considers, counterfactual and nomic, the relationship is not transitive. e may be dependent on c, and c on d, yet e may not be dependent on d. Causal connection, on the other hand, is defined to be transitive. A causal chain is a chain of events, each of which is dependent on a preceding link. One event is a cause of another if there is a causal chain linking the two, even if the second is not causally dependent on the first (1986[1973], p. 167). Thus, in the case above, d is a cause of e.

Having causal connections explicated in terms of a more primary relationship between events is extremely important. It resolves some major difficulties that stood in the way of formulating a relationship which always holds between one event and its effect. The regularities we discover in nature ground our beliefs about what conditions are necessary or sufficient for what. And there have been many attempts at theories of causation in terms of one or both of these relationships. But no account has been entirely satisfactory. Cases of pre-empting overdetermination, where there is an event which would have caused an effect if the actual cause had not occurred, have often been cited as counterexamples. Lewis shows how the actual cause in some such cases can be identified by tracing the chain of dependence (pp. 171–2). I believe that this tactic is very fruitful, and that it can be used to distinguish cases of causation from correlations due to a common cause. However to do this we need to take causal dependence as nomic, rather than counterfactual, dependence. Indeed I believe that an account in terms of chains of nomic dependence can handle other cases of preemption which Lewis later came to see as difficulties.

Clearly we will have trouble if we require that dependence holds between an event and the one that immediately precedes it, for there is no such unique event – however close two events may be, infinitely many others intervene. A dependence theory can accept an 'at-at' notion of a causal process. It is only necessary to detect for each event some sufficiently close preceding event such that there is dependence. The sequence of events that results is a causal chain, and the limit as more and more links are identified is a causal process.

There are some other features of a dependence theory of causation which Lewis points out and should be noted. In the first place it deals directly only with causation between pairs of events (p. 161). However there are good grounds for taking this relation to be central, and that accounts of other applications of the concept (for instance when we speak of the cause of something having a certain property) need to be built on it. He also makes it clear that he is not attempting to say what distinguishes one of the various causal factors operating as "*the* cause" (p. 162). We have already noted that this distinction is neither needed nor desirable in an explication of the concept as it is used in science. What is needed are the criteria by which any causal, or causally relevant, factor is to be identified as such. The most important respect in which Lewis's original theory is incomplete is that it deals only with deterministic changes (*ibid*). Lewis later proposed a way of extending his dependence theory to cover cases that are essentially probabilistic; this, and alternatives, will be considered later in the paper.

3.1 Nomic Dependence

It is very natural to interpret 'e depends on c' as saying that if the event c had been different then so would event e. Lewis adopts a subjunctive sentence similar to the foregoing as a satisfactory way of analysing the meaning of claims about causal dependence, for he is confident that counterfactuals are univocal statements with an unproblematic semantics. However, on the face of it, we learn about the dependence of an occurrence on various factors by discovering regularities in nature. Lewis, taking account of this, considers the thesis that a theory of dependence might be spelled out in terms of laws. However he dismisses it as unsatisfactory. He defines what he calls *nomic dependence* (hereafter abbreviated to 'n-dependence'). Conditionals are inevitably involved, but the only ones employed in the definition are material conditionals. So this definition expresses in extensional logic a relationship between facts and laws which gives promise of capturing what we mean when we say that an occurrence depends on certain factors. This should be an attractive option to those who are not happy about counterfactual conditionals as a basic conception. I will argue that Lewis's reasons for rejecting a theory based on n-dependence are not adequate. I will however modify his definition slightly, for reasons already indicated, by replacing 'laws' in the definition by 'projected generalizations' (I will pass over any inappropriateness in retaining this name in the face of the modification, for the notion I use is so very close to that defined by Lewis).

3.2 Nomic Dependence vs Counterfactual Dependence

Lewis's definitions of these concepts can be summed up as follows. Both require the notion of a family of statements each asserting the occurrence of a different event,

$\{X_1, X_2, \text{——}, X_n\}$ is a family of such statements, with X_i asserting that x_i occurred, if just one of the statements must be true. Assume this family and another, $\{Y_1, Y_2, \text{——}, Y_n\}$, of assertions about the occurrence of different but spatio-temporally related events. Now if $X_1 \Box\!\!\rightarrow Y_1, X_2 \Box\!\!\rightarrow Y_2, \text{——}, X_n \Box\!\!\rightarrow Y_n$, (where $X_i \Box\!\!\rightarrow Y_i$ asserts the counterfactual from X_i to Y_i), and if x_k and y_k $(1 \leq k \leq n)$ actually occur then y_k is *counterfactually dependent* on x_k.

Given the same two families of statements, then if the set of material conditionals, $\{X_1 \supset Y_1, X_2 \supset Y_2, \text{——}, X_n \supset Y_n\}$ are derivable from projected generalizations and conditions that obtain, then if x_k and y_k actually occur, y_k is *nomically dependent* on x_k.

Lewis points out that it is not possible to identify causal dependence with n-dependence. To do so would have perverse consequences, and would make it impossible to draw the important distinction between causal connections and correlations due to common cause. To make use of n-dependence in explicating causal dependence, it is necessary to formulate a further requirement which must also be applied. Lewis points to a requirement couched in terms of counterfactuals which would distin guish those cases of n-dependence which are causal dependencies.[4] So, believing that counterfactuals are involved anyway, he simply identifies causal dependence with counterfactual dependence, and makes no use of the concept of n-dependence. Lewis had no reason to look any further; however having doubts about counterfactuals, and believing that an explication of causation should indicate how we appraise causal claims via evidence from the actual world, I will explore the possibilities of n-dependence further, and seek a requirement which makes no appeal to counterfactuals yet selects those cases of n-dependence which are cases of causal dependence.

I propose that we take a sequence of events in the positive temporal direction as *prima facie* a causal chain, or process, if each link is n-dependent on an earlier link, and then specify criteria which rule out some *prima facie* cases. To see what these criteria should be, consider a case where we suspect that some event, say s, earlier than e and on which e is n-dependent is not a cause of e. s must be a reliable indicator of e. The question is how we might establish that s is nevertheless not a cause. We would be hard pressed to do so if s were the only event on which e were n-dependent. However if there should be a third event, say c, earlier than both s and e, on which e is also n-dependent, there may be some reason for taking the chain of causal dependence which leads to e to go via c rather than via s. Suppose that the laws which directly express the n-dependence of e on s (ND2) are derivable from more basic laws in virtue of which e and s are both n-dependent on c. In such a situation e is directly n-dependent on c in virtue of laws more basic than those that express its dependence on s. Since these latter laws are derivable from the former, e is also n-dependent on s in virtue of the more basic laws, but only via a chain of n-dependence which

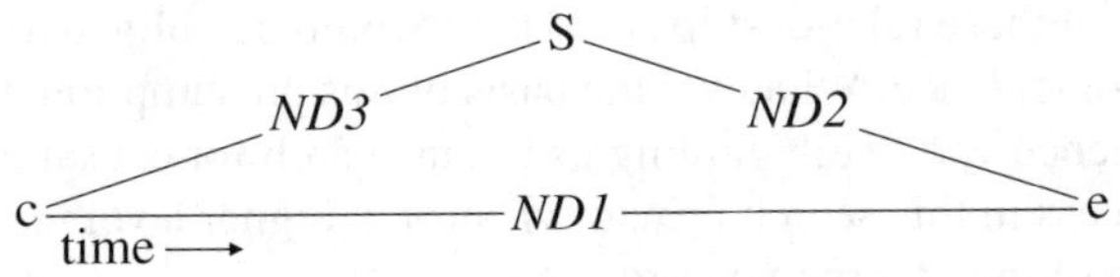

Figure 1: Take ND1 as the relation of n-dependence of event e on c, ND2 as that of e on s, and ND3 that of s on c.

includes c and in which some of the links have a temporal direction opposite to that of others. Thus the n-dependence of e on s in virtue of the more basic laws is less direct than its n-dependence on c. This kind of situation can be represented as in Figure 1. Here c is a common cause of s and e, and e is also n-dependent on s. Further, it can be shown that this latter dependence (ND2) is derivative from e and s being n-dependent on c (ND1 & ND3). (It would also be possible for c to be a common cause via chains of dependence, without s and e being n-dependent on c. In such cases the argument would be more complex – needing to take account of additional events between both c and s and c and e which are links in the two chains of dependence – but the same conclusion follows.) Given the definition of the relationship, the n-dependence of s on c (ND3) ensures that c occurs iff s occurs, so e is also n-dependent on s. This derivation of (ND2), the n-dependence of e on s, from laws more basic than those that directly express this relationship, shows that it is parasitic, *inter alia* on the n-dependence of e on c. Hence the former chain of dependence may be described as less direct than the latter, which consequently has a stronger claim than the former to be taken as the chain of causal dependence.

It is now possible to explicate a chain of causal dependence as follows: A chain of n-dependence which links events ordered in the forward temporal direction is a chain of causal dependence provided there is no other chain of n-dependence ending in the same event which is more direct. One chain of n-dependence a is more direct than another b if the facts and relationships that constitute a must be included as components of b for b to be subsumed under generalizations which are no less basic than those which cover a.

We may see the directness of a chain as determined by its route through events when the route is traced by laws which are no less basic than is necessary for the purpose. The simple chain from s to e is quite direct relative to a generalization according to which an e-type event occurs in such a context iff an s-type event does; but this relationship can be derived from more basic laws, in which case we must take account of the n-dependence of e on c.

We have seen that in at least some cases of a common cause the foregoing proposal can distinguish one event as having a better claim to be a cause than another. However two questions remain. Firstly, are all cases of n-dependence which are not cases of causal dependence cases of common cause? Secondly, can we be confident that in every case of common cause there will be the relationship between the laws involved that there is in the above case? In response to the first, if e is n-dependent on s, then there is a correlation between s-type events and e-type events. Given such a correlation we suspect that either one event causes the other or that they have a common cause. If neither of these relationships holds, we would be obliged to say that the correlation was a coincidence. Whatever the basis of this presumption, there is no doubt that it serves science very well, guiding us in our search for causal connections. The level of our success in this search provides a strong, higher level inductive argument for the principle of the common cause.

May not s and e have a common cause without the correlation between them being derivable from other laws in the way suggested in the example? This question immediately raises the one that is our prime concern: what is it for one event to have another as a cause? We have already noted that the ubiquity of causal idiom results in statements like "s did not bring e about" being given as the reason for concluding that s and e had a common cause. If we want to make any progress towards an explication of causation we need to consider what sort of answer could avoid anything that implies causal connections. I have argued that tracing dependence according to laws establishes a non-circular, *prima facie* case that there is a causal dependence. Such a case can be defeated if the n-dependence is parasitic on another. The notion of one dependence being parasitic on another draws on something that is crucial to science as we know it: that in many cases a generalization is in principle dispensable because everything entailed by it is entailed by a set of more basic principles. This shows how, in tracing chains of causal dependence we can distinguish some as derivative from others.

Examples of cases where we distinguish reliable signs or symptoms from causes, illustrate the point. Why do we believe that Koplik spots, which appear in a patient's mouth before any of the main symptoms of measles, are not a cause of the later symptoms? While avoiding causal idiom, we can point out that the generalization linking the spots to the later symptom is not basic, and that its derivation requires, *inter alia*, those more basic laws that link the initial infection to both the spots and later symptoms. Another classic example of a correlation which is not causal is that between the reading of a barometer and the weather.[5] Again, in believing that this is not a causal connection we believe that the generalization that there is such a correlation is derivable from other generalizations, which are consequently taken as more basic, about the way air pressure affects the instrument and also the weather. A generalization which is derivable is dispensable, so we only need the one from which it is derivable. (We may well believe that this in turn is derivable from yet more basic generalizations, but these further derivations are not necessary for the purpose in hand.) Once again ordering of generalizations according to derivability, and taking account of what conditions are needed to do so, reveals an asymmetry in dependence; the weather is only n-dependent on the instrument reading because, in part, of its dependence on the air pressure.

We may readily say what we take to be relevant in these cases by using counterfactuals: "if the human body had been just sufficiently different——", "if barometers worked differently so that——". Undoubtedly these locutions are used in this way. However, as we have noted earlier, there are objections to taking counterfactuals as unproblematic and as a suitable basis for the explication of an objective relationship. The foregoing argument for the reasonableness of an n-dependence theory of causation equally points to an unproblematic, objective relationship which underpins counterfactual idiom where this has some point.

4. COMPARISON AND ELABORATION

Since a proposition and its negation constitute a family, e will be counterfactually dependent on c provided (C □→E) and (not-C □→not-E). So on Lewis's account this will be the minimum requirement for causal dependence. The truth conditions for a counterfactual conditional are relationships of comparative similarity between possible worlds, however it seems fairly clear, and is acknowledged by Lewis, that our evidence for beliefs about such relationships must be what we learn about the actual world. So according to the counterfactual dependence theory, what we learn about the actual world leads to beliefs about relationships involving non-actual worlds, and it is these latter relationships which constitute causal dependence. However the way facts about our world confirm claims about degrees of similarity between various non-actual worlds cannot be described with any precision. This is especially clear in Lewis's argument against the acceptability of backtracking counterfactuals (1986, pp. 32 *et seq*). This points to the advantage of a theory according to which empirical evidence supports causal claims in a less devious way.

Granted that the counterfactual locution can play a useful role in characterizing dependence, the question is whether we can explicate a more precise relationship which grounds this way of speaking. According to the n-dependence theory, to establish that e is causally dependent on c we need to show how our generalized knowledge provides reason to expect something different to e in every context which differs from the actual one just in something different to c occurring in its place. This is done by deriving from established generalizations and a specification of the relevant context the set of material implications $\{C_1 \supset E_1, C_2 \supset E_2, \text{——} C_n \supset E_n\}$, where $\{C, C_1, \text{——} C_n\}$ and $\{E, E_1, \text{——} E_n\}$ are families of assertions about the occurrence of events (including, respectively, c and e). This requirement draws on our inductive knowledge and employs only extensional logic, so without circularity or invoking non-actual worlds captures the belief which grounds a counterfactual-dependence locution.

Consider now the counterfactual (C □→E). It must be true whenever both C and E are. So, on the counterfactual theory, that there is a causal dependence between two events which both occur depends entirely on the conditionals dealing with cases in which C is not the case. (This is commented on by Armstrong on the first page of his contribution to this collection.) Thus the theory invokes no positive connection between the cause and effect. For n-dependence, on the other hand, a material conditional linking cause and effect must not only be true but also be covered by some generalization that we accept.[6] In this respect the n-dependence explication captures an intuition that the counterfactual theory does not, requiring that the cause is both nomically sufficient and necessary for there to be causal dependence in the context which obtains.

4.1 The Individuation of Events

Although I have followed Lewis in taking causation as primarily a relation between events, a notion of events which differs significantly from his is possible and appropriate in the n-dependence theory. The counterfactual theory treats events as clearly delimited entities. This is necessary because that theory is concerned with what events in non-actual worlds are identical to the events under consideration. However problems emerge concerning such identities. I will argue that such problems can be avoided in the nomic dependence theory.

Paul Humphreys (1989, p. 24) adopts the following definition: "An *event* is a change in, or possession of, a property in a system on a trial [that is on a certain occasion]". This definition is well suited to the n-dependence theory. According to it we must identify a certain location to identify an event. However it would be impractical to equate an event with everything that is the case in a certain space-time region. What is relevant to the typical question that we ask about causal relations will occupy a region substantially larger than the minimum we can practically specify, yet such a region will also contain much that is not relevant to our question. Accordingly to identify an entity which enters into a causal relationship we must cite some property, or change of property, as well as a location. There are reasons, which we may ignore here, for referring to a concrete system rather than a region of space in specifying location, but so long as a location is identified, the assertion that some event occurred can be taken as an assertion about the character of what is the case, or what happens, at this location. Clearly a property can be characterized with more or less specificity. We may use a generic property to identify the effect which we wish to explain, but in establishing the cause take account of the more specific properties which constitute the generic one. Given a suitable location and property we will have successfully identified the effect in the actual world, and may then seek more information about it to establish its cause.

According to the n-dependence theory, causal connections consist in nomic relationships between facts in the actual world. Talk of other worlds may be used as a convenience, but is not essential to the theory; so whatever effectively identifies an event in the actual world succeeds in individuating it. However for the counterfactual theory there is always a question as to whether an event in another world which differs in some details from that in the actual world is nevertheless the same event. This is the source of a problem which Lewis explores via a number of examples, including ingenious chains and networks of neurons, each of which, on firing, stimulates or inhibits others from firing (1986, pp. 193–205). The main point can be made via a more mundane example which he also discusses. X shoots Y, but X was just one of seven men stalking Y, and if X had failed to fire, one of the others would have. Is Y dying from the bullet of one of the stalkers other than X, in another world, the same event as the death of Y in the actual world? If so then the counterfactual "if X had not fired, the death would not have occurred" is false, and the death is not causally dependent on X firing (nor on any of the chain of events that constitute the bullet moving from X's gun into Y's body). Lewis holds that common sense does not fully determine the standards for the identity of events, and considers whether the counterfactual theory could handle the foregoing problem by tightening these standards, so that very slight differences mean different events – taking events to be very fragile, in his terminology. However he recognizes awkward consequences from this proposal. Consider the case of a prisoner executed by a firing squad. Suppose one gentle soldier refrains from shooting, then high standards of fragility mean that no event in a world in which all the soldiers fired would be the same event as the actual death of the prisoner. So had the gentle soldier fired the actual death would not have occurred, and the actual death is causally dependent on the gentle soldier not firing. This certainly conflicts with what we would say in such a case. While Lewis does not entirely rule out the possibility of developing a satisfactory theory in terms of the fragility of events, he can see no way of doing so, and prefers to retain ordinary standards of fragility but re-define causal dependence so that it may be constituted by quasi-dependence as well as counterfactual dependence. This notion is defined so that Y's death is quasi-dependent on X's firing, on the grounds that in other sequences of the same intrinsic nature there would be counterfactual dependence (pp. 205–6). The modified theory has become somewhat complex. The truth conditions for counterfactuals are still invoked, but also judgements as to which processes have the same intrinsic character; this in turn requires that we disregard any irrelevant events (p. 205). To decide what is relevant we must surely take account of the nomic connections between events, and it would be much neater if nomic relations alone settled the matter.

In the original paper Lewis warns against taking that of which we seek the cause to be simply what fits a certain description; rather we must take it to be the event which is effectively picked out by that description. As we have seen, the rationale for this is that his theory requires the identification of events in non-actual worlds. If, however, we adopt a theory which establishes connections between events in the actual world by establishing how they are related with respect to regularities manifest in this same world, there is no difficulty in accepting that what we seek to account for is whatever meets a certain description, provided it is not ambiguous in the actual world. Certainly we cannot expect that the description we initially employ will tell us everything that is relevant to causal connections; these will be revealed in the investigation. There is no paradox even if we should discover that the original description which we take as identifying the effect needs to be corrected. Indeed if we discover an error in what we believed about part of the world, we had better be ready to correct it. (The coroner originally believes that the visible stab wound penetrated the heart, but the autopsy reveals that it did not, and that the victim drowned. This would not lead him to abandon the investigation just because he originally thought he was dealing with a stabbing.) I conclude that questions about what event is identified by a certain description involve nothing more than what description selects just those features of a part of the world which are relevant to the causal inquiry in hand.

4.2 The Depth of Causal Insight

We must recognize that there is a certain degree of openness in many causal claims. The n-dependence theory locates the source of such openness where it occurs. Suppose that it is claimed that a cause of a current occurring in a certain object is that the latter is made of copper. I think that our intuitive response would be that this is unsatisfactory, but that it does indicate something relevant. The n-dependence theory grounds this response, for the material implication (x is not copper) $\supset$ (x does not conduct electricity) cannot be derived from accepted generalizations together with any plausibly assumed conditions. However there is some disjunction of chemical kinds, say *K*, such that it includes everything that is a conductor, so there is a generalization in virtue of which x conducting electricity is n-dependent on its being *K*, which is consequently *prima facie* a cause of an object of this kind conducting electricity, and does not seem to be defeated by a different dependence being more direct. However if *K* remains no more than a disjunction of unrelated chemical kinds, we will feel that stating that an object is in the extension of *K* is a rather unsatisfactory account of a cause. This indicates that the n-dependence theory, indeed any theory of causation, needs to be complemented by taking account of what we may call "the depth of the causal insight" provided by an explanation. This will be determined by how basic the generalization employed is. Because of the similarity of behaviour of things of any of the different kinds that constitute *K*, we expect that there is some identifiable characteristic common to all elements of its extension, and that the generalization about *K* is derivative from a generalization invoking this characteristic. If this is so, then being a *K* is a cause of electricity being conducted, but citing *K* gives less information than could be given. Before we have discovered the common characteristic, we must make do with the account in terms of *K*, recognizing its incompleteness and its fallibility.

There is another feature of this example that calls for comment. Being a conductor is a qualitative concept employed where a quantitative one is needed. It covers a range of values of resistance with no precise limits. We should seek a causal account of the substance having a resistance of such and such magnitude rather than its being a conductor. So we need a further clause to the definition of depth of causal insight.

If there are different causal accounts of the same event under different descriptions, then an account which invokes more basic laws gives deeper causal insight, and so does an account in which the event is more precisely described.

To maximize the depth of causal insight we need descriptions in quantitative terms, where applicable, that are as precise as possible, and the laws employed to establish n-dependence need to be as basic as possible. Happily developments in theory which contribute to one of these desiderata typically also contribute to the other.

4.3 Basic Generalizations

According to the n-dependence theory, employing generalizations which are the most basic available plays a double role. It shows which chains of n-dependence are most direct and it maximizes the depth of causal insight which we achieve. However there is an important distinction. The former role is concerned with whether or not there is a relationship of causal dependence, and this is not so of the latter. Increasing the depth of causal insight is a desideratum, but achieving it is not a requisite for discovering causal relationships.

The capacity of the n-dependence theory to give univocal answers about causal connections depends on the extent to which a univocal ordering among generalizations can be achieved by science, and this is a contingent matter. I have indicated why I do not see this as a serious objection. Those who would have a concept which was necessarily applicable in all possible situations must consider whether we could make sense of claims about causal relations in a situation in which there were quite different orderings of relevant generalizations, each of which was equally well grounded. Our beliefs about which relationships are causal certainly correlate with beliefs that some generalizations are derivable from others. For instance, we believe that it is changes in the air pressure, rather than changes in the behaviour of the barometer, which cause a storm. We also believe that the generalization linking the storm to readings of the instrument depends on that which links it to atmospheric pressure. Once we agree that causal and counterfactual idiom may not be used as ultimate reasons for our causal claims, we see the significance of the connection between such beliefs.

5. PROBABILISTIC CAUSES

Physics tells us that some irreducible laws are non-deterministic. It is, of course, possible that the physicists are wrong; but it would be unwise for philosophers to commit themselves to a position that assumed that they are. So our account of causation must allow for events that are not determined. Nomic dependence, as so far defined, cannot be applied to such cases, for the required material implications cannot be derived. Nevertheless we certainly seek to cover non-determined occurrences by generalizations, albeit probabilistic ones; so it may be possible to extend the notion of n-dependence. Here it will not be possible to establish the dependence of e on c in the sense that some different event would have occurred if c had not, but we may hope to establish that the probability of e would not have been as it is if c had not occurred. Then we might say, in a weaker sense, that e depends on c. This extended notion of dependence will cover both deterministic and nondeterministic cases.

We should note that probabilities will be involved in very many cases where we are concerned with causes but have no reason to believe that they are probabilistic causes. In the typical case there are relevant factors that we cannot discover. In addition, we are often less than certain about the laws which apply. Consequently we can only speak of the probability that a certain factor is a cause. It is possible that in many such cases the causes are deterministic; Laplace thought that this was always so, and that probability merely reflected our ignorance. It is essential that cases where we can only achieve probable knowledge of causes are not confused with those cases where we have good grounds for believing that the only law-based relations relevant to causal connections are probabilistic.

5.1 Lewis on Probabilistic Causation

In one of the postscripts to his original article, Lewis (1986, pp. 175–84) proposes just such an extension to his dependence account of causation. He again takes the counterfactual language as irreducible. However, as with deterministic causation, it is possible to interpret what is expressed in counterfactual idiom as only a rough characterization to be replaced by a rigorous specification in terms of n-dependence. Where c and e both occur, Lewis holds that e is counterfactually, and so causally, dependent on c if the probability of e would not have been as high as it is if c had not been so. (He does not require that the final probability be high, only that it be increased from what it would have been without c.) Thus, for Lewis, as for many others, c is only a cause if its occurrence increases the probability of e; it must be positively relevant to e's occurrence. Against this, Salmon argues that any factor which is probabilistically relevant, either positively or negatively, to e should be taken as a cause (1984, pp. 192–202). A decision on this issue must be dependent on what we see as the point of a theory of causation. There is little doubt that our intuitions incline to the notion that a cause, if it does not ensure the effect, must at least make it more probable. However when we remember how recently there was a virtual consensus about determinism we can see that not too much reliance should be put on intuitions about nondeterministic cases. As I have argued in *Preliminaries* above, it is not a question of which theory is true, but of which proposal is best; and I believe that our prime concern should be with a concept of causation that will contribute most to scientific discourse.

If we should hold that an occurrence c, to be a cause, must increase the probability of e then we would be confronted with an awkward conclusion: in some cases we could then not allow a factor as a cause even although we knew that it occurred and that in doing so it influenced the probability of e. Suppose that we have background knowledge B and also know C, that c occurred, and that Pr(E/B) < Pr(E/B&C).[7] We then accept c as a cause of e. Suppose that we also know that the context of c included factor f, such that Pr(E/B&C&F) < Pr(E/B&C&not-F). According to Lewis's account f cannot be accepted as a cause of e; yet we know that f is relevant to the occurrence of e, and perhaps we know just how it changes the probability. Requiring that some factors which influence the occurrence of e should not be included among its causes has this consequence. Some factors which would have to be taken into account before the event occurred to make the best prediction would not be counted as causes. Identifying causal relations promises to add something to a mere subsumption of an event under laws, but a requirement that we ignore as causes some factors which are nomically relevant to e would seem to be doing the opposite. Further, if it is agreed, as Lewis proposes, that the selection of one of the causes as *the* cause is not important for a theory of causation, it seems appropriate to identify all relevant factors as causal. Again, if identifying causal connections is tracing chains of dependence it seems appropriate to include as a cause everything that influences the occurrence of each link. Such considerations indicate the superiority of accepting an explication which agrees with Salmon in identifying all factors which are probabilistically relevant.

5.2 Salmon on Statistical Relevance

In his earlier work on explanation, Salmon did not see causal connections as of central importance. Nevertheless his account of statistical relevance is very suggestive for a theory of probabilistic causes. He takes the probability of an event having a certain attribute which locates it in class *A* as the frequency of events of this class in a reference class which is defined in terms of just those factors which are statistically relevant to the event being in *A*. Consequently in his (1971) he holds that statistical explanation consists in specifying the characteristics a reference class, *R*, must have to achieve this. Employing the notion of event proposed in Section 4.1, we may take an event to be identified by its location in a system at a certain time. The extension of *A* and *R* will be such locations, and the location, l, which identifies the effect-event will be an element of both classes. Thus the probability of l being in the class *A* is the frequency with which elements of *R* are also elements of *A*. If it were possible to partition a proposed reference class so as to give a different probability of l being in *A*, some factor which was relevant to the occurrence of an *A* event would have been omitted from the definition of the original *R*. Consequently Salmon required that the reference class employed in satisfactory probabilistic explanation be homogeneous with respect to the attribute *A*. *R* must be homogeneous to the best of our knowledge for us to judge an explanation satisfactory. To be actually satisfactory it must possess an objective, rather than an epistemic, property. The notion of objective homogeneity is no simple matter, however Salmon in his (1984) devotes a chapter to its viability. If R should be objectively homogeneous but, as specified, not as wide as some other homogeneous reference class, then some factor has been employed in its definition which is not relevant to the probability of an A event. (As an example Salmon suggests explaining a person not getting pregnant by pointing out that this person is a man who takes the contraceptive pill.) If an explanation is to be free of information which is irrelevant, and so may mislead, the reference class must be the widest which is homogeneous with respect to the attribute to be explained. While the *Statistical Relevance* account of explanation does not directly speak of causes, it rests on the identification of a reference class defined by just those factors which are relevant to the probability that l is an element of A. We might very well say that it identifies all the factors which are causally relevant to the event constituted by l being of this kind.

The foregoing suggests that an SR explanation identifies causal factors by the very same criteria as n-dependence, provided we take n-dependence to cover the dependence of the probability of non-determined events; however this assumes that rational beliefs about probabilities are expressed in, or derivative from, projected generalizations that we accept. A belief about the frequency of an attribute in a reference class may not seem to involve projected generalizations, but as soon as we remember that we are not concerned with frequencies in a finite set, but with long-run frequencies defined in terms of limits, it is evident that they are. So the claims about frequency which must be used in applying SR are projected generalizations, and giving an SR explanation is the same as identifying the set of factors on which the probability of the explained event is n-dependent.

There is a further feature of the formal specification of SR which brings out this parallel. Salmon requires that complementary homogeneous reference classes be identified in a correct explanation (1971, pp. 76–7). Not only will R, of which l is an element, be identified but also other homogeneous reference classes, $R_1, R_2, \text{——}, R_n$ into which l might have fallen. For each R_i the frequency of A events is given, and it must be established that R_i is the largest class with that frequency. This specifies the probability of an A event at any location once we know something else about that location, namely into which of the possible reference classes it falls. Thus an SR explanation draws on our general probabilistic knowledge to exhibit the probability that l is of kind A in all the relevantly different contexts in which it could occur. SR can be expressed in terms of a set of material implications which are derivable from our generalized knowledge, each having this form:

$$x \varepsilon R_i \supset (\Pr(x \varepsilon A) = p_i) \text{————SRMI}$$

where x ranges over possible locations of events and R_i and A range over characteristics of locations.

It is necessary to add a restraint on the definition of R_i. The requirement that each R_i be objectively homogeneous with respect to A could be easily achieved by including membership of A in their definitions. If this were allowed we would always explain the character of an event by pointing out that it has that character, but we want an explanation of l being A in terms of some other feature of l; so a theory of explanation must rule out membership of *A* being part of the definition of R_i. If we take causation between events as always in the forward temporal direction, this will follow; what locates l as an element of R_i will be the relationship of l to another event which occurs just before l, and so before l has the characteristics which determine that it is an element of *A*. So the membership of an event-location in R_i will be definable in terms which are logically independent of those which determine that it is an element of *A*.

In dealing with deterministic causation we followed the common practice of taking causation to be a relationship between two events. It may seem that the foregoing proposals about probabilistic causation involve incompatible concepts: causes as factors rather than events. However this is no more than a verbal difference. A factor is something that is the case, some structure having or changing some property at some time; that is, an event. Suppose we take event c as a probabilistic cause of e. This description can be assimilated to the foregoing account of probabilistic causation thus. E is the fact that l is an element of *A*. c must then be a distinct event at location m different, but presumably reasonably close, to l, such that m being related as it is to l, and having a certain property is probabilistically relevant to l being *A*. The fact that c is an event which is distinct from e is no problem, for the properties which define *R* will include relational properties of l. Thus one of the properties which defines *R* will be that its elements are events spatio-temporally related in a certain way (in the way that e and c are related) to other events with certain properties (properties which make the event at location m of the kind that c is).[8]

5.3 Contributing and Counteracting Causes

Paul Humphreys (1989, p. 14 and *passim*) distinguishes between those factors which increase the probability of the effect and those which decrease it, calling the former contributing, and the latter, counteracting causes. Modifying the account of causation based on SR explanation in this way may seem to be only a small modification to bring it more into accord with our intuitions. However there are difficulties in holding that there is a real distinction here. As presented it suggests that the probability Pr(E/not-C) has the same status as Pr(E/C). I have already proposed to follow Humphreys in taking an event to be the occurrence or change of property at a certain location. 'not-C' will involve a very broad, generic property. If we seek to compare Pr(E/C) to a probability conditional on information of comparable specificity to C we should consider Pr(E/C_1), where C_1 specifies the character of l in much more detail than does not-C. However there are many different ways that l might be specified in detail which would all entail that it did not have the property asserted in C. Typically, the probability conditional on some of them will be greater than Pr(E/C), and some less than it.

Consider any case where the chance of an event depends on some variable magnitude. Suppose, for example, a set-up where a source of photons is directed through a slit in the general direction of a target, an area on a screen which is small in comparison to the area in which there is a non-zero probability of it arriving. The probability of a scintillation in the target due to the arrival of a photon is conditional on the angle between the line through the slit and the direction from slit to target, and it can be arranged so that this is variable. Now consider any angle for which the probability is non-zero. Is this relation between the two directions a contributing cause when there is, in fact, a scintillation? On Humphreys' definition an affirmative answer requires that Pr(E/C) > Pr(E/not-C) (taking E as asserting the occurrence of a scintillation and C specifying the angle). Assuming that a photon is emitted, Pr(E/not-C) can be calculated by integrating over the probabilities for all possible directions different to the actual one. Now if C is as detailed as we can make it, that is, if the direction is measured as precisely as possible, then Pr(E/C) will certainly be less than Pr(E/not-C), even if the direction is such as to maximize Pr(E/C). So considering quantitative causes, which are of crucial interest to science, it seems that no causal factor will be of the contributing kind if it is specified with sufficient precision. If we should seek to avoid this by basing the definition on a comparison with a causal factor of the same degree of specificity, the consequence is no more satisfactory: for there will be such a comparison for every possible angle, and some will have the actual direction of approach a contributing cause, and typically others will have it a counteracting cause.

This indicates very briefly the problems that I see with the distinction between contributing and counteracting causes. However what I have been able to say here indicates very little about Humphreys' careful analysis. An important feature of his approach is his denial that a causal explanation must be complete to be true (p. 9). His point is that we are rarely in a position to be confident that science has identified all relevant causal factors when it has surely made a serious contribution to understanding. He says that in such cases "we have good reason to suppose that what has been offered as an explanation is true and explanatorily informative." The trouble is that wherever we allow that there are causally relevant factors not taken into account, it remains possible that the causal structure is quite different to the way it seems on available information, indeed the situation may turn out to be deterministic. The assumption that as more and more explanatory information is acquired only the details, and not the general structure of the explanation, change is not justified. To deal with genuine cases of probabilistic causation, and to distinguish them from deterministic situations in which we nevertheless employ probabilities because of our lack of information, we must speak of objectively homogeneous reference classes. We may always be wrong about a class being or not being homogeneous, but if we are we may equally be wrong about the causal structure. An explication should say what needs to be the case for a certain causal relation to hold, irrespective of our being able to discover whether this is so.

The n-dependence explication of probabilistic causation will handle the example of the photon arriving at the target by the application of SRMI. From the irreducibly probabilistic theory that applies in such cases, we can derive the set of material implications required. These exhibit the different probabilities for each possible angle, and this shows how the probability of hitting the target is dependent on the actual angle between the line through the slit and that to the target. So this angle is a causally relevant factor. That, in most cases, there would have been a higher probability for some different angle, is not an issue.

5.4 Frequencies or Chances

Salmon, while favouring the frequency interpretation of objective probability, states that his SR account of explanation does not depend on it, and could be adapted to other theories of probability, including the propensity theory (1971, pp. 39–40 & 82). Indeed, in his (1984) he sees a positive role for propensity, arguing, however, that it cannot be accepted as the sole interpretation of probability (1984, pp. 204–5). The crucial point is that where we have beliefs about objective probability, they must ground rational opinion, or degrees of belief, about facts not yet directly known. Both frequency and objective chance interpretations of probability account for this; if either failed to do so, it would be inadequate. Thus although the probabilities have different meanings on the two interpretations, believing that an objective probability has a certain value has the same impact on expectations whichever is accepted. This indicates that Salmon is right, and that the requirements on probabilistic explanation expressed in his statistical relevance account could equally hold for a chance interpretation of probability. Certainly on the latter interpretation the material implications in SRMI will be derived from projected generalizations about chances, as is required by the nomic dependence theory.

I agree with Lewis, and many others, that a chance interpretation of objective probability is best. On the other hand Salmon's statistical relevance account of explanation has many virtues that can be readily adapted to express a dependence account of the causation of events which are not determined. It requires that all factors relevant to the probability of the effect be taken into account, and it does not invoke counterfactuals, resting instead on the instantiation of inductive generalizations. On both these counts I believe that Salmon is correct. Thus I see SRMI, with the probabilities interpreted as chances, as expressing the n-dependence criterion of which factors are causally relevant in cases of probabilistic causation.

5.5 The Thesis in Summary

It is now possible to summarize in general terms what is argued in this essay. In seeking to identify causal chains, or processes, we are obviously concerned with more than temporal order. The thesis is that we aim to trace chains of states or occurrences which are linked by generalizations expressing persisting patterns in nature. The linkage that needs to be established is n-dependence, which means that we seek to identify for each link those aspects of the situation preceding the effect which are nomically relevant to the chance of its occurrence. In deterministic cases, where the chance of the effect is unity conditional on some set of factors, it will, in general, be possible to discover more than one such set; in which case a further condition applies: that the set of factors would not be different if the n-dependence were established by generalizations which are more basic in the nomic systematization of generalizations. Where the n-dependence of an event is probabilistic, the factors are causal only if they jointly define a set of situations which is objectively homogeneous with respect to effect-events of the kind under which we seek to causally explain the effect. (That the house was destroyed may be causally explained by the dam collapsing; whereas that the house was destroyed with the west wall going first depends also on factors which directed the surge of water.) The homogeneity requirement ensures that there are no more basic regularities which if taken into account would have identified different factors as causally relevant. This explication of causation does not require that we have available what we believe to be a complete list of causally relevant factors before we can make a claim about causation, but it does mean that factors not taken into account can falsify such a claim. So a claim about a causal connection can be defeated by information that certain factors which were not known to be relevant actually are.

6. LAWS AND GENERALIZATIONS

The general statements employed in the explication proposed here are identified by their logical form, their being accepted as true on empirical grounds and on the further requirement that they must be grounded by inductive projection. The importance for science of distinguishing this kind of generalization was pointed out by Reichenbach in his (1976[1954]). He saw inductive verification, rather than mere truth, as essential for a general statement to be a law of nature (p. 12). For Reichenbach this requirement was but one of a set intended to explicate the concept of a law, and his other requirements are by no means redundant. It would not be plausible to suggest that all projected generalizations are laws. However I do not attempt an explication of laws in this paper, believing that it is sufficient to define n-dependence in terms of projected generalizations.

Let us consider some examples of generalizations that should not, by ordinary standards, be classified as laws, and see whether allowing the material conditionals which constitute nomic dependence to be derived from them has any paradoxical results. It will be convenient to use Reichenbach's requirements to suggest such generalizations. In the first place these requirements exclude as laws general conditionals with always-true consequents or always-false antecedents. These could not ground causal dependence, but since they are not projected generalizations no problem arises. Reichenbach's requirements also exclude as laws generalizations which misleadingly say less than has been established to be true. There would be no point in disallowing n-dependencies based on such generalizations, since they would hold in any case in virtue of the stronger generalizations which have also been established in these cases. Reichenbach, and many others, would exclude as laws generalizations which include proper names or references to particular space-time locations. Those general statements adopted just because they deal with a finite set, all members of which have been examined, will be excluded by this requirement. However these are also excluded by requiring that only projected generalizations be employed. A more likely source of trouble are those generalizations which are inductively adopted but which we suspect to have only a limited scope. For instance we may accept on inductive grounds a generalization about a biological species which we expect will eventually become extinct, and consequently have a finite extension. Further it is plausible to hold that the meaning of 'species' involves appeal to a much smaller finite set: those individuals with which we are acquainted and accept as of this kind, the remaining members being those individuals reproductively related. There will be similar generalizations in other parts of science, including physics, about structures which have come into being and remain fairly stable for a considerable time, for instance generalizations about the solar system. The character of such generalizations may well give reason to doubt that they should be classed as laws; the question is whether it is reasonable to accept claims of n-dependence based on them.

Science will seek for unlimited generalizations to subsume the kinds of limited ones we are considering, for those of the former kind maximize predictive power. Certainly this desideratum must be weighed against that of having well supported generalizations, but a policy of adopting only hypotheses with less scope than any competitor based on the same evidence, because it must have a higher probability, would defeat the very point of adopting hypotheses; it would result in never adopting any proposition with predictive content, since by this policy the bare statement of the evidence would be preferred to any generalization. Granted that induction is a rational policy, it will always require that we adopt an unlimited generalization unless there is a positive reason for adopting only a limited version of it. Consider the implications of this when we are confronted by some kind, like a species, which has the following characteristics. Firstly it is grounded on similarities over a variety of characteristics yet is not capable of precise definition, secondly we have reason to believe it has a finite extension, and thirdly various generalizations about this kind continue, for the most part, to be confirmed as more and more individuals are examined. The situation as described precludes our formulating and accepting a precise generalization with unlimited scope, but it does not prevent us from expecting that there is some such generalization which is true. Such examples are, in some respects, like the generalization about the disjunctive kind of substances which conduct electricity. Part of our dissatisfaction with basing causal claims on them is due to the lack of causal insight that they provide. We want to know what it is about a species in virtue of which all known members have some characteristic, and these days we expect that this will be the genetic constitution common to its members. While we have reason not to rest contented with the use of generalizations about species in making claims about causal dependence we have grounds to hope for a better situation as science progresses. In the mean time if we are interested in causal claims about living organisms it seems clear that we must take account of the species to which an individual belongs as one of the factors that is relevant to much of what happens. Even if generalizations such as those about species should not be accepted as laws, their on-going predictive success makes it reasonable to employ them in evaluating n-dependence claims until we have more basic generalizations to take their place.

I conclude that in general there is no reason to limit n-dependence to those relationships where the required material implications can be derived from what we unreservedly call 'laws'; it is enough that they be derived from projected generalizations that have not been supplanted in theoretical systematization. Two points are especially relevant here. Firstly, the ordering of generalizations with respect to which are more basic is to be employed to identify those cases of n-dependence which are causal dependence. Secondly it is acknowledged that it is not only important to identify causal dependencies, and hence causal connections, but also to do this in a way which maximizes the depth of causal insight. This will require that we seek out the most basic generalizations that can be employed to establish n-dependence. This theory recognizes the relevance of identifying how basic each generalization is compared to others. What it does not require is that a precise line be drawn between those projected generalizations which can be derived and those which are basic.

6.1 The Truth Conditions of Projected Generalizations

Projected generalizations have been defined by assertability conditions which determine when it is proper to apply the concept. However general statements make objective claims, so we might hope that a satisfactory explication should identify what needs to be objectively the case for a projected generalization to be true. An obvious suggestion is that what makes a projected generalization true is a set of facts of the kind which are relevant to its confirmation or falsification (knowing the assertability conditions ensures that we know what these are for any particular context). If we accept this then the truth condition of a projected generalization must have an infinity of elements; we might describe it as consisting of an infinite pattern in the facts. Some question the rationale of accepting beliefs about such infinite patterns. David Armstrong (1983, Ch. 4, Sec. 5) argues that on this interpretation induction must be an inference from one set – the instances that have been observed – to a completely disjoint set – the unobserved instances (p. 58). But unless the two sets are somehow unified the inference must be ungrounded. He concludes that in induction we accept something else, which explains what is observed and entails what is predicted (p. 55). This something extra is a connectedness in nature and if it exists we may take it to be what makes a law-statement true. Many others agree with Salmon that such a notion is unacceptable and contributes nothing to explanation. J.J.C. Smart argues that there is nothing problematic about taking laws to be infinite patterns, and that the rationality of induction rests on "simple brute facts on the cosmic scale" being "more believable than complex and idiosyncratic ones" (1985, p. 276). This means that there is nothing that is presently the case which makes a belief about a law true if it is true. Smart sees no problem in this; those who do will be inclined to some view like Armstrong's.

It is not possible to explore these views here, but one point needs to be made. It has usually been assumed that the issues raised concern laws, and that other generalizations are accidental truths whose truth conditions pose no problems. I have focused attention on projected generalizations, and indicated that there must be a question about their truth conditions. Drawing a line between laws and other projected generalizations is not simple. Of the examples we considered of generalizations which might not be allowed as laws the most interesting were those which we arrive at by induction, but which we expect to have a finite scope. Should we conclude that if a species is actually finite, a generalization about it will have a finite set of facts as truth conditions and so will be an accidental generalization? To do so would be to put such generalizations together with statements like "Every man in this room now either has a moustache or is taller than 1.75 meters". This would indeed be misleading. That there is a species with such and such a characteristic might be described as an accident, but that an individual born of parents of this species has it, is not. Further, a generalization about a species is not accepted because of an examination of all the elements of the set, but by an induction supported by background biological beliefs. Even although we cannot as yet specify it, we believe that there is an unlimited generalization about individuals with a certain kind of genetic composition which will mostly but not invariably be transmitted to progeny.

There is one kind of generalization which may seem to be both accidentally true yet not about a determinate, finite set. Reichenbach cited "all gold cubes are smaller than one cubic mile" (1976, p. 11) as a statement that may be true but should not be accepted as a law because we cannot verify its truth. There may be some sentences which just happen to be true but which we could have no grounds for believing. They would not be inductively confirmable, so not projected generalizations. However this sentence is hardly of this kind. To say that it may be true is rather an understatement. I imagine that we are all pretty confident that it is true, and this confidence surely rests in some way on induction. Is it an accidental truth, a projected generalization or both? The point is that there are various projected generalizations which bear on the likelihood that it is true: those about the way natural gold deposits come about, those about techniques for accumulating and casting gold, those about the kinds of things humans value and so on. However it is not itself a projected generalization; it has neither been accepted via an induction on the size of golden cubes, nor derived from a broader generalization that is inductively projected. Because beliefs about persisting regularities bear on its probability, I see it as misleading to say that this statement, if true, is accidentally true. However neither is it a law nor a projected generalization. If it should be cited in making predictions, one or more of the projected generalizations that give it some plausibility would be actually doing the predictive work. The suggestion that true generalizations are either laws or accidental truths is a serious over-simplification, and passes over a number of distinctions that need to be recognized.

I have argued that projected generalizations can be distinguished from other generalizations, that they do the predictive work of empirical knowledge and that they can be employed to define nomic dependence, which in turn can be used to explicate causal dependence. The systematizing of projected generalizations via the relationship of derivability is of immense importance in the practice of science; and this ordering is needed to select those cases of n-dependence which are also causal dependence. Thus the proposed explication of causal connections draws on distinctions that are already crucial to science. The concept of law has not been employed, and no explication of it has been offered. Nevertheless the concept of projected generalization, and the criteria by which they are ordered, take us quite some way along that path.

University of Melbourne

NOTES

[1] This leaves open the thesis, which I see as highly plausible, that in different contexts counterfactual locutions can be grounded very differently. For instance, some counterfactuals, such as "If he had not made a public apology he would not have had my vote" are grounded in a person's intentions, others, such as "If Jack were of lower rank than a sub-lieutenant he would not be a commissioned officer", in legal or verbal conventions. Other counterfactual locutions are grounded in our beliefs about causal connections. This thesis requires that our conception of causal connections, and of intentions and conventions, are prior to our understanding of counterfactuals.

[2] All emphases within quotes, unless otherwise indicated, are those of the author quoted.

[3] In this work I do not intend to take up the relationship between temporal and causal order. I will assume that the causes of an event are prior to, or no later than, that event.

[4] This requirement is that the laws and conditions in virtue of which the nomic dependence holds must be counterfactually independent of the antecedents of the derived material conditionals. Whether this requirement succeeds in distinguishing causes from symptoms depends on whether Lewis's antibacktracking thesis is accepted (see Clendinnen, 1992, p. 350).

[5] It is likely that the barometer example involves irreducibly probabilistic connections, and so is strictly not suitable to illustrate distinctions in deterministic contexts, however this example indicates others where an instrument, and the outcome of what is measured, are both covered by deterministic laws.

[6] A participant in the discussion of an earlier version of this paper pointed out that if the conditions in virtue of which n-dependence is considered are specified so as to include or entail E, the material condi-

tional C⊃E will be trivially true. This can be easily be avoided by requiring that C⊃E be derivable from laws and conditions but not from the conditions alone.

[7] Probability is here taken in the Credibility sense as a function of propositions. Capital letters are the assertions that the events named by the lowercase of the same letter occurred.

[8] It should be noted that the account of probabilistic causal relations between two events in Clendinnen (1992, p. 356) involved a serious blunder by having the effect event an element of the attribute class but not of the reference class. That account is superseded by the one in this paper.

REFERENCES

Armstrong, D.M. (1983), *What is a Law of Nature?*, Cambridge: Cambridge University Press

Carnap, R. (1950), *Logical Foundations of Probability*, Chicago: University of Chicago Press

Clendinnen, F.J. (1992), 'Nomic Dependence and Causation', *Philosophy of Science* **59**, 341–60

Clendinnen, F.J. (1996), 'Theorizing and Empirical Belief', in *Natural Kinds, Laws of Nature and Scientific Methodology*, Peter Riggs (ed.), Dordrecht: Kluwer, 63–92

Dowe, Phil (1992), 'Wesley Salmon's Process Theory of Causality and the Conserved Quantity Theory', *Philosophy of Science* **59**, 195–216

Humphreys, Paul W. (1989), *The Chances of Explanation*, Princeton: Princeton University Press

Lewis, David (1973), 'Causation', *Journal of Philosophy* **70**, 556–67, Reprinted in *Philosophical Papers*, Vol. II, by David Lewis, New York: Oxford University Press, 159–72 (References are to the reprint Lewis (1986).)

Reichenbach, Hans (1976), *Laws, Modalities and Counterfactuals*, Berkeley: University of California Press (second edn. of Reichenbach, 1954)

Reichenbach, Hans (1954), *Nomological Statements and Admissible Operations*, Amsterdam: North Holland

Salmon, Wesley C. (1967), *The Foundations of Scientific Inference*, Pittsburgh: University of Pittsburgh Press

Salmon, Wesley C. (ed.) (1971), *Statistical Explanation and Statistical Relevance*, Pittsburgh: University of Pittsburgh Press

Salmon, Wesley C. (1976), 'Foreword' to Reichenbach (1976)

Salmon, Wesley C. (1984), *Scientific Explanation and the Causal Structure of the World*, Princeton: Princeton University Press

Salmon, Wesley C. (1994), 'Causality Without Counterfactuals', *Philosophy of Science* **61**, 297–312

Smart, J.J.C. (1985), 'Laws of Nature and Cosmic Coincidences', *The Philosophical Quarterly* **35**, 272–280

JOHN D. COLLIER

CAUSATION IS THE TRANSFER OF INFORMATION

I. INTRODUCTION

Four general approaches to the metaphysics of causation are current in Australasian philosophy. One is a development of the regularity theory (attributed to Hume) that uses counterfactuals (Lewis, 1973; 1994). A second is based in the relations of universals, which determine laws, which in turn determine causal interactions of particulars (with the possible exception of singular causation, Armstrong, 1983). This broad approach goes back to Plato, and was also held in this century by Russell, who like Plato, but unlike the more recent version of Armstrong (1983), held there were no particulars as such, only universals. A third view, originating with Reichenbach and revived by Salmon (1984), holds that a causal process is one that can be marked. This view relies heavily on ideas about the transfer of information and the relation of information to probability, but it also needs uneliminable counterfactuals. The fourth view was developed recently by Dowe (1992) and Salmon (1994). It holds that a causal process involves the transfer of a non-zero valued conserved quantity. A considerable advantage of this approach over the others is that it requires neither counterfactuals nor abstracta like universals to explain causation.

The theory of causation offered here is a development of the mark approach that entails Dowe's conserved quantity approach. The basic idea is that causation is the transfer of a particular token of a quantity of information from one state of a system to another. Physical causation is a special case in which physical information instances are transferred from one state of a physical system to another. The approach can be interpreted as a Universals approach (depending on one's approach to mathematical objects and qualities), and it sheds some light on the nature of the regularity approach.[1] After motivating and describing this approach, I will sketch how it can be used to ground natural laws and how it relates to the four leading approaches, in particular how each can be conceived as a special case of my approach. Finally, I will show how my approach satisfies the requirements of Humean supervenience. The approach relies on concrete particulars and computational logic alone, and is the second stage of constructing a minimal metaphysics, started in (Collier, 1996a).

The approach is extraordinarily simple and intuitive, once the required technical apparatus is understood. The main problems are to give a precise and adequate account of information, and to avoid explicit reference to causation in the definition of information transfer. To satisfy the first requirement, the approach is based in computational information theory. It applies to all forms of causation, but requires a specific

H. Sankey (ed.), Causation and Laws of Nature, 215–245.

interpretation of information for each category of substance (assuming there is more than one). For the scientifically important case of physical causation I use Schrödinger's Negentropy Principle of Information (NPI). Causation can be represented as a computational process dynamically embodied in matter or whatever other "stuff" is involved, in which at least some initial information is retained in each stage of the process.[2] The second requirement, avoiding circularity, is achieved by defining causation in terms of the identity of information tokens.

2. THE ROLE OF FORM IN CAUSAL EXPLANATIONS

Suppose we want to ensure that someone is the person we believe them to be. We typically rely on distinguishing features such as their face, voice, fingerprints or DNA. These features are complex enough that they can distinguish a person from other people at any given time, and are stable enough that they reliably belong to the same person at different times (science fiction examples excepted). However, if the person should have a dopplegänger (a qualitatively identical counterpart), these indicators would not be enough for identification; we would need to know at least something of the spatiotemporal history of the particular instantiations of the qualities of the person we wish to identify. Sameness of person (or at least of their body) requires a causal connection between earlier stages and later stages. We can recognise this connection through identifying features and spatiotemporal continuity. The body transmits its own form from one spatiotemporal location to another. I will argue that not only is this sort of transmission an evidential basis for causal connection, but it can be used to define causal connection itself.

Central to my account is the propagation of form, as measured by information theoretic methods. This is not so foreign to traditional and contemporary views of causation as it might seem. By form, I mean the integrated determinate particular concrete qualities of any thing, of *any* kind.[3] Understanding the propagation of form is necessary for understanding contemporary science. If the reader finds this uncontroversial, I suggest they skip directly to §3.

Form includes the geometrised dynamics of grand cosmological theories like geometrodynamics (harking back to Platonic and Cartesian attempts to geometrise dynamics, Graves, 1971) and geometry and symmetry used to explain much of quantum particle physics (Feynman, 1965). It also includes the more common motions and forces of classical mechanics, as expressed in the Hamiltonian formulation with generalised coordinates.[4] Treatments of complex physical phenomena such as Bénard cell convection and other phenomena of fluid dynamics rely on knowledge of the form of the resulting convection cells to solve the equations of motion (Chandreshankar, 1961; Collier, Banerjee and Dyck, in press). In more highly nonlinear phenomena, standard mechanical techniques are much harder to apply, and even more knowledge of the idiosyncrasies of the form of particular phenomena are required to apply mechanical methods. Even in mathematics, qualitative formal changes have been invoked to explain "catastrophes" (Thom, 1975). Sudden changes are common in phase transitions in everyday complex phenomena like the weather, as well as in highly nonlinear physical, chemical, developmental, evolutionary, ecological, social and economic processes.

In biology, causal explanations in terms of the dynamics of form are especially common. Because of the complexity of macromolecules and their interactions, it seems likely that biologically oriented chemists will need to rely on the form of molecules indefinitely. This has led to information theoretic treatments of biochemical processes (Holzmüller, 1984; Küppers, 1990; Schneider, 1995). Even the reduction of population genetics to molecular genetics has failed to fulfill its promise because of many-many relations between the phenotypic traits of population biology and molecular genes. Although some molecular biology is now done with the aid of mechanics and quantum mechanics, these methods are limited when large interacting molecules are involved. In large scale biology (systematics, ecology, ontogeny and evolution), causal arguments usually concern some aspect of the transitions of form as I have defined it above (D'Arcy Thompson, 1942; Wiley, 1981; Brooks and Wiley, 1988; Ulanowicz, 1986).

The most dominant current view of cognition is the syntactic computational view, which bases cognitive processes on formal relations between thoughts. Whether or not the theory is true, it shows that psychologists are willing to take it for granted that form (viz., the syntax of representations) can be causal. Fodor (1968) argues that the physical embodiment of mental processes can vary widely, if the syntactic relations among ideas are functionally the same. To understand the embodiment of mind, if we accept that cognitive processes derive their formal relations from underlying dynamics, we need an account of the role of form and information in dynamics.[5]

Traditional linear and reductionist mechanical views of causation have had limited success in these emerging areas of study. Since the traditional views are well established, any adequate account of causation may initially seem counterintuitive. Causal studies using ideas of form, broadly construed as I have described it, have been more successful than mechanical approaches in the sciences of complex systems, but we need a precise account of the causal role of form to unify and normalise these studies. We need this because there is no hope that mechanical accounts will ever fully replace their currently less regarded competitors. The mechanical view is demonstrably too restrictive to deal with many kinds of possible systems that we are likely to encounter (see Collier and Hooker, 1999, for details).

The view I propose is not entirely without precedent. Except for its idealism, Leibniz' account of causation is in spirit the most developed precursor of the account I will give.[6] The case is complicated because of Leibniz' three level ontology (Gale, 1994). At the observable level Leibniz' physics was mechanical, however this dynamics was explained by the properties of monads, whose substantial form implied a primitive active force (encoded by the logico-mathematical structure of the form in a way similar to the compressed form of causal properties I will discuss later). This primitive active force produces the observable varieties of derivative active force through strictly logical and mathematical relations. At the metaphysical level, the substantial form is based in "clear perceptions" which are articulated in the structure of the substantial form of corporeal substance. A similar hierarchy exists for passive forces, which are similar to Hobbes' material cause.[7] The derivative passive force is a consequence at the corporeal level of Prime Matter, which is metaphysically based in the monad's confused perceptions (which, because unarticulated, cannot act as agent; see Christensen *et al.*, in preparation).

Leibniz expressed many of these ideas in 'On the elements of natural science' (ca. 1682–4, Leibniz, 1969, pp. 277–9). The following quotation illustrates the importance of form in Leibniz' philosophy:

> And the operation of a body cannot be understood adequately unless we know what its parts contribute; hence we cannot hope for the explanation of any corporeal phenomenon without taking up the arrangement of its parts. (Leibniz, 1969, p. 289)

An earlier version of this view can be found in the 1677 paper 'On the method of arriving at a true analysis of bodies and the causes of natural things'. The paper emphasises the importance of empirical observation, but the final paragraph makes clear the role of form in causal explanation:

> Analysis is of two kinds – one of bodies into various qualities, through phenomena or experiments, the other of sensible qualities into their causes or reasons, by ratiocination. So when undertaking accurate reasoning, we must seek the formal and universal qualities that are common to all hypotheses . . . If we combine these analyses with experiments, we shall discover in any substance whatever the cause of its properties. (Leibniz, 1969, pp. 175–76)

Again we see that for Leibniz, grouping observable phenomena by their qualities and changes in qualities is but a prelude to explanation in terms of substantial form. The full explanation from metaphysics to physics to phenomena should be *entirely* mathematical. I shall take advantage of the fact that mathematics is neutral to collapse Leibniz' three levels into one involving only concrete particulars. The first step is the quantification of form using recent developments in the logic of complexity.

3. QUANTIFICATION OF FORM VIA COMPLEXITY THEORY

A precise mathematical characterisation of form (more precisely, the common core of all possible conceptions of form) can be formulated in computational information theory (algorithmic complexity theory). This will provide the resources for a general account of causation as information transfer (whether physical or not) in §4. In §5 I will connect information to physical dynamics in an intuitive way through Schrödinger's Negentropy Principle of Information (NPI), which defines materially embodied information. Physical causation is defined in §6, using the resources of the previous three sections. A method of quantifying the hierarchical structure of a thing is given in §7, to distinguish between mere complexity and organised complexity. This is done through Charles Bennett's notion of logical depth. The notion of logical depth can be used to make sense of the account of laws as abstractions from particular cases of causation given in (Collier, 1996a) by showing how laws organise the superficial disorder of particular events and their relations. This completes the technical part of the chapter. The final sections look at some potential objections to the formal approach to causation, and the implications for the four current approaches to causation.

The quantification of form is a quantification of the complexity of a thing. Complexity has proven difficult to define. Different investigators, even in the same fields, use different notions. The Latin word means "to mutually entwine or pleat or weave together". In the clothing industry one fold (e.g. in a pleat) is a *simplex*, while multiple folds comprise a *complex.* The most fundamental type of complexity is *informational complexity*. It is fundamental in the sense that anything that is complex in any other way must also be informationally complex. A complex object requires more information to specify than a simple one. Even the sartorial origins of the word illustrate this relation: a complex pleat requires more information to specify than a simplex: one must specify at least that the folds are in a certain multiple, so a repeat specification is required in addition to the "produce fold" specifications. Further information might be required to specify any differences among the folds, and their relations to each other.

Two things of the same size or made from the same components might have very different informational complexities if one of them is more regular than the other. For example, a frame cube and a spatial structure composed of eight irregularly placed nodes with straight line connections between each node may encompass the same volume with the same number of components, but the regularity of the cube reduces the amount of information required to specify it. This information reduction results from the mutual constraints on values in the system implied by the regularities in the cube – all the sides, angles and nodes must be the same. This redundancy reduces the amount of information required in a program that draws the cube over that required by a program that draws the arbitrary eight node shape. Similarly, a sequence of 32 '7's requires a shorter program to produce than does an arbitrary sequence of decimal digits. The program merely needs to repeat the output of '7' 32 times, and 32 itself can be reduced to $2^{5,}$ indicating 5 doublings of an initial output of '7'. To take a less obvious case, any specific sequence of digits in the expansion of the transcendental number π=3.14159 . . . can be produced with a short program, despite the apparent randomness of expansions of p. The information required unambiguously to describe ordered and organised structures can be compressed due to the redundant information they contain; other structures cannot be so compressed. This is a property of the redundancy of the structures, not directly of any particular description of the structures, or language used for description.

The specification of the information content of a form or structure is analogous to an extended game of "twenty questions", in which each question is answered yes or no to identify some target. Each accurate answer makes a distinction[8] corresponding to some difference between the thing in question and at least one other object. The answers to the questions encode the distinct structure of the target of the questions. Every determinate aspect of the form of any thing is included in its encoding. Thus, the encoding from questions and target perfectly represents the form of the target. Nothing else is left to encode, and the form can be recovered without loss from the encoding by examining the questions and decoding the answers (assuming the questions to be well formed, and the answers to be accurate). Such an encoding is an isomorphic map of the form of an entity like an object, property or system onto a string in which each entry is a "yes" or a "no", or a "1" or a "0". This string is an object to which computational complexity theory (a branch of mathematics) can be applied. The method is analogous to the use of complex numbers (the Cauchy-Riemann technique) to solve certain difficult problems in mathematical physics. The form is first converted to a tractable encoding, certain results can be derived, and then these can be applied to the original form in the knowledge that the form can be recovered with the inverse function. There is no implication that forms *are* strings of 1s and 0s any more than that the physical systems to which complex analysis of energy or other relations is applied really involve imaginary numbers.

Let s be mapped isomorphically onto some binary string σ_s (i.e. so that s and only s can be recovered from the inverse mapping), then the informational complexity of s is the length in bits of the shortest self-delimiting computer program on a reference universal Turing machine that produces σ_s, minus any computational overhead required to run the program, i.e. $C_I = \text{length}(\sigma_s) - \mathbf{O}(1)$.[9] The first (positive) part of this measure is often called *algorithmic complexity*, or Kolmogorov complexity. The second part of the measure, $\mathbf{O}(1)$, is a constant (order of magnitude 1) representing the computational overhead required to produce the string σ_s. This is the complexity of the program that computes σ_s. It is machine dependent, but can be reduced to an arbitrarily small value, mitigating the machine dependence.[10] I deduct it to define the informational complexity to get a machine independent measure that is directly numerically comparable to Shannon information, permitting identification of algorithmic complexity and combinatorial and probabilistic measures of information.[11] The resulting value of the informational complexity is the information in the original thing, a measure of its form. Nothing additional is needed to specify the form of anything. Consequently, I propose that the information, as measured by complexity theory, *is* the form measured, despite disparate approaches to form in differing sciences and philosophies. Nothing determinate remains to specify. Any proposed further distinctions that go beyond this are distinctions without a difference, to use a Scholastic saw. The language of information theory is as precise a language as we can have. Once all distinctions are made, there is nothing else we could say about something that gives any more information about it.

All noncomputable strings are algorithmically random (Li and Vitànyi, 1990). They cannot be compressed, by definition; so they contain no detectable overall order, and cannot be distinguished from random strings by any effective statistical test. This notion of randomness can be generalised to finite strings with the notion of *effective randomness*: a string is effectively random if it cannot be compressed.[12] Random strings do not contain information in earlier parts of the sequence that determines later members of the sequence in any way (or else they could be compressed).[13] Thus any system or process whose trajectory cannot be specified in a way that can be compressed is dynamically disorganised and effectively random. Such a system or process can have specific consequences, but cannot control anything, since these effects are indistinguishable from random by any effective procedure: no pattern (form) can be generated except by chance.

Algorithmic information theory can be used to quantitatively examine relations of information, and thus of form. The assignment of an information value to a system, state, object or property is similar to the assignment of an energy to a state of a system, and allows us to talk unambiguously of both the form and its value.[14] We can then compare the form of two states, and of the transfer of form between states. In addition, and unlike for energy (whose relations also require dynamical laws), there are necessary relations between information instances that depend on whether a second instance is a theorem of the theory comprising the first instance and computation theory. Except for noncomputable cases, this relation is equivalent to there being a Turing type computation from the first information to the second. There are several relations of note: the information I_A contained in A contains the information in B iff I_B is logically entailed by I_A, and vice versa. This implies that the information in A is equivalent to the information in B if and only if each contains the other. The information in B given the information in A, and their mutual information can be expressed in a similar way. These relations are all standard in algorithmic complexity theory (Li and Vitànyi, 1993, pp. 87ff). They allow us to talk about changes of form from one state of a system to another (processes), and between states of different systems (interactions). A relation between the information in A and that in B is not causal unless A and B have mutual information (they must be correlated). Sufficiency is not guaranteed, since the formal content of A and B can overlap by chance.

4. CAUSATION IS THE TRANSFER OF INFORMATION

The previous section placed the existence of mutual information as a necessary condition on the dynamics of form, but left open the possibility of chance coincidence of forms. An account of information as transfer of form must rule this possibility out non-circularly. I do this by stipulating that the information transferred must be *the same particular information*. Thus, we have the first definition of a causal process:

> P is a causal process in system S from time t_0 to t_1 iff some particular part of the form of S involved in stages of P is preserved from t_0 to t_1.

The preservation of form, here, implies that the form preserved is the identical form, not only in value and logical structure, but in fact. This definition is temporally symmetrical, in keeping with the temporal symmetry of fundamental laws of mechanics and quantum mechanics. Temporal asymmetry is an additional condition that will be considered in §6.1. The definition is similar to Salmon's PCI (1984, p. 155) according to which a process that transmits its own structure can propagate a causal influence from one spacetime locale to another. The main difference is that Salmon defines the transmission of structure as the capacity to carry a mark. This notion requires an irreducible counterfactual formulation (Kitcher, 1989; Dowe, 1992), as Salmon has admitted (1994). This violates my methodological assumption that all fundamental concepts should refer only to logic, mathematics and concrete particulars (see footnote 2 above).

Another serious problem for the mark approach is that some causal processes cannot be marked. An electron, for example, has too few fundamental properties to be easily marked (though polarisation is a possibility). Other fundamental particles (photons and neutrinos, for example) cannot be marked without changing them into other particles. Equilibrium statistical mechanical systems cannot be marked without changing them from equilibrium. The mark will disappear relatively quickly through dissipation unless the mark puts the system quite far from equilibrium, which changes the dynamical properties of the system substantially. Since such basic dynamical processes as fundamental particle propagation and equilibrium processes are clearly causal, but cannot be marked, the mark method fails for reasons independent of any problem of the counterfactual dependence of the approach. In addition, it is unclear how nonphysical processes (if such exist, like the thought processes of God or other immaterial minds) could be marked. I avoid both the problem of unmarkable causal processes and the problem of counterfactual dependence by dropping marking entirely in favour of the transfer of form in general, rather than just of marks. The ability to mark a causal process remains important as an intuition leading to the definition above, however.

Critics might complain that my minimised definition of causation is wildly circular, since the identity of preserved form entails a causal connection. I am glad they share this intuition with me, since an adequate analysis of causation should correspond to intuitive ideas about causation. Any suspicion that causation has been surreptitiously imported in the above definition must involve the preservation of information, since form in a state of a system has been defined with purely logical notions. Preservation, though, I have stipulated to be identity, which is also a logical notion.[15] There is no direct reference to causation in the definition. This is perhaps more clear in the following variant:

> P is a causal process in system S from time t_0 to t_1 iff some particular part of the information of S involved in stages of P is identical at t_0 and t_1.

This may seem like a trick, and indeed it would come to very little unless there is a way to determine the identity of information over time without using causal notions explicitly. This is an epistemological problem, which I defer until later. It turns out that there are simple methods for many interesting cases. From a strictly ontological view, the above definition is all that is needed, though the metaphysics of identity will depend on the substantial categories involved. Information tokens are temporal particulars. In physics with spatio-temporal locality, they are space-time "worms" (see §6.1).

The notion of transfer of information is useful:

> Information I is transferred from t_0 to t_1 iff the same (particular) information exists at t_0 and t_1.

The definition of causal process can then be revised to:

> P is a causal process in system S from time t_0 to t_1 iff some part of the information of S involved in stages of P is transferred from t_0 to t_1.[16]

Interactive causation can now be defined easily:

> F is a causal interaction between S_1 and S_2 iff F involves the transfer of information from S_1 to S_2, and/or vice versa.

This allows a straightforward definition of causal forks, which are central to discussions of common cause and temporal asymmetry (Salmon, 1984):

> F is an interactive fork iff F is a causal interaction, and F has distinct past branches and distinct future branches.

and,

> F is a conjunctive fork iff F is a causal interaction, and F has one distinct past branch and multiple distinct future branches, or vice versa.

Interactive forks are X-shaped, being open to the future and past, while conjunctive forks are Y-shaped, being open in only one temporal direction. The probability relations Reichenbach used to define conjunctive forks follow from these definitions, the mathematics of conditional information, temporal asymmetry, and the probabilities derived from the information from the mathematical relation between informational complexity and probability-based definitions of information (justified by the definability of randomness within complexity theory), as do the probabilities for interactive forks. There is no room to prove this here, since apart from the mathematics we need a satisfactory account of the individuation and identity of dynamical processes that is beyond the scope of this chapter. It should be obvious, though, given that a causal process preserves information, that a past common cause (shared information) makes later correlation more certain than a present correlation of two events makes later interaction probable, though the reasons for this are not presently transparent by any means (Horwich, 1988). Likewise, an interactive fork gives information about both past and future probabilities, because the identity of the information in the interaction restricts the possibilities at both of the open ends of the forks.

Pure interactions between independent processes are rare, if not nonexistent. Interaction through potential fields (like gravity) occurs among all bodies continuously. If gravity and other fields are involved in the dynamics of interacting systems, enlarging the system to include all interactions is better than to talk of interacting systems. This is standard practice in much of modern physics, for example, when using the Hamiltonian formulation of Newtonian mechanics.

According to the information theoretic definition of causality, the necessity of causal relations follows easily, since the informational relations are computational. The information transferred must be in the effect and it must be in the cause, therefore the relevant information is entailed by both the cause and the effect. Furthermore, the existence of the identical information (token) in both the cause and effect is both a necessary and a sufficient condition for causation. We can think of a causal process as a computation (though perhaps not a Turing computation or equivalent) in which the information in the initial state determines information in the final state. The effect, inasmuch as it is determined, is necessitated by the cause, and the cause must contain the determined information in the effect. Although the causal relation is necessary, its conditions are contingent, so it is necessary only in the sense that given the relata it cannot be false that it holds, not that it must hold (see Collier, 1996a for more on this form of necessitation, and its role in explaining the necessity of natural kinds and laws). Note that the only necessity needed to explain causal necessity is logical entailment. This is one great advantage of the information theoretic approach to causation, since it avoids direct appeals to modalities. Counterfactual causal reasoning fixes some counterfactual conditions in distinction to the actual conditions either implicitly or explicitly through either context or conventions of language. Counterfactual causal reasoning is thus grounded in hypothetical variations of actual conditions.

Locality, both spatial and temporal, is a common constraint on causation. Hume's "constant conjunction" is usually interpreted this way. While it is unclear how causation could be propagated nonlocally, some recent approaches to the interpretation of quantum mechanics (e.g. Bohm, 1980) permit something like nonlocal causation by allowing the same information (in Bohm's case "the implicate order") to appear in spatially disparate places with no spatially continuous connection. Temporally nonlocal causation is even more difficult to understand, but following its suggestion to me (by C.B. Martin) I have been able to see no way to rule it out. Like spatially nonlocal causation, temporally nonlocal causation is possible only if the same information is transferred from one time to another without the information existing at all times in between. Any problems in applying this idea are purely epistemological: we need to know it is the same information, and not an independent chance or otherwise determined convergence. Resolving these problems, however, requires an appropriate notion of information and identity for the appropriate metaphysical category.

The epistemological problems are diminished immensely if temporal locality is required. If there is a sequence of temporal stages between the start and end of a candidate causal process for which there is no stage at which the apparently transferred information does not exist, the candidate process is temporally local. All other things being equal, it is far more parsimonious to assume that the identical information exists at each stage of a candidate local process than that the information at each stage arises independently. The odds against a candidate local causal process being noncausal (i.e. apparently but not actually transferring the identical information) are astronomical. The main exception is an independent common cause, as in epiphenomena like Leibniz' universal harmony. There are difficulties distinguishing epiphenomena from direct causal phenomena, but in many cases intervention or further knowledge can provide the information needed to make the distinction. For example, we can tell that the apparent flow of lights on a theatre marquee is not causal by examining the circuitry. The null hypothesis, though, despite these possibilities, would be that candidate causal processes are causal processes. Lacking other information, that hypothesis is always the most parsimonious and the most probable. Unfortunately, it can't be shown conclusively that any apparent causal process is really causal, but this sort of problem is to be expected of contingent hypotheses. The important thing to note is that (ignoring pragmatic considerations) any talk of causation can be eliminated in favour of talk of the transfer of the same information throughout the apparent process.

It is interesting to note that my approach to causation permits an effectively random system to be a cause. A large random system will have ordered parts, and an infinite random system will have ordered parts of arbitrarily large size (see footnote 13 above). If the universe originated as an infinite random system, as suggested by David Layzer (1990), then ordered random fluctuations would be expected, and our observable world could be caused by a particularly large fluctuation that later differentiates through phase transitions into the variety that we observe today. This cosmological theory requires the pre-existence of a random "stuff" with the capability of self interaction. No intelligence or pre-existing order is required to explain the causal origin of the order and organisation in the observable world. This is contrary to the views of my rationalist predecessors like Aristotle, Descartes and Leibniz.

So far, this account of causation has very little flesh; it is just a formal framework. This will be remedied in the next two sections in which I apply the framework to physical causation.

5. THE NEGENTROPY PRINCIPLE OF INFORMATION

To connect information theory to physical causation, it is useful to define the notions of order and disorder in a system in terms of informational complexity. The idea of disorder is connected to the idea of entropy, which has its origins in thermodynamics, but is now largely explained via statistical mechanics. The statistical notion of entropy has allowed the extension of the idea in a number of directions, directions that do not always sit happily with each other. In particular, the entropy in mathematical communications theory (Shannon and Weaver, 1949), identified with information, should not be confused with physical entropy (though they are not completely unrelated). Incompatibilities between formal mathematical conceptions of entropy and the thermodynamic entropy of physics have the potential to cause much confusion over what applications of the ideas of entropy and information are proper (e.g. Wicken, 1987; Brooks *et al.*, 1986).

To prevent such problems I adopt the interpretive heuristic known as NPI, according to which the information in a specific state of a physical system is a measure of the capacity of the system in that state to do work (Schrödinger, 1944; Brillouin, 1962, p. 153), where work is defined as the application of a force in a specific direction, through a specific distance.[17] Work capacity is the ability to control a physical process, and is thus closely related to causality. Nevertheless, it is a state variable of a system, and involves no external relations, especially to effects. So the concept of work capacity is not explicitly causal (though the concept of work is).[18] Through the connection with work, NPI ties information, and so complexity and order, to dynamics. NPI implies that *physical information* (Brillouin, 1962)[19] has the opposite sign to physical entropy, and represents the difference between the maximal possible entropy of the system (its entropy after all constraints internal to the system have been removed and the system has fully relaxed, i.e. has gone to equilibrium) and the actual entropy, i.e.,

$$\text{NPI: } I_P = H_{MAX} - H_{ACT}$$

where the environment of the system and the set of external constraints on the system are presumed to be constant. The actual entropy, H_{ACT}, is a specific physical value that can in principle be measured directly (Atkins, 1994), while the maximal entropy, H_{MAX}, of the system is also unique, since it is a fundamental theorem of thermodynamics that the order of removal of constraints does not affect the value of the state variables at equilibrium (Kestin, 1968). This implies that the equilibrium state contains no trace of the history of the system, but is determined entirely by synchronic boundary conditions. Physical information, then, is a unique and dynamically fundamental measure of the amount of form, order or regularity in a state of physical system. Its value is non-zero only if the system is not at equilibrium with its environment. It is a measure of the deviation of the system from that equilibrium. It is important to remember that NPI is not a formal or operational definition and, given the current proliferation of formalisms for entropy and information, it needs to be interpreted as appropriate for a given formalism and for a given physical system and its environment.[20]

On the other hand, NPI is an implicit definition, since it determines how terms like entropy and information are to be used in a physical context. As in mathematics, central definitions in empirical theory should be supported with an existence proof. This is done by showing that violating the definition would violate any known or theoretically projected observations (Mach, 1960, pp. 264ff). If we assume NPI, then reliable production or reproduction of one bit of information requires a degradation of at least $kT\ln 2$ exergy (available energy), where k is Boltzmann's constant in a purely numerical form (Brillouin, 1962, p. 3), and T is temperature measured in energy units. This relation must hold, or Maxwell's demon will come to haunt us, and the Second Law of Thermodynamics will come tumbling down. NPI, then, reflects the strongly confirmed intuitions of physicists and engineers that the physical form of things cannot be used in some tricky way to control effectively random physical processes. There are strong reasons to believe that this is logically impossible in a world restricted to physical causation (Collier, 1990b). I will return to this later in §6.1. NPI is empirically justified; we know, for example, that violation of NPI, which would amount to using information to reduce the entropy of an isolated system, violates our most common experiences of physical systems. NPI implies that a bit of information can be identified with the minute but non-negligible physical value $k\ln 2$ and that its transfer from one system or part of a system to another will require the *transfer* of at least $kT\ln 2$ exergy (see Brillouin, 1962 for details). This gives us a quantitative physical measure of form that is directly related to exergy and entropy, the central concepts in non-equilibrium processes. These relations allow us to study complexity changes in physical processes, and permit principled extensions of the concepts of entropy and information.[21]

NPI can be motivated more directly from information theory. This might be useful to those who find themselves on the wrong side of C.P. Snow's two cultures, the divide being the understanding of entropy. Entropy cannot be explained simply without loss of content[22], but the following explanation will give the main details, though it will give no idea of how to apply the ideas (unlike the way I introduced NPI above, which rigorously connects information to known physical principles and their common applications). H_{MAX} represents a possible state of the system in which there is no internal structure except for random fluctuations. All possible microstates of the system are equally likely. There is no physical information within the system, and it cannot do any work internally, since it is statistically uniform except for random fluctuations, which, because of their random nature, cannot be harnessed for any purpose from within the system. The actual entropy, however, except for systems in equilibrium, permits internal work, since there is energy available in the nonuniformities that can be used to guide other energy. The information equivalent to this ordered energy is just that we would obtain with a perfect "game of twenty questions" that determines the information gap between the information of the macrostate and the information of the microstate, and hence the probability distribution of microstates. It therefore represents the form (nonrandom component) of the system, according to the definitions of §3. This justifies the connection between form and capacity for work. Any other consideration of dynamics and physical information will have to be consistent with this connection (however subtle) between dynamics and form, i.e. any physical system must satisfy NPI.

There are two ways that entropy is significant in physical systems, sorting and energy availability, though they are really extremes of one set of principles. To take a simple example of sorting, imagine that we start with a container of m "red" and n "white" molecules in an ideal gas at equilibrium, S_0, and it ends in a state, S_1, in which all the red molecules are on the right side of the container, and the white molecules are on the left side, so that we could move a frictionless screen into the container to separate completely the red and white molecules without doing any additional work. The entropy of S_0 is $-\Sigma P_0 k\ln P_0$, and the entropy of S_1 is $-\Sigma P_1 k\ln P_1$, where P_0 is the inverse of the number of complexions in the initial state, and P_1 is the inverse of the number of complexions in the final state. Simplifying again, assume the m = n = 1.[23] Then the entropy of the final state is obviously 0, since there is only one possibility, in which the red molecule is on the right, and the white molecule is on the left, so $P_1 = 1$. The entropy of the initial state is higher: both molecules can be either on the right or the left, or there can be a red on the left or a red on the right, giving four distinct possibilities, and $P_0 = .25$. If we know that the system is in S_1, we have 2 bits more information than if we knew merely that it was in S_0. For example, we might have the information that no two molecules are on the same side, and that a red molecule is on the right, requiring two binary discriminations. To slide the screen in at an appropriate time, we need the information that the system is in S_1, i.e. we need the information difference between S_0 and S_1. This is exactly equivalent to the minimum entropy produced in a physical process that moves the system from S_0 to S_1, as can be seen by setting k to 1, and using base 2 logarithms to get the entropy in bits. To move the system from S_0 to S_1, then, requires at least 2T work. This is a very small amount; the actual work input would be larger to cover any energy stored and/or dissipated. Alternatively, a system in S_1 can do at most 2T work before it has dissipated all its available energy from this source. Putting this in other words, the system can make at most two binary distinctions, as can be seen by reversing the process.[24] These two bits measure the maximal controlling potential of the system: implemented as a controller, controlling either itself or another system, the system could function as at most two binary switches. Calculating the physical information for each case from the definition above, $I_P(S_0) = 0$, while $I_P(S_1) = 2$. As it should, the difference gives us the amount of information lost or gained in going from one state to the other. A number of years ago it was confirmed that the entropy production of the kidneys above what could be attributed to the basal metabolism of its cells, could be attributed to the entropy produced in sorting molecules for elimination. Presumably, more subtle measurements would also confirm a physical realisation of the molecule example.

The relations between information and energetic work capacity are somewhat subtle, since they involve the correct application of NPI, which is not yet a canonical part of physics.[25] The physical information in a given system state, its capacity to do work, breaks into two components, one that is not constrained by the cohesion in the system, and one that is. The former, called *intropy*, ι, is defined by $\Delta\iota = \Delta(\text{exergy})/T$, so that $\int T\Delta\iota$ measures the available energy to do work, while the latter, called *enformation*, ε, measures the structural constraints internal to the system that can guide energy to do work (Collier, 1990a). Enformation determines the additional energy that would be obtained in a system S if all cohesive constraints on S were released. Intropy measures the ordered energy that is not controlled by cohesive system processes, i.e. by system laws, it is unconstrained and so free to do work. For this reason, though ordered, both intropy and exergy are system statistical properties in this sense: their condition cannot be computed from the cohesive or constrained system state, the cohesive state information determines the micro state underlying the intropy only up to an ensemble of intropy-equivalent microstates. There is another system statistical quantity, entropy, S, but it is completely disordered or random, it cannot be finitely computed from any finite system information.[26] Entropy is expressed by equiprobable states, and so appears as heat which has no capacity to do work; $\Delta S = \Delta Q/T$, where

Q is heat, and $\int T\Delta S$ measures heat. Enformation is required for work to be accomplished, since unguided energy cannot do work.[27] Intropy is required for work in dissipative systems, to balance dissipation (S production).

Consider, for example, a system S with heat Q as its only unconstrained energy. If S is at equilibrium then the only enformation is the existence of a system temperature (not that it is of some specific value T), for only that follows from the system constraints, and $\iota = 0$ and Q is entropic since Q cannot do work on S. If S nomicly maintains an internal temperature gradient G then G is enformation for S since it cannot be released to do work without first altering the cohesive structures of S. If G is unconstrained by S then G expresses intropy in S since G is an ordering of the heat energy and work can be done in S because of G. (In fact S will dissipate G, creating entropy, until equilibrium is reached.) Further, note that if S, even if at internal equilibrium with $G = 0$, is made part of a larger system S_s where it is in contact with another subsystem P of S_s at a lower temperature, then there is now a new temperature gradient G_s unconstrained by S_s so S will do work on P with heat flowing between them until equilibrium is reached ($G_s = 0$) at some intermediate temperature; hence G_s is intropic in S_s even though S has no intropy and S's temperature, which serves in part to determine G_s, is enformation in S.[28] These analyses carry over to all other physical forms of energy.

The main difference between intropy and enformation is the spatial and temporal scale of the dynamical processes that underlie them.[29] The dynamics underlying intropy have a scale smaller than that of the whole system, and involve no long term or spatially extended constraints, except those that govern the system as a whole, which in turn constitute the system enformation. The intropy of a system S is by definition equal to the difference between S's actual entropy and its maximal entropy when exergy has been fully dissipated (given enformation invariant, i.e. S's constraints remaining unchanged, and environment invariant); so, $\iota = I_P = H_{MAX}(S) - H_{ACT}(S)$, all at constant environment and constraints. The enformation is just the additional information equal to the difference between $H_{ACT}(S)$ and the entropy of the set of system components that result when the constraints on S are fully dissipated and S comes to equilibrium with its environment (assumed to remain otherwise invariant); $\varepsilon = I_E = H_{MAX}(S_E) - H_{MAX}(S)$. Note that $I_P(S) = \iota + \varepsilon = H_{MAX}(S_E) - H_{ACT}(S)$ as required by NPI. This is perhaps more clear with an example. A steam engine has an intropy determined by the thermodynamic potential generated in its steam generator, due to the temperature and pressure differences between the generator and the condenser. Unless exergy is applied to the generator, the intropy drops as the engine does work, and the generator and condenser temperatures and pressures gradually equilibrate with each other. The enformation of the engine is its structural design, which guides the steam and the piston the steam pushes to do work. The design confines the steam in a regular way over time and place. If the engine rusts into unrecoverable waste, its enformation is completely gone (as is its intropy, which can no longer be contained), and it has become one with its supersystem, i.e. its surroundings. Such is life.

As noted, NPI allows us to divide a physical system into a regular, ordered part, represented by the physical information of the system, i.e. $\iota + \varepsilon$, and a random, disordered part, represented by the system entropy. The *orderedness* of the system is its information content divided by the equilibrium (i.e. maximal) entropy, i.e.; O =

I_P/H_{MAX}, while the disorderedness is the actual entropy divided by the equilibrium entropy, i.e. $D = H_{ACT}/H_{MAX}$ (Layzer, 1975; Landsberg, 1984); it follows from NPI that $O+D = 1$. The informational complexity of the information in the system, $C_I(I_P)$, is equal to the information required to distinguish the macrostate of the system from other macrostates of the system, and from those of all other systems made from the same components.[30] The mathematical relations between statistical entropy and algorithmic information (Kolmogorov, 1965, 1968; Li and Vitányi, 1993) ensure that $C_I(I_P) = H_{MAX} - H_{ACT}$, so $C_I(I_P) = I_P$. This is so since the physical information of a system determines its regularity and this regularity can be neither more nor less informationally complex than is required to specify the regularity. (The informational complexity of the disordered part is equal to the entropy of the system, i.e. $C_I(H_{MAX} - I_P) = C_I(H_{ACT}) = H_{ACT}$ and since $O = I_P/H_{MAX}$, the ordered content of $S = H_{MAX}O = I_p$ as required.) These identities allow us to use the resources of algorithmic complexity theory to discuss physical information, in particular to apply computation theory to the regularities of physical systems. This move has always been implicit in the use of deductive reasoning to make physical predictions, and should be non-controversial. The main achievement here is to tie together explicitly computational and causal reasoning within a common mathematical language (see also Landauer, 1961, 1987; Bennett, 1988).[31] It is important to note, however, that NPI can be stated entirely in terms of state descriptions and relations between state descriptions, and involves no explicit importation of causal notions.

6. PHYSICAL CAUSATION

The analysis of causation in §4 is very abstract and perhaps hard to comprehend. In this section I use NPI to give an account of physical causation, the target of most contemporary metaphysical accounts of causation. My account divides into ontological and epistemological issues.

6.1 Ontological Issues

The mark approach fails because of its dependence on counterfactuals, and the inability of some obviously causal processes to be marked (see §4). This problem can be overcome if we take the form of the states of a physical process to itself be a mark, where the information in the mark is given by the methods of §5. The mark approach is attractive, since we can make a recognisable mark, and then check it later. A paradigmatic example is signing or sealing across the flap of an envelope so we, or someone else, can check that it is the original envelope, and that it has not been tampered with. Modern computer security methods using open and private keys are directly analogous, and much more secure. Unfortunately, many causal processes are too simple to mark at all, let alone permit the complex mathematical methods of open key security. Security and recognition, however, are more epistemological problems than ontological ones, and I will postpone this issue until section §6.2. For now I will concentrate on the ontology of the transfer of physical form.

Information preserved in physical causation will have constrained (enformational) and may have unconstrained (intropic) components. For example, a steam locomotive retains its structure as it moves along the tracks, but it also burns fuel for its intropy, part of which is converted into motion. It is only the enformation that is essential to a dynamical process, since the intropy is statistical and its microscopic basis is possibly chaotically variable, whereas the enformation guides the dynamical process, and constitutes the structure of the system at a given time. Therefore we might try:

> P is a physical causal process in system S from time t_0 to t_1 iff some part of the enformation in S is transferred from t_0 to t_1.

We may or may not want to add locality requirements. Familiar cases of physical causality are both temporally and spatially local.

Unfortunately, pseudoprocesses like the passing of a beam of laser light across the face of the moon satisfy this definition, but the causal process involved is actually a good deal more complicated. NPI can help us here. First, though, it helps to invoke Russell's "at-at" theory of causal propagation (Salmon, 1984, pp. 147ff) to ensure locality:

> P is a causal process in system S from time t_0 to t_1 iff some part of the enformation in S is identical from t_0 to t_1, and at all times between t_0 and t_1.

As mentioned in §4, the at-at approach to locality makes the information token a spacetime "worm". Locality disallows causation over a temporal gap, but it is very much in tune with the intuitions of physicists and other scientists that all physical causation is local. The main exception might arise in quantum mechanics on Bohm's implicate order approach, which is controversial, and nonetheless requires locality of a different sort through the enfolding of the universe. The above definition can be revised, if necessary, to take into account this different sort of locality. Of course the intuitions of physicists may be false, but at this time they are our best experts on how to interpret causality and cognate terms.

The pseudoprocess problem is then the underlying problem for the information transfer theory, as it is for the mark approach. I attack this problem by invoking NPI explicitly:

> P is a physical causal process in system S from time t_0 to t_1 iff some part of the enformation in S is transferred from t_0 to t_1, and at all times between t_0 and t_1, all consistent with NPI.

Consistency with NPI is a fairly strong constraint. It requires that causal processes be consistent with entropy changes in the processes. This is enough to rule out the flashlight beam across the moon pseudoprocess, since the information in the spot comes from nowhere, and goes to nowhere, if the movement is all there is to the process. This violates not only the Second Law of Thermodynamics, but also strong physical intuitions that embodied order cannot just appear and disappear. Quantum mechanics and the emergence of dissipative structures seem to violate this intuition, but on closer study symmetry requirements in quantum mechanics and the influence of the order in microscopic fluctuations ensure that no new information is generated.

The Second Law itself has an interesting status. Although reversible systems are possible in nature (apparently reversible systems can be designed in the laboratory as thermodynamic branch systems, but in fact they obey the Second Law when it is properly interpreted), it is impossible for any physical device or physical intervention to control the overall direction of the entropy gradient because to do so is computationally impossible (Bennett, 1987; Collier, 1990b). Reversal of the normal increase in entropy requires very special conditions that can be detected by examining the history, boundary conditions and dynamics of the system. Consistency with the Second Law is not merely an empirical requirement; it is closer to a logical constraint, and holds for abstract computational systems as much as for physical systems (Landauer, 1961, 1987; Bennett, 1988; Li and Vitànyi, 1993). Processes can be distinguished from pseudoprocesses, then, by their consistency with the Second Law, if we take care to ensure that special conditions allowing spontaneous reversal of entropy increase do not hold. It is possible (though highly unlikely) that a pseudoprocess could by chance mimic a real process with respect to the constraints of NPI, but experimental intervention could detect this mimicry to a high degree of probability.

NPI ensures that if information is not lost, the causal process is temporally symmetrical, and there is no internally defined temporal direction. If dissipation occurs, however, the information in the final state is less than in the initial state and the initial state cannot be recovered from the final state. Consequently, dissipative causal processes are temporally directed. The complete nature of dissipation is not yet completely understood (Sklar, 1986, 1993), but we know that it occurs regularly. I cannot give a complete account of chance causation here, but I will give a brief sketch. If the information in the effect cannot be computed from the information in the cause, the system is not predictable, even though it may be deterministic in the sense that the same cause would produce the same effect. In either deterministic or indeterministic cases with this sort of informational gap between cause and effect, the probability of the effect can be computed by the informational size of the gap by using the standard relations between information and probability. Perhaps the most interesting case is deterministic chance, which at first appears to be an oxymoron. Consider a coin toss. Suppose that the coin's side is narrow compared with the roughness of the surface on which it lands, so it comes up either heads or tails. Suppose further that its trajectory takes it through a chaotic region in the phase space of the coin toss in which the head and tail attractor basins are arbitrarily close to each other (the definition of a chaotic region). The path of the coin to its eventual end in one of the attractors in this case cannot be computed with any finite resources (by any effective procedure). This means that the attractor the coin ends up in is irreducibly statistical, in the sense that there is no effective statistical procedure that could distinguish the attractor selected (however much it is determined) from a chance occurrence (see end of section §3). The actual odds can be computed by the size of the information gaps in prediction of each of the outcomes, since some of the form can be tracked (e.g. the coin keeps its shape). If the coin has the right form (i.e. it is balanced and symmetrical), the odds will be roughly 50–0 for heads or tails. Note that no counterfactuals are required for this account of chance, nor is the chance in any way subjective.

Some readers might resist the idea of a deterministic system being a chance system, but since no effective statistical procedure can distinguish a chaotic system from a chance system, the difference is certainly beyond our reach. The decision to call systems like the coin toss chance systems is somewhat arbitrary, but it is consistent with anything we could know about statistics or probability. The distinction between unpredictable systems and indeterministic systems forces us to choose which to call intrinsically chancy. Since chance has long been associated with limits on predictability, even by those like Hume who considered it to be a purely subjective matter, I believe that the association of chance with intrinsic unpredictability rather than with indeterminism is justified. The difference between chance deterministic systems and chance indeterministic systems, then, is that the information gap in the former is only in the computability of the information transferred, while in the latter the gap is in information transferred. Deterministic systems are entirely causal, but indeterministic systems are not. A completely random system might still be completely determined. Our universe might be such a system (see §4 above), showing only local, but not global order beyond the constraints of logic.

6.2 Epistemological Issues

One problem with the information theoretic approach is that it requires precise assessments of the quantity of form. This is difficult even for simple molecules, though techniques are being developed using informational complexity (Holzmüller, 1984; Küppers, 1990; Schneider, 1995; Yagil, 1993a, 1993b, 1995). A further problem is that the maximally compressed form of an arbitrary string is not computable in general, though again, methods have been developed for special cases, and approximation methods have been developed that work for a wide range of cases (Rissanen, 1989; Wallace and Freeman, 1987). This problem does not affect the metaphysical explanation of causation in terms of information transfer, however.

Perhaps a more serious problem is determining the identity of information in a system trajectory. For example, apparent causation might really be Leibnizian preestablished harmony. We might not be able to tell the difference, but the information flow would be different if God or some demon is the cause of the apparent direct causation. This situation does not violate the informational metaphysics, however, since the information flow in the preestablished harmony case would be from God to individual monads, with the form originating in God.[32] The problem of the intervention of gods in a specific causal process is just a special case of the Leibniz case, and can be handled the same way, as long as they are subject to the constraints of logic. If they are not, and the effects of their actions bear no determinate relation to the cause, the effects are chance, and can be handled as such.

What appears to be a causal process from the information theoretic point of view might be a chance sequence, or contain a chance element that breaks the causal chain. For example, at one instant the causal chain might end indeterministically, and a new chain might start at the next instant, also indeterministically, where the form types involved are identical to the types if the chain were unbroken. Phil Dowe's chapter in this volume deals with the identity across time issue fairly effectively by showing that other approaches to causation also suffer from the problem. I see no conclusive way around it. I think we just have to live with this possibility. On the other hand, if locality holds, and NPI is applied, we can reduce the probability that what appears to be a transfer of the same information is actually a chance configuration to a minuscule consideration, as argued in §3. The lack of certainty should not be philosophically bothersome, since we cannot be certain of contingencies in any case.

7. ORGANISATION, LOGICAL DEPTH AND CAUSAL LAWS

One last technical idea will be useful for connecting causality to causal laws. The redundancy in a system (physically, its I_P) can be decomposed into orders n based on the number of components, n, required to detect the redundancy of order n (Shannon and Weaver 1949).[33] Complex chaotic (and nearly chaotic) conservative systems, e.g. a steel ball pendulum swung over a pair of magnets under frictionless conditions, typically show relatively little low order redundancy, but a significant amount of high order redundancy, while living systems typically show significant redundancies at both low and high orders (Christensen *et al.*, in preparation). It is ultimately an empirical matter just how local and global redundancies interrelate to lower and higher order redundancies in particular classes of systems, though usually higher order redundancies will also have large temporal or physical scale, or both.

In many of these cases the higher order redundancy is hidden or buried, in the sense that it is not evident from inspecting small parts of the system or local segments of the dynamic trajectory of the system. Nevertheless, it can be seen in the overall structure of the system, and/or in the statistics of its trajectory. For example, the trajectory of a chaotic system is locally chaotic, but it is (probably) confined to spatially restricted attractor basins. Because the information in such systems involves large numbers of components considered together without any possibility of simplification to logically additive combinations of subsystems (the systems are nonlinear), computation of the surface form from the maximally compressed form (typically an equation) requires many individual steps, i.e. it has considerable *logical depth* (Bennett, 1985; Li and Vitányi 1990, 238). Bennett has proposed that logical depth, a measure of buried redundancy, is a suitable measure of the organisation in a system.

Formally, logical depth is a measure of the least computation time (in number of computational steps) required to compute an uncompressed string from its maximally compressed form.[34] Physically, the logical depth of a system places a lower limit on how quickly the system can form from disassembled resources.[35] Organisation requires complex large scale correlations in the diverse local dynamics of a system. This, in turn, requires considerable high order redundancy, and a relatively lower low order redundancy. This implies a high degree of physically manifested logical depth. Whether or not organisation requires anything else is somewhat unclear right now. For present purposes, high level redundancy implied by logical depth will be a more important consideration than organisation or dynamical time, since it will be shown to explicate natural laws as described in (Collier, 1996a). A deep system is not maximally complex, because of the buried redundancy (more internally ordered than a gas), but it is not maximally ordered either, because of its surface complexity (less ordered than a crystal).

Logical depth requires correlation (redundancy), but is silent about dynamics. No dynamical interconnections among the parts of the system are implied, because of the formal nature of the concept (which, like all purely formal concepts, ignores dynamics). Logical depth needs to be supplemented with a dynamical account of depth, within the context of NPI. How to do this is not presently entirely clear (because the dynamical grounding of logical depth requires a way to physically quantify the notion of computational time, or, equivalently, of a computational step, and how to do this properly is not clear). But when we *do* observe organisation we can reasonably infer that it is the result of a dynamical process that can produce depth. The most likely source of the complex connections in an organised system is an historically long dynamical process. Bennett recognised this in the following conjecture:

A structure is deep, if it is superficially random but subtly redundant, in other words, if almost all its algorithmic probability is contributed by slow-running programs .... A priori the most probable explanation of 'organized information' such as the sequence of bases in a naturally occurring DNA molecule is that it is the product of an extremely long biological process. (Bennett, 1985; quoted in Li and Vitányi, 1990, p. 238)

However we should also note that higher order redundancy could arise accidentally as an epiphenomenon (a mere correlation), but then it would not be based on a cohesive structure (cf. Collier, 1988) and so its emergence can't be controlled and it will not persist.

Entrenchment is physically embodied depth *per se*, with no direct implications concerning the historical origins of the depth. *Canalisation*, on the other hand is entrenchment resulting from a deep historical process (and also describes the process). Bennett's conjecture is, then, that cases of entrenchment are, most likely, cases of canalisation. This is an empirical claim. Natural laws are usually taken to be entrenched, but not canalised. Future studies in cosmology may prove this wrong. On the information theoretic account, the same historical origin for the same forms in different systems, including law-like behaviour, is an attractive hypothesis. In any case, logical depth implies high order redundancy, whether it took a long time to form or not. This high order redundancy is a measure of organisation. Natural laws are at the maximal level (or levels, if specificity of information is not linearly ordered) that there is redundancy within a system (cf. Collier, 1996a), and are specified by this redundancy (information) and the inverse mapping function. As such, they serve as constraints on the behaviour of any system. They are thus abstractions from concrete particulars. System laws are not always true scientific laws, which must be general. This can be assured by taking as the system the physical world. This is a standard scientific practice, according to which a purported scientific law that fails under some conditions is thereby shown not to be a law after all.

The information theoretic approach to causation can be used, then, to interpret natural laws in the minimalist metaphysics described in (Collier, 1996a), according to which laws are relations between natural kinds, which are in turn the least determinate classes related by the mathematical relation required to ensure particular instances of the laws hold. These classes are defined in terms of their information, and the mathematical relation is computational consequence, ensuring necessity, given the existence of the particular informational structures (i.e. forms).[36] The laws turn out to be computationally deep, in the sense that the phenomena obeying the laws show high order redundancy, and the computation of the surface phenomena is relatively long (Collier and Hooker, 1999, also Collier, 1996b). The explication of causation, laws and counterfactuals, then, requires only logic with identity (computation theory) and particular concrete circumstances. This is the sort of metaphysics the logical empiricists were looking for, but they made the mistake of relying too heavily on natural language and phenomenal classifications (i.e., they put epistemology before ontology). Of course computation theory was poorly developed before their program was undermined by their mistakes, so they had no way to recover from those mistakes.[37]

8. INFORMATION THEORETIC CAUSATION AND OTHER APPROACHES

Some aspects of the information theoretic approach to causation can be clarified by comparing it with other accounts of causation. I will deal with each of the major current approaches in turn. Not surprisingly, as a minimalist approach, my approach can be interpreted as a version of each of the others with suitable additional assumptions.

8.1 The Regularity Approach

The regularity approach to causation is widely supported by philosophers of science, since it seems to represent well how scientists actually establish correlations and causal influence through controlled experiments using statistical methods (Giere, 1984). Information content (compressibility and depth) is a measure of regularity. It is more reliable than the constant conjunction approach: (1) constant conjunction fails for accidental generalisations, whereas the information transfer model does not because it requires computational necessitation, and (2) samples of chaotic systems appear irregular, thus unlawlike, but have a simple generating function that can often be recovered by computational methods.[38] For systems in chaotic regions, random sampling of data will give results indistinguishable from chance events, even though the generating function for the data points can be quite simple. Minor deviations in initial or boundary conditions can lead to wildly different behaviour, so experiments are not repeatable. Constant conjunction as a means to decide regularity is unworkable. Time series analysis can improve the chances of finding the generating function, especially if the basic dynamics of the system can be guessed from analogy to more tractable systems. There is still a problem of determining the dimensionality of the phase space of the system, and wrong guesses can lead investigators far astray. Testing guesses with computer models is helpful, but the mathematics of chaos ensures that there is no reliable method for finding the generating function of a given time series: the problem is computationally intractable.

The alternative Humean approach uses counterfactuals (Lewis, 1973). This presents problems of its own. Any attempt to distinguish laws from accidental generalisations using counterfactuals without going beyond particulars by using a possible worlds ontology is plagued by the lack of a computable similarity metric across deterministically chaotic worlds. The phenomena in such worlds might as well be related by chance, since by any effective statistical procedure, their relations are chance. This objection is not telling, however, except for verificationists. There is a deeper problem for anyone who is not a verificationist, or anyone who is a metaphysical realist. It seems we can imagine two worlds, one with deterministic chaotic generators, and one produced solely by chance, which are nonetheless identical in the form of all their particulars. Either these worlds are distinguished by the separate existence of laws, which undermines the reason for inferring possible worlds, or else the two worlds must be the same. This latter assumption seems to me to be arbitrarily verificationist, especially given that unrestricted verificationism directly undermines the possible worlds approach, and is also contrary to metaphysical realism. If one is willing to swallow these consequences, then I can see no objection. The same argument can be applied to chance and nonchance worlds that are not chaotic, which is perhaps more telling.

These problems are also telling against any attempt to reduce causation to probability, since it is question begging to infer common cause from probability considerations if causation is defined in terms of probability relations. Probability considerations alone cannot distinguish between a world with chance regularities and one in which the regularities are caused. On the informational approach, there are no such problems. The metaphysical distinction between chance correlations and causal correlations depends on the identity of information. The sensible hypothesis, on any reliable evidence that there is a possibility that the world is not a chance world, would be that the world has natural causal laws that preserve information and determine the probabilities. But that hypothesis could be wrong.

8.2 Universals and Natural Kinds

A distinction might be considered to be the only required universal. I see no great advantage in making distinction a universal, but it is certainly a natural kind in the sense of (Collier, 1996a). The issue of whether or not it is a universal ultimately depends on the ontological status of mathematical objects. Other natural kinds are forms that can be analysed in terms of their informational structure, perhaps in terms of their distinctions from other natural kinds. In any case, the information theoretic approach to causation does not need to invoke either universals or natural kinds except as abstractions from particular systems and their particular properties. If mathematical objects are universals, then so are natural kinds, but so are a lot of other forms as well. Invoking universals seems to have no additional explanatory value in the case of causation, since all possible distinctions are already presupposed by the information theoretic account. [39]

8.3 The Conserved Quantity Approach

Salmon has dropped the mark approach, and has adopted Dowe's conserved quantity approach (Dowe, 1992; Salmon, 1994). The idea is that causal processes, unlike pseudocausal processes (like the spot of a flashlight crossing the face of the moon), involve non-zero conserved quantities. The problem with this approach is that it does not allow for causation in dissipative systems, which predominate in this world. It is possible that dissipative processes can be reduced to underlying conservative approaches, but how to do this is not presently known (Sklar, 1986, 1993, pp. 297ff). Energy and momentum are conserved in dissipative processes in our world (so far as we know). Nevertheless, it seems to be possible to have physical systems in which not even energy and momentum are conserved (e.g. through the spontaneous disappearance of energy/momentum from the physical world as it dissipates, or through the appearance of matter through Hoyle's empirically refuted but seemingly possible continuous creation). Dissipative systems of this sort still preserve information, even if they do not conserve the quantity of information or any other non-zero physical quantity.

The information approach and the conserved quantity approach are equivalent in conservative causal processes and in causal interactions because conservation laws are equivalent to informational symmetries (cf. Collier, 1996b and references therein). In any conservative process, no information is lost, and no information is gained. However, the quantity of information is not necessarily conserved in causal processes (by definition it is not conserved in dissipative processes), though some information is preserved in any causal process. I suppose that focusing on the preserved information in the process as the conserved quantity might make the two approaches identical, but this seems a bit stretched to me. In our world, there are conserved quantities in all causal processes (energy/momentum and charge/parity/time, so far as we know), but the information theoretic approach will work in worlds in which there are no conservation laws as well, if such worlds are indeed possible. At the very least, the information theoretic approach and the conserved quantity approach mutually support each other in worlds in which all causal processes involve some conserved quantity. The main practical advantages that I see for the informational approach is that it explains the longstanding importance of form in accounts of causation, and it does not rule out dissipative worlds. Theoretically, the approach gives a deeper insight into the nature of conservation as a form of symmetry.

8.4 Humean Supervenience

Humean supervenience requires that causation, natural kinds and natural laws are supervenient on particulars. It is satisfied trivially by the information theoretic approach that I have proposed. All that is required for my general account is the particular information in particular things and their computational relations, and a natural way to specify the information in things that is epistemologically accessible. For physical things, NPI provides the last condition. Although NPI cannot be completely specified right now, if ever, the practices of scientists and engineers allow us to use information as unambiguously as any concept in science. I cannot say exactly what NPI means, but I can show how it is used. Unfortunately, it is my experience that the learning process takes many weeks at least. It takes much longer if the student keeps asking for explanations in English. Information theory is the most precise language we can have. Asking for a clearer natural language explanation is pointless. To paraphrase Wittgenstein, we must learn by practice what cannot be said.

9. CONCLUSION

The identification of causation with information transfer permits a minimalist metaphysics using only computational logic and the identity through time of contingent particulars. It also helps to account for the persistence of causal explanations involving form from the beginnings of science to the present. It needs no possible worlds or universals, so it is ontologically parsimonious. Necessitation arises naturally from the computational basis of the approach: causal processes can be thought of as analogue computations that can, when recursively definable, be mapped onto digital computations for tractability. There are some epistemological difficulties with the approach, but it shares these difficulties with more elaborate approaches. The more elaborate approaches can be seen as special cases of the information theoretic approach involving at least one further methodological or empirical assumption, so it can be seen as the core of the other current approaches. Furthermore, it is immediately compatible with modern computational techniques, and thus can be applied directly using conventional methods that have been developed by physicists. NPI and the quantitative methods of Statistical Mechanics permit the quantification of the strength of causal interactions. I doubt these advantages can be gained by any other philosophical approach to causation.

University of Newcastle

ACKNOWLEDGEMENTS

This work was undertaken with the generous support of Cliff Hooker and the Complex Dynamical Systems Group at the University of Newcastle, financially, intellectually and emotionally. I would also like to thank Howard Sankey for his encouragement in producing this chapter. Penetrating questions by David Armstrong, Neil Thomason, John Clendinnen and Tim O'Meara, who were present at the first public presentation of the ideas in this chapter have greatly improved the final version. Comments by Gad Yagil on some ideas contained in this chapter helped me to clarify some central issues concerning NPI. Ric Arthur and Jonathan D.H. Smith made helpful suggestions on historical background and mathematical details, respectively. A very audience interactive version of this paper given at the Department of Philosophy at Newcastle University helped me to clear up three central sticking points of an early version of this chapter. John Wright's acute questions were especially helpful. Malcolm Forster and Steve Savitt made some useful suggestions on presentation. Finally, I would like to thank two anonymous referees, some of whose suggestions I adopted. Other of the referees' suggestions helped me to see that I had not expressed my line of argument concerning the significance of form as clearly as I might have. I hope that the present version dissolves the basis of these suggestions.

NOTES

1 Jack Smart suggested to me that my approach might be a regularity approach in the wider sense that includes his own account of causation. On my account all detectable causation involves compressible relations between cause and effect (§4 below). Inasmuch as, given both compressibility and all other evidence being equal, it is almost always more parsimonious to assume identity of information token rather than coincidence (and never the opposite), compressibility almost always justifies the inference to causation. If meaning is determined by verification conditions (which I doubt) then my theory is indistinguishable from a regularity theory in Smart's wide sense, since the exceptions are not decidable on the basis of any evidence (see §8.1 below for further discussion).

2 Causal connection is necessary in the same way that a computation or deduction is necessary, but it is not necessary in the sense that it is impossible for things to be otherwise. The necessity depends on contingent conditions analogous to the premises of a valid argument (see §4 below). I proposed this kind of necessity for laws in (Collier, 1996a). Something that is necessary in this way cannot be false or other than it is, but is contingently true; i.e. it is contingent that it is. This theory of causation fills out the uninterpreted use of 'causation' in concrete particular instances in (Collier, 1996a), and is part of a project to produce a minimal metaphysics depending on logic, mathematics and contingent concrete particulars alone.

[3] I will give a more precise, mathematical characterisation of form in §3 below. My definition of form may seem very broad. It is. Naturally, if there are any exceptions, my account fails for that sort of case, but I believe there are none, nor can there be.

[4] On generalised coordinates, see (Goldstein, 1980). On the embedding of classical mechanics as well as more recent non-mechanical physics in the dynamics of form, see (Collier and Hooker, 1999)

[5] This argument is developed in (Collier, 1990a) and (Christensen *et al*, in preparation).

[6] Some other possible precursors are Plato, Aristotle, the geometric forms of the atomists' atoms, Descartes' geometric view of dynamics, and Spinoza's theory of perception. I mention them mostly to avoid being accused of thinking myself especially original. These are failed attempts that imported unnecessary metaphysical elements to fill gaps in the accounts that disappear with a proper understanding of computational processes and their relation to physical processes. My position differs from Wittgenstein's position in the Tractatus (1961, see 2.0 to 2.063 especially) in using computational logic broadly construed. I also differ with Wittgenstein on 2.062, in which he says that a state of affairs cannot be inferred from another state of affairs (see §4 below). States of affairs which are unanalysable distinctions or differences may be the only exception, and might satisfy the requirements for Wittgenstein's elementary propositions (for reasons to think not, see Bell and Demopoulos, 1996).

[7] Although Hobbes attributed causation to the mechanical collisions of contiguous bodies, at least one of which is in motion, he believed that causation was necessary. If the total cause is present, then the effect must occur; if the effect is present, then the total cause must have existed (Hobbes, 1839: 9.3). The total cause is made of both the efficient cause (being accidents in the agent) and the material cause (being accidents in the patient). Form played no role for Hobbes' view of causation, except in the geometry of the accidental motions of the agent and patient. On the other hand, despite this extreme mechanism, there is no cause unless the geometries are precisely correct to necessitate the effect.

[8] The logic of distinctions has been worked out by George Spencer Brown (1969) and is provably equivalent to the propositional calculus (Banaschewski, 1977). This is the basis of the binary (Boolean) logic of conventional computers.

[9] On the original definition, length(σ_s) = min{|p|: pÎ{0,1}* & M(p) = σ_s} = min{|p|: pÎ{0,1}* & f(p) = s}, |p| being the length of p, which is a binary string (i.e. pÎ{0,1}*, the set of all strings formed from the elements 1 and 0), and M being a specific Turing machine, and f being the decoding function to recover σ_s from p and then s from σ_s. This definition requires an O(logn) correction for a number of standard information theoretic functions. The newer definition, now standard, sets length(σ_s) to be the input of the shortest program to produce σ_s for a self-delimited reference universal Turing machine. This approach avoids O(logn) corrections in most cases, and also makes the relation between complexity and randomness more direct (Li and Vitányi, 1990).

[10] For a technical review of the logic of algorithmic complexity and related concepts, see (Li and Vitànyi, 1990 and 1993). The complexity of a program is itself a matter for algorithmic complexity theory. Since a universal Turing machine can duplicate each program on any other Turing machine M, there is a partial recursive function f_0 for which the algorithmic complexity is less than or equal to the algorithmic complexity, plus a constant involving the computational overhead of duplicating the particular M, calculated using any other f. This is called the Invariance Theorem, a fundamental result of algorithmic complexity theory (for discussion of this, to some, counterintuitive theorem, see Li and Vitànyi, 1993, pp. 90–5). Since there is a clear sense in which f_0 is optimal, the Invariance Theorem justifies ignoring the language dependence of length(σ_s), and this is now common practice for theoretical work. String maps of highly complex structures can be computed, in general, with the same computational overhead as those of simple structures (the computational overhead is nearly constant), so for complex structures (large CI) the negative component of informational complexity is negligible. Furthermore, in comparisons of algorithmic complexity, the overhead drops out except for a very small part required to make comparisons of complexity (even this drops out in comparisons of comparisons of complexity), so the relative algorithmic complexity is almost a direct measure of the relative informational complexity, especially for large CI.

[11] The more operational approach that retains the constant achieves only correspondence in the infinite limit, which is the only case in which the computational overhead, being a constant, is infinitesimal in proportion and is therefore strictly negligible (Kolmogorov, 1968; Li and Vitànyi, 1990).

[12] Since it is possible to change an effectively random string into a compressible string with the change of one digit and yet, intuitively, the change of one digit should not affect whether a string is random, randomness of finite strings of length n is loosely defined as incompressibility within O(logn) (Li and Vitányi,

1990: 201). By far the greatest proportion of strings are random and in the infinite case the set of non-random strings has measure 1. It is also worth noting that there are infinite binary strings whose frequency of 1s in the long run is .5, even though the strings are compressible, e.g. an alternation of 1s and 0s. These strings cannot be distinguished by any effective statistical procedure (see above). If probability requires randomness, probability is not identical to frequency in the long run. It seems unreasonable, e.g. to assign .5 to probability of a 1 at a given point in the sequence because the frequency of 1s in the long run is .5, if the chance of getting a 1 at any point in the sequence can be determined exactly to be 1 or 0.

13 The converse is not true. Arbitrarily long substrings of non-computable strings (and, for that matter, incompressible finite strings) can be highly ordered, and therefore computable, but the location and length of these highly ordered sub-strings cannot be predicted from earlier or later elements in the string. In general, the incompressibility of a string does not imply the incompressibility of its substrings.

14 As with assigning energy values to real systems, assigning information values for practical purposes is not always easy.

15 The nature of identity is not important here, as long as identicals are indiscernable, i.e. if a=b, then there is no way in which a is distinct from b, i.e. they contain the same information.

16 It is tempting to define a cause as the origin of the information in a causal process. Quite aside from problems of which end of a causal process to look for the origin, the usual continuity of causal processes makes this notion poorly defined. Our usual conception(s) of cause has(ve) a pragmatic character that defies simple analysis because of the explanatory and utilitarian goals it (they) presuppose(s). Nonetheless, I am confident that my minimalist notion of causal process is presupposed by both vulgar and scientific uses of the term 'cause'. Transfer of information is necessary for causation, and is sufficient except for pragmatic concerns.

17 Work has dimensions of energy in standard mechanics, and thus has no direction. However, since it is the result of a force applied through a distance, it must be directed. Surely, undirected force is useless. However, this changes the units of work, since energy is not a vector. Interestingly, Schrödinger (1944) considered exergy as a measure of physical information, but rejected it because people were easily confused about energy concepts. This is remarkable, since exergy and entropy do not have the same dimensions.

18 Though work capacity is a dispositional concept, it is defined through NPI in terms of the state variables of a system, which can be understood categorically. The causal power of a system is determined by its work capacity. The details are relatively simple, but are beyond the scope of this paper, since explaining them requires clearing up some common misconceptions about statistical mechanics.

19 Brillouin (1962, p. 152) refers to physical information as bound information but in the light of my distinction between intropy and enformation (see below), I avoid this term (since in one obvious sense intropy, being unconstrained by the system, is not bound). Brillouin defines bound information as a special case of free information, which is abstract, and takes no regard of the physical significance of possible cases. Bound information occurs when the possible cases can be regarded as the complexions of a single physical system.

20 With respect to the need to interpret the principle in relation to the system and environment under consideration, the situation is exactly paralleled by that for energy and momentum. By referring information to the system environment the need to define some absolute reference point where all constraints of any kind are relaxed, which is not obviously a well defined condition is avoided. Just as there are very different formulae for all the forms of potential energy in different systems, so too are there for forms of entropy and information. The 0th Law of Thermodynamics suggests an absolute measure of entropy, but in practice the "freezing out" of complex order in the hierarchy of energy levels precludes strict application of this "law", except to ideal gases. For the 0th Law to apply, all degrees of freedom of a system must be equally flexible. This is very unlikely to be true in any real physical system (see also Yagil, 1993b).

21 It is worth noting at this point that logical processes, such as computations, obey the Second Law as well, in the sense that a computation can produce only as much information as it starts with, and generally will produce less. There are theoretically possible reversible computers, but they produce vast amounts of waste stored bits if they compute practical results. Consequently, arguments concerning the dynamics of physical complexity also apply to any sort of creature governed by logic. This places some limits on the role of gods as counterexamples to causal analyses, unless the gods act inconsistently with logic. We might as well just assume uncaused events in these supposed counterexamples (see §6.2).

22 Many bright students have taken more than one full course on the subject at university without coming to understand entropy properly.

[23] This is not quite as simple as Szillard's case (see Brillouin, 1962, pp. 176ff), which uses only one molecule!

[24] NPI is assumed throughout, as is the impossibility of a Maxwellian demon (Brillouin, 1962; Bennett, 1982; Collier, 1990b). Szillard's original argument makes the connection to work more obvious by using a molecule pushing on a cylinder in a piston, but the more general arguments by Bennett and Collier examine (in different ways) the computational problem the demon is supposed to solve. The connection to work is implied by thermodynamics and NPI. Szillard used thermodynamics explicitly, but NPI only implicitly, which meant that his exorcism of the demon could not be general. Denbigh and Denbigh (1985) argue that information is not required for the exorcism, since thermodynamics can be used in each instance. It seems to have escaped them that proving this requires something at least as strong as NPI. The problem of Maxwell's demon is especially important because it forces us to be explicit about the relations between control and physical activity. A demon that could store information in some non-physical form could perform its sorting job, though at the expense of producing waste (unusable) information in this storage.

[25] That would require the equivalent of the acceptance of the ideas in this section.

[26] One obvious information basis to consider is a complete microscopic description of a system. However, behind this statement lies the vexed issue of a principled resolution of the relations between mechanics and thermodynamics that respects the irreversibility of the latter despite the reversibility of the former. While the analysis offered here represents a small step toward greater clarity about this complex issue, I do not pursue it here.

[27] Archimedes' lever with which he could move the world, like any other machine, must have a specific form: it must be rigid, it must be long enough, there must be a fixed fulcrum, and there must be a force applied in the right direction. If any of these are lacking, the lever would not work. No amount of energy applied without regard to the form in which it is applied can do work, except by accident.

[28] There is nothing arbitrary about these system-relative distinctions; each is grounded in the system dynamics. Relational properties, like intropy, entropy and enformation, necessarily produce relativised applications across relationally distinct contexts, e.g. S and S_s here, and it is an error (albeit a common one) to equate this to relativism, which is the absence of any principled basis for distinguishing conflicting claims across contexts.

[29] All enformation except perhaps the enformation in some fundamental particles, like protons, will eventually decay, which means that at some temporal scale all, or at least most, enformation behaves as intropy. The scale is set by natural properties of the system in question. Specifically, the extent of the cohesion of the system implies a natural scale (Collier, 1988; Collier and Hooker, 1999, Christensen *et al.*, in preparation).

[30] A complete physical specification would amount to a maximally efficient physical procedure for preparing the system, S, in the macrostate in question from raw resources, R (Collier, 1990a). Furthermore, the procedure should be self-delimiting (it finishes when S is assembled, and only when S is assembled). The information content of this specification is just IP plus any intropy that must be dissipated in the process. The latter is called the thermodynamic depth of the state of the system, and is equal to $H_{ACT}(R) - H_{ACT}(S)$ if there are no practical restrictions on possible physical processes. The algorithmic complexity analogue of thermodynamic depth is the complexity decrease between the initial and final states of a computation (through memory erasure). This quantity is often ignored in algorithmic complexity theory, but see (Bennett, 1985; Collier, 1990b; also Fredkin and Toffoli, 1982), who would hold that the analogy is a physical identity.

[31] There is one further terminological issue concerning physical information that should be noted. By NPI, the disordered part of the system does not contain information (because it cannot contribute to work), but the information required to specify the complete microstate of the system is equal to the information in the macrostate plus the information required to specify the disordered part. Layzer (1975; 1990) speaks of the information required to specify the disordered part as the "microinformation" of microstates, as if the information were actually in the microstate. This information can do work only if it is somehow expressed macroscopically. For this reason, I prefer to regard unexpressed microinformation as a form of potential information (Gatlin, 1972; Collier, 1986; Brooks and Wiley, 1988). Expressed information is sometimes called stored information (Gatlin, 1972; Brooks and Wiley, 1988). Potential information can also be directly expressed as intropy, e.g. in the Brownian motion of a particle, as opposed to at the expense of enformation, e.g. when micro fluctuations disrupt structure. Although expression as intropy is physically possible, it cannot be physically controlled (Collier, 1990b). Control of this process would imply the existence of a Maxwellian

demon. In dissipative structures, especially those formed in systems with multiple attractors, in which the branch system followed in a phase change is determined by random fluctuations, potential information can be expressed macroscopically at the expense of dissipation outside the macroscopic structure.

32 It seems to me that Leibniz had something like this in mind, but it is unclear to me how the generation of new information could be possible without God suffering from the waste problem of the computational version of Maxwell's demon. God could solve the problem by storing huge amounts of waste storage someplace otherwhere, but it would certainly complicate the metaphysics. I believe my one levelled approach is more parsimonious.

33 This is a strictly mathematical decomposition. Physical decomposability is not required. A level of organisation is a dynamically grounded real structural feature of a complex system which occurs when (and only when) cohesive structures emerge and operate to create organisation (Collier, 1988). The same level may manifest or support many different orders of organisation and the same order of organisation may be manifested or supported at many different organisational levels.

34 Some adjustments are required to the definition to get a reasonable value of depth for finite strings. We want to rule out cases in which the most compressed program to produce a string is slow, but a slightly longer program can produce the string much more quickly. To accommodate this problem, the depth is defined relative to a significance level s, so that the depth of a string at significance level s is the time required to compute the string by a program no more than s bits longer than the minimal program. A second refinement, depth of a sequence relative to the depth of the length of the sequence, is required to eliminate another artefact of the definition of depth. All sequences of n 0s are intuitively equally trivial, however the depth of each string depends on the depth of n itself. The additional depth due to sequence of 0s is small. The depth of a sequence of n 0s relative to the depth of the length of the sequence itself is always small. This relative depth correctly indicates the triviality of sequences of the same symbol.

35 Since computation is a formal concept, while time is a dynamical concept, it isn't completely clear how we can get a dynamical measure of computation time. Generally, the minimal assembly time of a system will be less than the expected assembly time for assembly through random collisions, which we can compute from physical and chemical principles. Maximally complex systems are an exception, since they can be produced only by comparing randomly produced structures with a non-compressible template.

36 For an explanation of how this supports counterfactuals, see (Collier, 1996a) and §4 above.

37 See (Collier, 1990a) for a discussion of the inadequacy of Carnap's attempt to determine the information in a proposition.

38 It is worth noting that our solar system, the epitome of regularity in classical physics, is stable for only relatively short periods. Over longer periods it is difficult to predict its evolution (physicist Philip Morrison calls this the "Poincaré Shuffle"). In a world with infinite time, it is mathematically impossible to predict the evolution of the solar system. On the other hand, dissipative processes like tidal dissipation probably explain the regularity that we observe in the solar system. A world in which all processes are in a chaotic regime would need to lack such dissipative processes that produce regularity.

39 There is an ingenious but not entirely convincing argument by Russell that nominalists are committed to at least one universal, similarity. I take it that all distinctions are particular, and depend only on the existence of distinct particulars.

REFERENCES

Armstrong, D.M. (1983), *What is a Law of Nature?*, Cambridge: Cambridge University Press

Atkins, P.W. (1994), *The Second Law: Energy, Chaos, and Form*, New York: Scientific American Library

Banaschewski, B. (1977), 'On G. Spencer Brown's Laws of Form', *Notre Dame Journal of Formal Logic* **18**, 507–9

Bell, John L. and William Demopoulos (1996), 'Elementary Propositions and Independence', *Notre Dame Journal of Formal Logic* **37**, 112–24

Bennett, C.H. (1982), 'The Thermodynamics of Computation: A Review', *International Review of Theoretical Physics* **21**, 905–40. Reprinted in Lef and Rex (eds) *Maxwell's Demon*

Bennett, C. H. (1985), 'Dissipation, Information, Computational Complexity and the Definition of Organization', in D. Pines (ed.), *Emerging Syntheses In Science. Proceedings of the Founding Workshops of the Santa Fe Institute*, 297–313

Bennett, C.H. (1987), 'Demons, Engines and the Second Law', *Scientific American*, 257, no. 5, 108–16
Bennett, C.H. (1988), 'Notes on the History of Reversible Computation', *IBM Journal of Research and Development* **32**, 16–23. Reprinted in Lef and Rex (eds) *Maxwell's Demon*
Bohm, David (1980), *Wholeness and the Implicate Order*, London: Routledge & Kegan Paul
Brillouin, L. (1962), *Science and Information Theory*, second edition, New York: Academic Press
Brooks, Daniel R. and Edward O. Wiley (1988), *Evolution as Entropy: Toward a Unified Theory of Biology, 2nd edition*, Chicago: University of Chicago Press
Brooks, D.R., E.O. Wiley and John D. Collier (1986), 'Definitions of Terms and the Essence of Theories: A Response to J.S. Wicken', *Systematic Zoology* **35**, 640–7
Chandrasekhar, S. (1961), *Hydrodynamic and Hydromagnetic Stability*, Oxford: Clarendon Press.
Christensen, W.D., Collier, John D. and Hooker, C.A. (in preparation), 'Autonomy, Adaptiveness and Anticipation: Towards Foundations for Life and Intelligence in Complex, Adaptive, Self-organising Systems'
Collier, John D. (1988), 'Supervenience and Reduction in Biological Hierarchies', in M. Matthen and B. Linsky (eds), *Philosophy and Biology: Canadian Journal of Philosophy Supplementary Volume* **14**, 209–34
Collier, John D. (1990a), 'Intrinsic Information', in Philip Hanson (ed.), *Information, Language and Cognition: Vancouver Studies in Cognitive Science, Vol. 1.*, Oxford: University of Oxford Press, 390–409
Collier, John D. (1990b), 'Two Faces of Maxwell's Demon Reveal the Nature of Irreversibility', *Studies in the History and Philosophy of Science* **21**, 257–68
Collier, John D. (1993), 'Out of Equilibrium: New Approaches to Biological and Social Change', *Biology and Philosophy* **8**, 445–56
Collier, John D. (1996a), 'On the Necessity of Natural Kinds', in Peter Riggs (ed.), *Natural Kinds, Laws of Nature and Scientific Reasoning*, Dordrecht: Kluwer, 1–10
Collier, John D. (1996b), 'Information Originates in Symmetry Breaking', *Symmetry: Culture and Science* **7**, 247–56
Collier, John D., S. Banerjee and Len Dyck (in press), 'A Non-equilibrium Perspective Linking Development and Evolution', in John D. Collier and Douglas Siege Causey (eds), *Between Order and Chaos: Studies in Non-Equilibrium Biology*
Collier, John D., E. O. Wiley and D.R. Brooks (in press), 'Bridging the Gap Between Pattern and Process', in John D. Collier and Douglas Siege Causey (eds), *Between Order and Chaos: Studies in Non-Equilibrium Biology*
Collier, John D. and C.A. Hooker (1999), 'Complexly Organised Dynamical Systems', *Open Systems and Information Dynamics*
Collier, John D. and Scott Muller (submitted), 'Emergence in Natural Hierarchies'
Collier, John D. and Douglas Siegel-Causey (in press), *Between Order and Chaos: Studies in Non-Equilibrium Biology*
Denbigh, K.G. and J.S. Denbigh (1985), *Entropy in Relation to Incomplete Knowledge*, Cambridge: Cambridge University Press
Dowe, P. (1992), 'Wesley Salmon's Process Theory of Causality and the Conserved Quantity Theory', *Philosophy of Science* **59**, 195–216
Fodor, Jerry A. (1968), *Psychological Explanation: An Introduction to the Philosophy of Psychology*, New York: Random House
Feynman, Richard P. (1965), *The Character of Physical Law*, Cambridge, Mass.: MIT Press
Fredkin E. and T. Toffoli (1982), *International Journal of Theoretical Physics* **21**, 219
Gale, George (1994), 'The Physical Theory of Leibniz', in Roger Woolhouse (ed.), *G. W. Leibniz: Critical Assessments*, London: Routledge & Kegan Paul, 227–39
Gatlin, Lyla L. (1972), *Information Theory and the Living System*, New York: Columbia University Press
Giere, Ronald N. (1984), *Understanding Scientific Reasoning*, 2nd ed. New York: Holt, Rinehart, and Winston
Goldstein, Herbert (1980), *Classical Mechanics, 2nd ed.* Reading, Mass.: Addison-Wesley
Graves, John Cowperthwaite (1971), *The Conceptual Foundations of Contemporary Relativity Theory*, Cambridge, Mass.: MIT Press
Hobbes, Thomas (1839), *Collected Works, Volume 1*, William Molesworth (ed.), London: John Bohn

Holzmüller, Werner (1984), *Information in Biological Systems: The Role of Macromolecules*, translated by Manfred Hecker, Cambridge: Cambridge University Press

Horwich, Paul (1988), *Asymmetries in Time*, Cambridge, Mass.: MIT Press

Kestin, Joseph (1968), *A Course in Thermodynamics*, Waltham, Mass.: Blaisdell

Kitcher, P. (1989), 'Explanatory Unification and the Causal Structure of the World', in P. Kitcher and W.C. Salmon (eds) *Minnesota Studies in the Philosophy of Science, Vol. 13 Scientific Explanation*, Minneapolis: University of Minnesota Press, 410–505

Kolmogorov, A.N. (1965), 'Three Approaches to the Quantitative Definition of Information', *Problems of Information Transmission* **1**, 1–7

Kolmogorov, A.N. (1968), 'Logical Basis for Information Theory and Probability Theory', *IEEE Transactions on Information Theory* **14**, 662–4

Küppers, Bernd-Olaf (1990), *Information and the Origin of Life*, Cambridge: MIT Press

Landauer, Rolf (1961), 'Irreversibility and Heat Generation in the Computing Process', *IBM Journal of Research and Development* **5**, 183–91. Reprinted in Lef and Rex (eds) *Maxwell's Demon*

Landauer, Rolf (1987), 'Computation: A Fundamental Physical View', *Physica Scripta* **35**, 88–95. Reprinted in Lef and Rex (eds) *Maxwell's Demon*

Landsberg, P.T. (1984), 'Can Entropy and 'Order' Increase Together?', *Physics Letters* **102A**, 171–3

Layzer, D. (1975), 'The Arrow of Time', *Scientific American* **233**, 56–69

Layzer, David (1990), *Cosmogenesis: the Growth of Order in the Universe*, New York: Oxford University Press

Lef, Harvey S. and Andrew F. Rex (1990), *Maxwell's Demon: Entropy, Information, Computing*, Princeton: Princeton University Press

Leibniz, W.G. (1969), *The Yale Leibniz*, translated by G.H.R. Parkinson, New Haven: Yale University Press

Lewis, David (1973), 'Causation', *Journal of Philosophy* **70**, 556–67

Lewis, David (1994), 'Chance and Credence: Humean Supervenience Debugged', *Mind* **103**, 473–90

Li, Ming and Paul Vitànyi (1990), 'Kolmogorov Complexity and its Applications', in *Handbook of Theoretical Computer Science*, edited by J. van Leeuwen. Dordrecht: Elsevier

Li, Ming and Paul Vitànyi (1993), *An Introduction to Kolmogorov Complexity and its Applications, 2nd edition*, New York: Springer-Verlag

Mach, Ernst (1960), *The Science of Mechanics*, Lasalle: Open Court

Rissanen, Jorma (1989), *Stochastic Complexity in Statistical Inquiry*, Teaneck, NJ: World Scientific

Russell, Bertrand (1913), 'On the Notion of Cause', *Proceedings of the Aristotelian Society, New Series* **13**, 1–26

Salmon, Wesley C. (1984), *Scientific Explanation and the Causal Structure of the World*, Princeton: Princeton University Press

Salmon, Wesley C. (1994), 'Causality Without Counterfactuals', *Philosophy of Science* **61**, 297–312

Schneider, T.S. (1995), 'An Equation for the Second Law of Thermodynamics', WWW: http://www-lmmb.ncifcrf.gov/~toms/paper/secondlaw/index.html.

Schrödinger, Irwin (1944), *What is Life?*, reprinted in *What is Life? And Mind and Matter*, Cambridge: Cambridge University Press

Shannon, C.E. and Weaver, W. (1949), *The Mathematical Theory of Communication*, Urbana: University of Illinois Press

Sklar, Larry (1986), 'The Elusive Object of Desire', in Arthur Fine and Peter Machamer (eds), *PSA 1986: Proceedings of the 1986 Bienneial Meeting of the Philosophy of Science Association*, volume 2. East Lansing: Philosophy of Science Association, 209–25,reprinted in Steven F. Savitt (ed.) *Time's Arrows Today* Cambridge: Cambridge University Press, 209–25

Sklar, Larry (1993), *Physics and Chance*, Cambridge: Cambridge University Press

Spencer-Brown, G. (1969), *Laws of Form*, London: Allen & Unwin

Thom, René (1975), *Structural Stability and Morphogenesis*, Reading, Mass.: W.A. Benjamin

Thompson, D'Arcy Wentworth (1942), *On Growth and Form, 2nd ed.*, Cambridge: Cambridge University Press

Ulanowicz, R.E. (1986), *Growth and Development: Ecosystems Phenomenology*, New York: Springer Verlag

Wallace, C.S. and P.R. Freeman (1987), 'Estimation and Inference by Compact Coding', *Journal of the Royal Statistical Society, Series B, Methodology* **49**, 240–65

Wicken, Jeffrey S. (1987), *Evolution, Thermodynamics and Information: Extending the Darwinian Paradigm*, New York: Oxford University Press

Wiley, E.O. (1981), *Phylogenetics: The Theory and Practice of Phylogenetic Systematics*, New York: Wiley-Interscience

Wittgenstein, Ludwig (1961), *Tractatus Logico-Philosophicus*, translated by D.F. Pears and B.F McGuiness, London: Routledge & Kegan Paul

Yagil, Gad (1993a), 'Complexity Analysis of a Protein Molecule', in J. Demongeot, and V. Capesso (eds), *Mathematics Applied to Biology and Medicine*, Winnipeg: Wuerz Publishing, 305–13

Yagil, Gad (1993b), 'On the Structural Complexity of Templated Systems', in L. Nadel and D. Stein (eds) *1992 Lectures in Complex Systems*, Reading, Mass.: Addison-Wesley

Yagil, Gad (1995), 'Complexity Analysis of a Self-Organizing vs. a Template-Directed System', in F. Moran, A. Moreno, J.J. Morleo, and P. Chacón (eds.) *Advances in Artificial Life*, New York: Springer, 179–87

PHIL DOWE

GOOD CONNECTIONS: CAUSATION AND CAUSAL PROCESSES*

I. INTRODUCTION

Clearly the concept of a 'causal process' has something to do with the concept 'causation'. Two events, or facts, or states of affairs are connected by the relation causation when the first is the cause of the second and the second is the effect of the first. A causal process, such as an apple decaying, or an atom decaying, or a billiard ball moving across a table, is typically distinguished from a pseudo process, such as a shadow moving or a spot of light moving across a wall. A pseudo process, as distinct from a causal process, does not contribute to the causal structure of the world (Dowe, 1996; Salmon, 1984).

But what is the conceptual connection between 'a causal process' and 'causation'? The natural answer is that causal processes are to be analysed or explained in terms of causation. For example, the distinction between a causal process and a pseudo process can be explained by the fact that two distinct temporal parts of a causal process are related by the relation of causation, whereas two distinct temporal parts of a pseudo process will not be connected by the relation of causation. Thus, according to this view, causal processes are analysed in terms of the relation of causation.

But some philosophers, such as Bertrand Russell and Wesley Salmon, have taken the explanation to go in the other direction. These philosophers have argued that causation needs to be analysed, or explained, in terms of causal processes. For example, Salmon writes,

> Focusing upon processes rather than events has, in my opinion, enormous heuristic (if not systematic) value. As John Venn said in 1866, "substitute for the time honoured 'chain of causation' so often introduced into discussions upon this subject, the phrase 'rope of causation', and see what a very different aspect the question will wear". (1984, p. 183)

I take it that we should read Salmon as saying here that there might be "systematic value" in analysing or explaining causation in terms of causal processes. After offering his 'at-at theory' of causal processes, Salmon comments,

> It is tempting, of course, to try to reduce causal processes to chains of events, indeed people frequently speak of causal chains . . . The point of the 'at-at' theory, it seems to me, is to show that no such question about the causal relations among the constituents of the process need arise. (1984, pp. 156–7)

The motivation for taking this approach seems to be a frustration with philosophical attempts to say what causation is, in particular certain frustrations that arise from the

H. Sankey (ed.), Causation and Laws of Nature, 247–263.

Humean tradition of accounting for causation in terms of regularities. By focusing on causal processes rather than the relation between events, Salmon hopes to circumvent these difficulties. Bertrand Russell had adopted the same approach, as we shall see in the next section, although his motivation was somewhat different.

In this paper I will attempt to defend this approach by defending a non-causal analysis of causal processes. A 'non-causal analysis' is an account of causal processes which does not appeal to the concept of causation; which is required or else the account of causation in terms of causal processes would be circular. This defence comes in two parts. Firstly, I defend a non-causal theory of causal processes – the conserved quantity theory – against some objections which focus on the question of identity through time. These objections are dealt with in sections 2–5. Secondly, I answer an objection to the claim that causal processes are adequate to explain or account for causation. This objection, which concerns so-called 'preventings and omissions', is dealt with in section 6.

2. RUSSELL ON CAUSAL LINES AND IDENTITY

Bertrand Russell's 1912/13 paper, 'On the Notion of Cause', is famous for the quote,

> The law of causality, I believe, like much that passes muster among philosophers, is a relic of a bygone age, surviving, like the monarchy, only because it is erroneously supposed to do no harm. (1912, p. 1)

In that paper Russell argued that the philosopher's concept of causation involving, as it does, the law of universal determinism that every event has a cause and the associated concept of causation as a relation between events, is "otiose" and in modern science is replaced by the concept of causal laws understood in terms of functional relations, where these causal laws are not necessarily deterministic.

In a later book written in 1948, entitled *Human Knowledge*, Bertrand Russell outlines a similar view but does so in language which is much more flattering to causation. He still holds that the philosophical idea of causation should be seen as a primitive version of the scientific idea of causal laws. However, his emphasis now is on certain postulates of causation which he takes to be fundamental to scientific (inductive) inference. A key postulate concerns the idea of causal lines or, in our terminology, causal processes. Russell's 1948 view is that causal lines replace the primitive notion of causation in the scientific view of the world, and not only replace but also explain the extent to which the primitive notion, causation, is correct. He writes,

> The concept "cause", as it occurs in the works of most philosophers, is one which is apparently not used in any advanced science. But the concepts that are used have been developed from the primitive concept (which is that prevalent among philosophers), and the primitive concept, as I shall try to show, still has importance as the source of approximate generalisations and pre-scientific inductions, and as a concept which is valid when suitably limited. (1948, p. 471)

Russell also says, "When two events belong to one causal line the earlier may be said to "cause" the later. In this way laws of the form "A causes B" may preserve a certain validity." (1948, p. 334). So Russell can be seen, in his 1948 book, as proposing the view that within limits causal lines, or causal processes, may be taken to analyse causation where causation is understood as the philosopher's concept of a relation between events.

So what is a causal line? Russell writes,

I call a series of events a "causal line" if, given some of them, we can infer something about the others without having to know anything about the environment. (1948, p. 333)

A causal line may always be regarded as a persistence of something, a person, a table, a photon, or what not. Throughout a given causal line, there may be constancy of quality, constancy of structure, or gradual changes in either, but not sudden change of any considerable magnitude. (1948, pp. 475–7)

So the trajectory through time of something is a causal line if it doesn't change too much, and if it persists in isolation from other things. A series of events which display this kind of similarity display what Russell calls 'quasi-permanence'.

Bertrand Russell also distinguishes two kinds of causal laws, both of which are to be found in fundamental science. One kind of causal law concerns change. This can be understood in terms of functional relations. Another kind of causal law concerns persistence, and Newton's first law of motion which says that a body will move in a straight line unless acted on by a force, is offered as an example (1948, p. 327). Here we wish to focus on the second kind, that is on causation as persistence, for reasons that will become clear.

Wesley Salmon has urged a number of objections against Russell's theory of causal lines (1984, pp. 140–5). The first objection is that Russell's theory is couched in epistemic terms rather than ontological terms, yet causation is itself an ontic matter not an epistemic matter. The second objection is that Russell's theory of a causal line does not enable the distinction between pseudo and causal processes to be made, yet to delineate causal from pseudo processes is a key issue which needs to be addressed by any theory of causal processes. As Reichenbach argued (1958, pp. 147–9), as he reflected on the implications of Einstein's special theory of relativity, science requires that we distinguish between causal and pseudo processes. Reichenbach noticed that the central principle that nothing travels faster than the speed of light is 'violated' by certain processes. For example, a spot of light moving along a wall is capable of moving faster than the speed of light.[1] Other examples include shadows, and the point of intersection of two rulers.[2] Such pseudo processes (as I call them[3]) do not violate special relativity, Reichenbach argued, simply because they are not causal processes, and the principle that nothing travels faster than the speed of light applies only to causal processes. Thus special relativity demands a distinction between causal and pseudo processes. But Russell's theory doesn't explain this distinction, because both causal processes and pseudo processes display constancy of structure and quality; and both licence inferences of the sort Russell has in mind. For example, the phase velocity of a wave packet is a pseudo process but the group velocity is a causal process; yet both licence reliable predictions.

Turning now to the question of identity through time, in *Human Knowledge* we find that Bertrand Russell sees that there is an important connection between causal process and identity. Russell writes,

The concept of more or less permanent physical object in its common-sense form involves "substance", and when "substance" is rejected we have to find some other way of defining the identity of a physical object at different times. I think this must be done by means of the concept "causal line". (1948, p. 333)

So the concept of a causal line can be used to explain the identity through time of an object or a person. Elsewhere Russell writes,

The law of quasi-permanence as I intend it . . . is designed to explain the success of the common-sense notion of "things" and the physical notion of "matter" (in classical physics) . . . a "thing" or a piece of matter is not to be regarded as a single persistent substantial entity, but as a string of events having a certain kind of causal connection with each other. This kind is what I call "quasi-permanence". The causal law that I suggest may be enunciated as follows: "Given an event at a certain time, then at any slightly earlier or slightly later time there is, at some neighbouring place, a closely similar event". I do not assert that this happens always, but only that it happens very often – sufficiently often to give a high probability to an induction confirming it in a particular case. When "substance" is abandoned, the identity, for commonsense, of a thing or a person at different times must be explained as consisting in what may be called a "causal line". (1948, pp. 475–7)

Russell's causal theory of identity asserts that the identity over time of an object or a person consists in the different temporal parts of that person being all part of the one causal line. This is the causal theory of identity couched in terms of causal processes or lines. A causal line in turn is understood by way of an inference which is licensed by the law of quasi permanence. With this background, we now turn to a contemporary version of the non-causal account of causal processes.

3. CONSERVED QUANTITY THEORY OF CAUSAL PROCESSES

The conserved quantity theory of causal processes (Dowe, 1992; 1995) is an example of a 'non-causal' theory of causal processes. Following (Salmon, 1984), this approach takes the key task to be to distinguish causal from pseudo processes. According to the conserved quantity theory, a causal process is a world line of an object which possesses a conserved quantity, where a process is a world line of an object. A causal process is one which possesses a conserved quantity, and as far as we can know, this can be any of the quantities that contemporary science tells us are universally conserved, for example, mass-energy, linear momentum or charge. The definition of an object is taken liberally as anything which is recognised in the ontology of science or of commonsense. This will include objects such as spots and shadows. A causal interaction is defined as the intersection of world lines which involves exchange of a conserved quantity.

A further proposal that I want to consider here is a thesis that I will call Strict Identity (SI): an object is wholly existent at each and every instant in its history (Dowe, 1995). One implication of this requirement is that the various time slices of the process each represent one and the same object at different times, thus the object has an identity over time.[4] I am not saying that the CQ theory needs SI. But it needs *some* thesis of identity, for there is important work to be done. Perhaps there is a weaker account of identity that will do that work; I simply show that SI does do the job. To see what work needs doing, we turn to the time wise gerrymander.

One kind of world line that doesn't qualify as a process is the time wise gerrymander. A time wise gerrymander is an alleged object which refers to different objects at different times, for example, in a billiards game suppose that at time t_1 a cue ball is struck which subsequently collides with the red ball at t_2, and which in turn hits the black ball at t_3, which in turn rolls into the pocket at t_4. Then one could define the alleged object, x, such that

x is the cue ball at $t_1 \leq t_2$,
x is the red ball at $t_2 \leq t_3$,
x is the black ball at $t_3 \leq t_4$.

The world line of this alleged object is a time like worm in space-time. In addition, the object has spatiotemporal continuity and a certain constancy of structure, a certain similarity over time, although it does change its colour. But, according to the CQ theory, plus SI, such a world line does not represent a process, let alone a causal process, simply for the reason that this is not the world line of an object since objects must be wholly existent at each instant in their history. The three phases of x are not phases of a single object.

SI counters the use of timewise gerrymanders as counterexamples to the conserved quantity theory. For example, take a spot of light moving along a wall, perhaps generated by a single spotlight some distance away. According to the conserved quantity theory, this spot is not a causal process because the object which is represented by the world line is not the kind of object to which a conserved quantity can be ascribed (Dowe, 1995). Spots of light have certain properties, such as size or speed, but they do not have conserved quantities such as energy or momentum.

Salmon has given a variation of this example as an objection to the conserved quantity theory (1994). According to Salmon, individual spots of light do manifest energy and therefore we can think of a process defined as "the segment of wall currently illuminated". This object can be represented by a world line in space-time and it manifests a conserved quantity, namely energy, and therefore, according to the conserved quantity theory, it ought to count as a causal process. Yet clearly it is a pseudo process. However, such an alleged object turns out to be a time wise gerrymander of the same kind as the series of billiard balls defined above. The object which moves, the spot, doesn't possess energy, so its worldline is not a causal process. The series of segments of the wall, each of which do possess energy, is not wholly existent at each time, so it is not the worldline of an object. It too is a time wise gerrymander (Dowe, 1995).

A similar example has been given by Christopher Read Hitchcock (1995) involving a charged shadow. Suppose we have a charged plate and a shadow moving across that plate, then at each instant the region of the plate covered by the shadow does manifest a conserved quantity, namely charge. However, this example can be ruled out for the same reason as Salmon's example, because the charged shadow also is a time wise gerrymander. The object which has charge is the segment of the plate, the object which moves and therefore makes up the process, or constitutes the parts of the world line in question, is the shadow. The shadow and the section of the plate are not one and the same thing; the section of the plate possesses a conserved quantity, the shadow does not have a conserved quantity. Shadows, like spots of light, do not possess quantities such as charge or energy.

Is it simply arbitrary to count as genuine objects non-causal objects such as spots and shadows while ruling out the 'objects' which constitute timewise gerrymanders?

Not at all. Properties attributed to timewise gerrymanders are in fact predicated on a series of *different* individuals, whereas properties attributed to pseudo processes at different times are predicated on the same individual.

For example, energy, attributed to the series of illuminated segments of wall is strictly speaking predicated on different segments at different times. On the other hand, the shape of a spot is predicated on the one individual – the spot – at all times.

So the conserved quantity theory, together with SI, presupposes a concept of identity through time as follows. A process is a world line of an object. In order to know what is a process one needs to know that the different parts of the world line one has drawn represent one and the same object, hence it is necessary that one has already at hand a clear concept of identity through time. Thus, the conserved quantity theory takes identity through time as in one sense primitive in the theory. Identity through time is part of what goes to make up the definition of a causal process, therefore we can say that a causal process is defined, or analysed, or explained in terms of a notion of identity through time.

Thus we have a non-causal theory of causal processes, a non-causal theory which appeals to conserved quantities, space-time points and also the identity through time of an object. Appeal to causation is not necessary to explain the difference between causal and pseudo processes, but appeal to identity through time is necessary. This is in agreement with Russell in that it finds an explanatory connection between causal processes and identity, but it is in disagreement with Russell over the direction of that explanatory connection. Russell explains identity in terms of causal processes, here we explain causal processes in terms of identity.

There are two objections which arise rather naturally in response to the above conserved quantity theory. The first objection has been raised by a number of philosophers, including Wesley Salmon. In a recent paper Salmon says, "I have offered a concept of causal transmission analysed in terms of the "at-at" theory for which Dowe has traded an unanalysed concept of genidentity. This is not, I think, an advantageous exchange" (1997, p. 468). Salmon comments on his own revised version, that "it yields a criterion that is impeccably empirical, and thus it provides an acceptable answer to the fundamental problem Hume raised about causality" (1997, p. 468).

So Salmon's objection to taking identity over time as primitive in a theory of causal processes is just that it violates the empiricist's stricture that one should not invoke empirically inaccessible elements as unanalysed or primitive in a philosophical theory.

There are three points to be made in reply to this objection. Firstly, I admit that there might be a need to explain what identity is. I don't want to be committed to a realist account of identity – 'realist' in the sense of Tooley (1987) – where identity is postulated to be an unanalysable primitive. I prefer to leave the options open. SI is one way to understand identity, but if there is some other way, apart from appealing to causation, then that's fine.

Secondly, therefore I take the position that the question of identity is open. But philosophy is an ongoing enterprise where advances always raise new questions. The task of explaining or analysing does not have to reach a terminus point in order for there to be progress. In this respect, philosophy is like science, which, as Musgrave once pointed out (1977) continually provides explanations, but never final explanations.

Thirdly, and more directly, Salmon's position may be open to an (ad hominem) symmetric complaint, namely, can the concept of a 'causal process' be explained without reference to identity? If it can't then the notion of a causal process is an unanalysed primitive in the theory, and by Salmon's lights that is unacceptable. Thus one question I want to focus on is "can causal processes be explained without reference to identity?" If the answer is no, then things are even, if I don't explain identity, then equally the alternative doesn't explain causal processes.[5] If this was so then

Salmon's charge would be deflected, although we would all agree that there is a need for further work. Whether it is so remains to be seen.

The second objection has been raised by Max Kistler of the Centre for Research in Applied Epistemology, Ecole Polytechnique in Paris. Kistler points out that the above account cannot explain why objects persist through time. He writes, referring not only to the conserved quantity theory but also Salmon's mark-transmission theory, "Such theories cannot *explain why* causal processes can transmit marks or why objects persist through time. They just have to take these as primitive facts" (Kistler, 1998).[6] The point seems to be that it is detrimental to an account of causation that it rules out the causal theory of identity, for the causal theory of identity can explain why objects persist through time. We have seen Russell's causal theory of identity, but the approach has a longer history than that. Hume, for example, offered a causal theory of identity when he said that the self identity of an object through time can be analysed in terms of spatiotemporal continuity, relations of resemblance and causation. We note that the causal theory of identity has today reached a status of widespread popularity,[7] although it has its critics.[8] Thus the objection to the conserved quantity theory is simply that it rules out a very plausible and widely held account of identity, offering none in its place.

In reply to this charge I make three comments. Firstly, it isn't quite true to say that I offer no account of identity. In fact if I am proposing SI then that doesn't require further explanation – the object is the same thing at different times analogously to the way the same property is instantiated in different individuals. Secondly, this argument is also open to a symmetric charge. If my approach doesn't explain identity, then we can ask whether the alternative offers an explanation of causal processes. The first question is the one just discussed, "Can we explain causal processes without appeal to identity?"

Thirdly, Kistler's charge prompts a second question, which challenges the causal theory of identity more directly: "Can identity really be explained in terms of causal processes?"

If the answer to both these is no, then appeal to the identity theory of causal processes is exonerated. If the answer to the first question is no, then we have established the symmetric charge against Kistler and Salmon, because if I don't explain identity, he in turn doesn't explain causal processes. If so, Kistler's charge is deflected. If the answer to the second question is no, then even if we had a decent non-identity account of causal processes, now that the causal theory is ruled out, any adequate explanation of identity is available to me as much as to my opponent. Again, and independently, Kistler's charge is deflected.

In this paper I want to argue that both these questions should be answered 'no'. We cannot account for causal processes except in terms of identity; and in any case causal processes do not provide an adequate account of identity. In the next section, "Salmon and Non-interacting Intersections", I look at Salmon's most recent attempt to provide an account of causal processes, and I show that it fails to avoid appeal to identity. In section 5. "Two Problems for the Causal Theory of Identity", I will raise some difficulties for the attempt to explain identity in terms of causal processes. If I am right on both these points, then these objections to the conserved quantity theory have been successfully deflected.

4. SALMON AND NON-INTERACTING INTERSECTIONS

Over the years Wesley Salmon has defended a number of theories of causal processes but most recently he has come out in defence of the conserved quantity theory (Salmon, 1997). He wishes, however, to avoid making any assumptions about identity through time. He proposes the following definition, "A causal process is the world line of an object that transmits a non-zero amount of a conserved quantity at each moment of its history (each space-time point of its trajectory)" (Salmon, 1997, p. 468). The concept of transmission is to be understood by the following definition.

> A process transmits a conserved quantity between A and B (A^1B) if and only if it possesses [a fixed amount of] this quantity at A and at B and at every stage of the process between A and B without any interaction in the open interval (A, B) that involves an exchange of that particular conserved quantity. (Salmon, 1997, p. 463)

Here, instead of an assumption about identity, Salmon appeals to a special kind of regularity which involves the possession of a fixed amount of a conserved quantity at every space-time point of the process which he calls the at-at theory.

According to Salmon, this also avoids the difficulties that Russell's account faces, as outlined above. Salmon's account is purely ontic, not epistemic, and further it does draw the distinction between pseudo and causal processes. A pseudo process is one which does not transmit a fixed amount of a conserved quantity.

This account also solves the problem of the moving spot of light, Salmon claims, because the spot does not manifest the energy *in the absence* of interactions, but rather, because of a continual set of interactions as the beam of light interacts with the wall.

But Salmon's account faces a number of difficulties. In particular, I want to argue that Salmon's account does not avoid the presumption of identity. The problem I want to focus on involves non-interacting intersections. For example, suppose we have two particles which are able to pass through each other without having any effect on each other. Then, on Salmon's theory, we are not able to tell whether we have a case where two causal processes pass through each other or whether we have a causal interaction where two particles collide and exchange quantities on colliding. But the assumption of identity over time determines which of these two alternatives is in fact the case.

Suppose we have two objects exactly similar, except that one has conserved quantity q=n, while the other has q=m (m^1n). Suppose, as before, they pass (or appear to pass) straight through each other, so that their worldlines intersect. After the intersection there is no change to the non-specified properties of either. Is this a causal interaction?

Both Salmon and I agree that a causal interaction involves an exchange of a conserved quantity. (See Dowe, 1992, 1995; Salmon, 1994, 1997). However, Salmon's account of a causal process fails to decide unambiguously whether our case is a causal interaction. One alternative has it that object *a* (q=m) and object *b* (q=n) pass each other by unchanged: no causal interaction. But another alternative says that there is a causal interaction where *a* (q=m) changes direction and amount of q (to q=n) while *b* also changes direction and amount of q (to q=m). This involves an exchange of a conserved quantity, and so qualifies as a causal interaction. Disambiguation is achieved by the notion of identity through time if it is the case that *a* is the same object before and after the intersection, then there is no causal interaction.

This is not such an uncommon scenario. A stray neutrino passes through my body.

Am I still myself, unaffected by the event, or am I now the thing that used to be the neutrino, having been radically transformed by the experience? Other non-interacting intersections include radio waves entering a building and light coming through the glass window.

Salmon's theory requires, in a causal process, that a conserved quantity is possessed continuously, in the absence of causal interaction. But in non-interacting intersections it would be indeterminate as to whether we have a causal process.

So, Salmon's theory does involve a notion of identity through time. Continuous possession does not avoid the problem. This casts serious doubt over the prospects of a non-identity theory of causal processes. At least we may say, on a pairwise consideration of Salmon's and my theory, I have answered Salmon's claim to have uncovered a reason to prefer his to mine.

5. TWO ARGUMENTS AGAINST THE CAUSAL THEORY OF IDENTITY

As mentioned above, the causal theory of identity through time has become quite popular in recent years. But I think that there remains a number of more specific considerations that tell against it which have been overlooked, and in this discussion I will present two of those considerations. Note that I am concerned with the causal theory of identity just within the context of considerations of causal processes.

The first argument against the causal theory of identity appeals to the idea of a pseudo process. Pseudo processes have identity over time without causation. Consider the spotlight spot moving across a wall. Its temporal parts are not causally related – it is not a causal process. But its temporal parts are related by the identity relation – the spot at time t_1 is the *same* spot that exists at a later time t_2. If there were two spots moving around in a chaotic fashion we might start to wonder which is which, thereby presupposing identity through time. To give another example, my shadow is always my shadow – it has identity through time, but it is not a causal process. Therefore since you can have identity without causation, the causal theory of identity is false.

There are two replies available to the defender of the causal theory, although neither is entirely satisfactory. The first reply is to deny that pseudo processes display identity through time. If we speak and think of the spot or shadow as being the 'same' shadow then that is only a manner of speaking that is not grounded in reality.

Consider an example intended to support this line of thinking. Suppose in my invisible pen I have 'magic ink' which disappears as soon as it hits the page.[9] Then if I draw a line across the page it would look as if an ink spot is moving across the page. But what we're really seeing is a succession of many different ink spots. So, here we have a pseudo process – although unlike the kinds of pseudo processes considered above, the ink spot cannot travel faster than the speed of light. The claim is, this pseudo process does not exhibit identity over time. We might be tempted to think of the moving ink spot as one thing but really it is a succession of different things.

But this example trades on an ambiguity. One sense of 'spot' (call it 'sense *a*') refers to the globules of ink which land on the page. Such spots do not move across the page. Another sense of 'spot' ('sense *b*') refers to the blue thing that moves across the page; and it's this second sense which is claimed, in the above argument, to exhibit identity over time. Exactly the same distinction is evident in the case of the light spot.

There are two senses of 'spot': sense *a* refers to a particular aggregate of photons which impinges on the wall at a particular time: such a spot does not move along the wall. Sense *b* refers to the spot which travels across the wall. The only difference between the ink case and the light case is that in common talk 'an ink spot' normally has sense *a*, whereas a 'spot of light' normally has sense *b*. In any case, the point is that the fact that we have a succession of sense *a* spots not related by identity does not mean that sense *b* spots do not exhibit identity over time.

Common ways of thinking and speaking grant to pseudo processes identity over time. But perhaps this first reply simply asserts that common ways of thinking are mistaken about that identity, and that sense *b* spots are not genuine objects which move across walls and display identity over time.

But this eliminativism runs into problems with scientific cases of pseudo processes. Here I am thinking in particular of the example of the water ripples moving at the so-called phase velocity. (See Dowe, 1992, p. 212). Such ripples can move faster than the speed of light, but they cannot be used to transfer information, as they are pseudo processes. But they figure in the scientific description of the world, so they need to be regarded as genuine items with spacetime trajectories and therefore with identity over time. *Something* moves at the phase velocity. Therefore, quite apart from its intuitive awkwardness, this eliminativism faces quite serious difficulties with scientific cases.

The second reply available to the proponent of the causal theory is to distinguish grades of identity in the following way. Top grade identity always involves causation. But there is also a lower grade of identity such as that displayed by pseudo processes, which does not directly involve causation. Then we can say that spots and shadows have identity in a sense, although not in the same sense as do billiard balls and persons. So there are two types of identity – the causal and the non-casual.

This may at first appear completely *ad hoc*, but in fact it can be given a credible basis. A pseudo process depends in an ontological sense on a causal process or on causal processes: the shadow is the shadow *of* the car, and the movement of the shadow is dependent on the movement of the car. In fact, all characteristics of pseudo processes depend on characteristics of causal processes in the sense that there can be no difference in the characteristics of a pseudo process without there being some difference in the characteristics of some causal process. So low grade identity can be understood the following way: an object has low-grade identity over time only if it is dependent in the appropriate way on an object (or objects) which has high-grade identity over time. In other words, a process can have identity over time without causation only if it depends in the appropriate way on a process which does involve causation.[10] However, while this reply is not entirely *ad hoc*, it does come at a cost, in that it introduces further complication, and as such represents a loss of theoretical simplicity and unity. On this view there are two kinds of identity, whereas before we thought that there was only one.

The second argument against the causal theory of identity is that the move of adding causation to other identity conditions such as continuity and similarity fails to achieve what it is commonly thought to achieve. This second argument is best presented as a response to an argument by Armstrong for the causal theory, against a particular non-causal conception of identity through time, *viz.*, a straight similarity-continuity account, such as Russell's.

Armstrong's argument involves two gods who are creating and destroying things independently. By coincidence, one destroys a certain item just at the same instant that the other creates an exact replica of that item at the same location (1980, p. 76).[11] It would seem to everyone that the same object still exists, but everyone would be wrong. It is really a different thing. Similarity and continuity are satisfied, so Armstrong concludes that we need to appeal to a causal theory instead. Later stages are not directly causally connected to earlier stages, which explains why they aren't the same thing.

However, we can modify the example in such a way that it undermines this move. Suppose we have a 'random creator' – a being located in space which is able to create beings at will, instantaneously, out of nothing, but it has no control over the nature or location of its creations. Conceptually, this is not far removed from Armstrong's gods instantaneously creating things. Suppose at time t this being creates by chance a replica of itself, and at the same instant one of Armstrong's gods annihilates this being, leaving just its replica. Suppose also that the replica appears at the very location vacated by the original being. To all appearances the being simply continued to exist uninterrupted, since there is spatiotemporal continuity and constancy of properties. But, as before, appearances are wrong; the being is no longer the same being. But now we have a further complication, because the earlier being was the cause of the existence of the later being.[12] So now we have spatiotemporal continuity and similarity *and* causation without identity. Therefore this common motivation for the causal theory of identity fails – the addition of causation does not solve the problem identified by Armstrong. This does not prove that identity doesn't require causation, because the answer may be to find a further condition to add to causation. But it does remove a powerful argument for the conclusion that identity needs causation. Whatever additional condition one finds might be available to me, too, for it may do the trick without the causal condition. Without arguments such as Armstrong's, the charge – that the identity theory of causal processes fails because it rules out the causal theory of identity – certainly loses its sting.

A range of objections to the example of the replicator will fail simply because they apply equally to Armstrong's two gods. For example, objections to creation ex nihilo, or to the presumption of discrete time, both apply to Armstrong's argument as much as to mine. Since my argument is presented simply to undermine Armstrong's, such objections are of no relevance here.

6. PREVENTINGS AND OMISSIONS

We return, finally, to the connection between causation and causal processes. If causation is to be analysed in terms of causal processes, then for A to be the cause of B it is a necessary condition that A and B are linked by a causal process or a set of causal processes and interactions. (For example, Dowe, 1992, 1995; Salmon, 1984, 1994). 'Preventings' and 'omissions' are types of counterexamples to this condition. To use an example due to Brian Ellis, pulling down the blind causes the room to be dark, but there is no set of causal processes and interactions linking the cause to the effect. On the contrary, pulling down the blind prevents the process which transmits energy from the sun to the room. To make the difficulty more perspicuous, one could consider the version where John's pulling down the blind caused Mary to stumble into the chair (because she couldn't see properly). In this section I sketch a strategy for overcoming these counterexamples.[13] Conceptual analysis shows that the difficulty is an illusion.

Other examples of this general type of counterexample have been given in the literature on causation and perception. Ehring gives an example of turning a switch to turn off a light. (Ehring, 1986, p. 251). Flicking the switch causes the light to go off, but energy is not transmitted from the switch to the light going out. Rather, the flicking of the switch stops the flow of energy through the switch to the globe, "eliminating transference". Ehring offers this case as a counterexample to the transference theories of Aronson and Fair. It is equally a problem for process theories because the cause (flicking the switch) and the effect (the light going out) are not linked by a set of causal processes and interactions. In general, whenever a causal process is interrupted, we have an instance of a *preventing*.

Another example is considered by Goldman (1977, p. 281), as a difficulty for the causal theory of perception. Someone staring at the sky sees a black patch, which turns out to be a black hole. Goldman's intuition is that she saw the black hole. The black hole is the cause of the visual experience, but there is no energy transfer from the black hole to the perceiver. Similar cases occur whenever we see black objects. Goldman also considers a related case (1977, pp. 282–3). An electrical failure causes Big Ben to fail to deliver its usual midnight chime. A man who lives next door jumps out of bed, shouting "what was that?" The failure of Big Ben to chime caused the man to wake up, but there was no energy transfer from the clock to the man's ear. Big Ben causes the effect by an act of *omission*.[14] Other examples of causing by omission are not hard to think of. The parent's inattention was the negligent cause of the child's death. The fieldsman's lack of judgment was to blame for the ball running to the boundary.

Brian Ellis, who offers an analysis of causation in terms of the action of forces, takes, as a solution to this difficulty, the line that ordinary talk is just mistaken when it takes these cases as cases of causation, since there is no action of forces (1990, p. 66). Call this 'Ellis-eliminativism'. I think this is close, but not quite right.

I suggest that the way to handle preventings and omissions is by redescription, in order to reveal the truthmakers for the relevant facts. The sentences describing preventings and omissions are true, properly understood, but they do not concern straight causation in the actual world, contrary to first impressions.[15]

The first hurdle is negative events. Negative events are non-happenings – certain things not happening, and negative facts are things not being a certain way. In everyday language negative events sometimes appear in causal descriptions; for example, 'not washing your hands caused the spread of bacteria'. Omissions are negative events, as in the example just given.

Sometimes apparently positive events are really negative events; for example, 'missing selection' means 'not being selected'. Some of the examples given above fall into this category. 'The room being dark' means 'the room not being light', and if we wish to describe things in terms of quantities such as energy, the latter is the preferred description. Thus, on redescription we can see that a case like 'pulling down the blind causes the room to be dark' is really a case of preventing: 'pulling down the blind prevents the room from being light'. In fact, both omissions and preventings involve negative events, and redescription of these problematic cases will reveal, firstly, that there are negative events or facts involved in a prima facie causal story.

Next, we turn to *preventings*, which are more straightforward than omissions. As the most basic form of this type of case, consider three billiard balls shown in Figure 1(a), where A is the motion of the first ball, B the motion of the second, and C the motion of the third; and call the interaction between the first two balls AB, and the interaction between the first and third balls AC. Suppose we have the claim '*C prevents collision AB*'. As we have seen, 'preventing x' is equivalent to 'causing not x'. So the preventing can be written as:

I C caused not AB.

I claim that all preventings should be written this way; for example, 'pulling down the blind caused the room to not be light', 'flicking the switch caused the light to not be on'. But, even if this is granted, we still face the problem that C and not AB are not linked by an appropriate set of processes and interactions. In addition we have the puzzle that non-events are effects of actual causes. How can this be?

I suggest that the apparently straightforward causal claim I is actually substantially more complex than it looks. Since it is in part a claim about what does not happen, it can be redescribed, partly in terms of what would happen under non-actual conditions. In fact, the appropriate redescription of preventings always yields at least two propositions: a causal claim about the actual world, and a counterfactual claim, i.e. a claim about causation in a merely possible world. In our case the two propositions are:

Ia C caused interaction AC, and
Ib If AC had not occurred, A would have caused interaction AB.

Ib is a counterfactual conditional. Now, I wish to make no commitments here on how

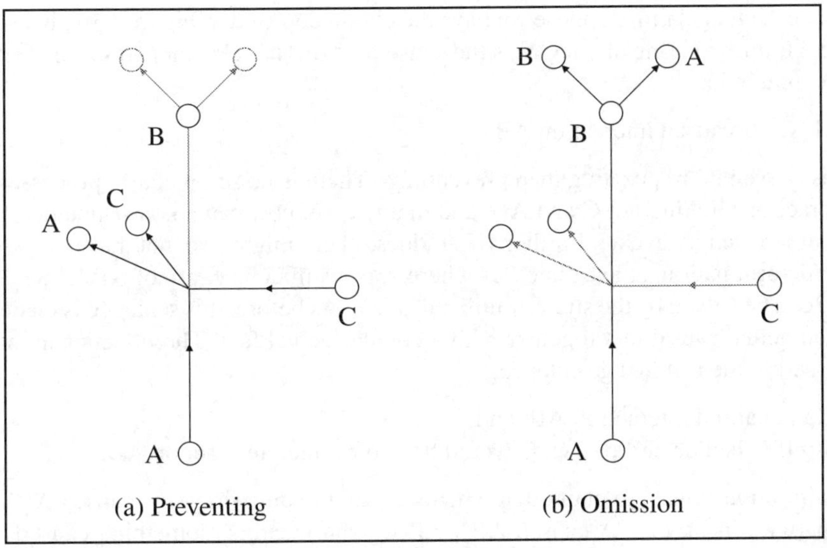

Figure 1: Preventing and Omission.

counterfactuals are to be analysed or explained. My account should work for any theory of counterfactuals.[16] However, leaving aside technicalities, Ib can be pictured as being about causation in a merely possible world, W*, a world just like ours up until the time of AC (just like ours with respect to particular matters of fact up until the time of AC, with respect to laws of nature, and with respect to any higher order facts that there may be), but where AC doesn't occur. At W* we have the straightforward causal claim, replacing Ib:

Ib* At W*: A caused interaction AB.

Ib* is a straightforward causal statement involving causal processes. In short, statement I is successfully redescribed as a statement about causation in the actual world, together with a statement about causation in a nearby merely possible world. If both these statement are true in their respective worlds, then statement I is true. However, I is not about direct causation between C and B in the actual world. In this sense, Ellis-eliminativism is right. 'Pulling down the blind causes the room to be dark' is not a straightforward causal claim in the sense of 'pulling up the blind causes the room to be light'. Yet the judgement is true, properly understood, so eliminativism is in that sense wrong.

So, in general, 'C prevented B' will translate to something of the form 'C caused interaction AC; and if AC had not occurred, then A would have caused B'. For example, what makes true 'pulling down the blind causes the room to be dark' is 'pulling down the blind causes the light to be reflected; and if the light were not reflected then the room would be light'.

This is too crude. It can't be part of what we mean when we say "C prevented B" that A rather than something else was the cause of B . To sophisticate the account we should replace 'A' with 'some event'. But the crude model should suffice to indicate how preventings and omissions are to be handled.

Omissions are a little more complex, because they involve a nested or second order counterfactual claim. Suppose we have the claim, depicted in Figure 1(b), that someone's failure to bring about C was the cause of AB. The relevant part of this fact can be expressed as:

II Not C caused interaction AB.

This is even more puzzling than preventings. There is no set of causal processes and interactions linking not C and AB, and in any case, how can a non-happening bring about an actual event? Further, if it does, then might we not have a case of overdetermination, at least in cases where A is a sufficient cause of AB? These puzzles can be solved by the strategy utilized above. As before, this sentence is made true by an actual causal fact together with a counterfactual fact. These facts may be expressed in the following sentences:

IIa A caused interaction AB, and
IIb If C had occurred, then C would have prevented interaction AB.

Again, I don't mean to claim that when we say 'the omission of C caused AB', that in part we mean that 'A caused AB'. All we can mean is 'something caused AB'.

However, the actual truthmaker of our claim, which is what we are interested in here, is the fact that A caused AB.

IIb is a claim about world W′, which is a world like the actual world except that C does occur, causing interaction AC, and hence interaction AB does not occur. Thus, IIb can be replaced by:

IIb′At W′: C prevented AB.

But notice that since IIb′ involves a preventing, rather than a straight causal claim, we still have the problem that there is no set of causal processes and interactions linking the cause and the effect. So the preventing at world W′ needs to be understood along the lines sketched above for preventings. Since preventings themselves involve a counterfactual, we have a nested counterfactual, a counterfactual truth *at* world W′. Thus IIb′ should be redescribed in terms of two propositions:

IIb′a At W′: C caused interaction AC, and
IIb′b At W′: If AC had not occurred, A would have caused AB,

and IIb′b can be written as

IIb′b′ At W″: A caused AB, where W″ is just like W′ up to the time of AC, but AC doesn't occur.

This doesn't violate the requirement that causes are connected by causal processes. (W″ is not the actual world because at W″ C occurs, whereas in the actual world C does not occur.)

In general, we can say the omission, 'not C', causes an event, 'interaction AB', just if A causes interaction AB and if C had occurred, C would have caused AC and AB would not have happened; and had C occurred but AC not, then A would have caused AB. So preventings and omissions can be treated as veiled counterfactual claims. To understand the true causal story, we need to make judgments about counterfactual situations, as well as about the actual situation. This involves recognising the veiled counterfactual claim. If this is right then we have answered the objection to the process account of causation.

How good an answer is this? Recall that this is supposed to be conceptual analysis which dissolves an apparent philosophical puzzle by clarification. While I will readily grant that this is no more than a rough outline of an answer, I won't accept that the approach is wildly implausible. To ease any worries about this, allow me to end with a short dialogue.

You say that the father's inattention was the cause of the child's accident. Surely you don't mean he literally made the child run into the path of the car, or the car to hit the child. Rather, you mean his failure to guard the child was the cause in the sense that if he had guarded the child, the accident would not have happened. You don't mean that he literally caused the accident; you mean it was possible for him to have prevented it.

Yes, that's what I mean. And you, when you talk of a scenario where the father stopped the accident, you don't mean that he bore any literal causal connection to a real thing called an accident. You mean that had he not acted the way he did, some

circumstances would have brought about the accident.
Yes, that's exactly what I mean.

University of Tasmania

NOTES

* I would like to thank David Armstrong, David Lewis, Hugh Mellor and an anonymous referee for comments which have led to improvements to this paper.

[1] One just needs a powerful enough light and a wall sufficiently large and sufficiently distant.

[2] See Salmon's clear exposition in his (1984, pp. 141–4). For the debate as to how to characterize distinguish causal from pseudo processes see Dowe (1992; 1995), Salmon (1984; 1994).

[3] Reichenbach called them "unreal sequences" (1958, pp. 147–9).

[4] One objection commonly raised to SI is the problem of 'temporary intrinsics', in Lewis' phrase (1986, pp. 202–4). If an individual can have contradictory intrinsic properties at different times, how can it exhibit strict identity? In reply, I would say that such properties are time indexed, and that there is no contradiction in having one intrinsic property at one time and another at a different time. One way to explicate this, mentioned (but not endorsed) by David Lewis is to treat these properties as disguised relations–relations to times.

[5] David Lewis points out (conversation) that things aren't equal. We can state my claim without using identity by using a property of totality: there is a thing and the whole of it is located at t1 and the whole of it is located at t2. In any case, if we are really talking about strict identity then that doesn't require further explanation (defense maybe, but not explanation) since nothing is clearer than self identity (Lewis, 1986, pp. 192–3.)

[6] pp. 25–6 draft.

[7] For example, Armstrong (1980); Lewis (1983, ch 5); Nozick (1981); Parfitt (1984); Shoemaker (1984, ch 11); Swoyer (1984). "Virtually all leading contributors to the current literature on personal identity . . . subscribe to it", according to Kolak (1987, p. 339).

[8] For example, Ehring (1991); Hirsch (1982, pp. 218–25); Kolak and Martin (1987).

[9] My thanks to an anonymous referee for this example.

[10] My formulation of this response was spurred by some comments by Michael Tooley (personal communication) on my first argument.

[11] In that version the 'items' were Richard Taylor and his twin.

[12] However, there is no relevant transfer of information from one to the other, in the sense that the state of the random creator is not the reason the replica has the state it does.

[13] This is equally a difficulty for the 'transference theories' of Aronson (1971) and Fair (1979), and for the 'force theories' of Bigelow (1988), Bigelow and Pargetter (1990) and Heathcote (1989).

[14] Michael Tye also considers this example (1982, p. 234) concluding that causation should not be analysed in terms of energy transmission.

[15] To my knowledge, the only other philosopher to defend in print anything like this intuition is David Fair (1979, pp. 246–7). However, I do not accept Fair's view that such truths are to understood as claims about 'plausible possible worlds', because that idea is subjective (relative to human beliefs), and I consider facts about preventings and omissions to be objective.

[16] The only restriction is that whatever are the truthmakers of our counterfactuals, they had better not include actual negative events. When I say my account should work for any theory of counterfactuals I mean this quite generally. The nature of states of affairs that I take to be the truthmakers of omissions depends on your theory of counterfactuals – if it involves modal realism then those states of affairs are existing nonactual states of affairs, if one holds an "actualist" account of counterfactuals such as Armstrong's combinatorial theory (1989), then those states of affairs are actual existing states of affairs.

REFERENCES

Armstrong, D.M. (1980), 'Identity Through Time', in P. van Inwagen (Eds.), *Time and Cause*, Dordrecht: Reidel, 67–78

Armstrong, D.M. (1989), *A Combinatorial Theory of Possibility*, Cambridge: Cambridge University Press

Aronson, J. (1971), "On the Grammar of 'Cause'", *Synthese* **22**, 414–30

Bigelow, J., Ellis, B., & Pargetter, R. (1988), 'Forces', *Philosophy of Science* **55**, 614–30

Bigelow, J., & Pargetter, R. (1990), 'Metaphysics of Causation', *Erkenntnis* **33**, 89–119

Dowe, P. (1992), 'Wesley Salmon's Process Theory of Causality and the Conserved Quantity Theory', *Philosophy of Science* **59**, 195–216

Dowe, P. (1995), 'Causality and Conserved Quantities: A Reply to Salmon', *Philosophy of Science* **62**, 321–33

Dowe, P. (1996), 'Backwards Causation and the Direction of Causal Processes', *Mind* **105**, 227–48

Ehring, D. (1986), 'The Transference Theory of Causation', *Synthese* **67**, 249–58

Ehring, D. (1991), 'Motion, Causation, and the Causal Theory of Identity', *Australasian Journal of Philosophy* **69**, 180–94

Ellis, B. (1990), *Truth and Objectivity*, Oxford: Blackwell

Fair, D. (1979), 'Causation and the Flow of Energy', *Erkenntnis* **14**, 219–50

Goldman, A.I. (1977), 'Perceptual Objects', *Synthese* **35**, 257–84

Heathcote, A. (1989), 'A Theory of Causality: Causality = Interaction (as Defined by a Suitable Quantum Field Theory)', *Erkenntnis* **31**, 77–108

Hirsch, E. (1982), *The Concept of Identity*, New York: Oxford University Press

Hitchcock, C. (1995), 'Salmon on Explanatory Relevance', *Philosophy of Science* **62**, 304–20

Kistler, M. (1998), 'In Defense of the Transference Theory of Causation', *Erkenntnis* **48**, 1–24

Kolak, D., & Martin, R. (1987), 'Personal Identity and Causality: Becoming Unglued', *American Philosophical Quarterly* **24**, 339–47

Lewis, D. (1983), *Philosophical Papers Volume 1*, Cambridge: Cambridge University Press

Lewis, D. (1986), *On the Plurality of Worlds*, Oxford: Basil Blackwell

Musgrave, A. (1977), 'Explanation, Description and Scientific Realism', *Scientia* **112**, 727–42

Nozick, R. (1981), *Philosophical Explanations*, Cambridge, Mass.: Harvard University Press

Parfitt, D. (1984), *Reasons and Persons*, Oxford: Clarendon

Reichenbach, H. (1958), *The Philosophy of Space and Time*, New York: Dover

Russell, B. (1913), 'On the Notion of Cause', *Proceedings of the Aristotelian Society* **13**, 1–26

Russell, B. (1948), *Human Knowledge*, New York: Simon and Schuster

Salmon, W. (1984), *Scientific Explanation and the Causal Structure of the World*, Princeton: Princeton University Press

Salmon, W. (1994), 'Causality Without Counterfactuals', *Philosophy of Science* **61**, 297–312

Salmon, W. (1997), 'Causality and Explanation: A Reply to Two Critiques', *Philosophy of Science* **64**, 461–77

Shoemaker, S. (1984), *Identity, Cause and Mind*, Cambridge: Cambridge University Press

Swoyer, C. (1984), 'Causation and Identity', In P. French, T. Uehling, & H. Wettstein (Eds.), *Midwest Studies in Philosophy IX*, Minneapolis: University of Minnesota Press, 593–622

Tooley, M. (1987), *Causation: A Realist Approach*, Oxford: Clarendon Press

Tye, M. (1982), 'A Causal Analysis of Seeing', *Philosophy and Phenomenological Research* **42**, 311–25

KEVIN B. KORB

PROBABILISTIC CAUSAL STRUCTURE[1]

"The word 'cause' is so inextricably bound up with misleading associations as to make its complete extrusion from the philosophical vocabulary desirable."—B. Russell (1912)

1. THE HUMEAN WORLD

The concept of *causation* is critical to our understanding of how we can understand our world, because causal relations just are the rules by which our world operates. Bertrand Russell supposed early in this century that the concept could be done away with since the product of science appeared in every case to be no more nor less than a set of functional relations – an axiomatized theory in which no primitive predicate of the form 'Cause(X, Y)' appears. He later abandoned this view and with excellent reason: even if causation does not figure as a subject matter of any science,[2] it figures *essentially* as a predicate in the metatheory of science and particularly in any plausible accounting of how scientists may come to learn theories that are true of the world. The unhappy consequence of this fact is that in order to come to an understanding of epistemology – how anyone can learn anything about the world in which one finds oneself – we must come to some understanding of the nature of causation, an understanding which has proved highly elusive. Here I shall review some of the recent philosophical literature on the subject of causation, with special emphasis on what appears to be clearly the most promising avenue of analysis: the probabilistic conception of causality. In the end I will offer a criterion of causality which improves upon all those reviewed and which may serve to clarify the role of controlled experimentation in science and which may also serve in attempts to automate causal reasoning within an artificial intelligence and, in particular, to automate reasoning to causal models from data (causal inference) and reasoning with causal models during prediction or planning.

David Hume presented a deflationary account of causation. As an empiricist, believing that all justifiable synthetic knowledge is constructed out of sense impressions, he held that for causal relations between events to be known they would have to either be directly perceptible or be logical constructions out of what is directly perceptible. "Looking around" revealed to Hume that causal connections are *never* directly perceptible: although the consumption of bread and the subsequent state of satiety are perceptible, no connection between the two is; further, what follows a "known" cause may be imagined to be any state of affairs whatsoever and not just its putative effect. Indeed, according to Hume, there can be no way even in principle to distinguish

H. Sankey (ed.), Causation and Laws of Nature, 265–311.

universal regularities from universal laws – for example, there is allegedly no way to distinguish the assertion that all unsupported objects of a certain class fall to the ground from the assertion that unsupported objects in that class near the earth are *attracted* by the earth according to Galileo's law of free fall. Hume admitted that we appear to be compelled, given the idea of a cause, to think of its putative effect; but his claim was that this is a fact about human psychology, not of the world known. Hence, Hume proposed the following definition of causality in Book I of his *Treatise*:

> We may define a CAUSE to be an object precedent and contiguous to another, and where all the objects resembling the former are placed in like relations of precedency and contiguity to those objects that resemble the latter. (Hume, 1739)

Thus,

(D1) C *causes* E iff (if and only if)

1. C precedes E
2. C and E are contiguous
3. All similar C's are so related to some similar E

Each of these conditions has been disputed at some point or other, but Hume's account of causality has nonetheless set the stage for every subsequent discussion of causality, at least until very recent times. The majority of philosophers, until perhaps fifteen years ago (indeed, perhaps to the present), have been Humeans in the sense that they have believed that the concept of causal relation is non-primitive and, in particular, that it should be defined in terms of regular associations between events classified as causes and other events classified as effects. And the determinism that underwrites the Humean view has been itself well-nigh universal. Similarly widespread is the empiricist, reductionist view that non-mathematical concepts must be coextensive with some set of perceptibles or fully analyzable into the language of perceptibles, even while the logical positivist program to produce such analyses (such as Carnap's *Der logische Aufbau der Welt*, 1928) has been well and truly rejected as unachievable, not just in practice, but in principle (e.g., by Carnap himself, 1936/37).

Here I present what has been the minority view, that causality is a primitive relation, not fully reducible to empirical relations more congenial to a positivist metaphysician. Objections to Humean metaphysics have always been present, indeed they have largely fueled the discussion of causality as Humeans have twisted and turned in their attempts to find a set of conditions which somehow avoid ridiculous implications while capturing a reasonable class of what are apparently causal relationships.[3] Already Thomas Reid in the eighteenth century pointed out that Hume's definition failed to distinguish causal regularities from "spurious" cases: for example, night precedes day, night and day are contiguous, and these are (or could be) always true – hence, on Hume's definition above night causes day! Of couse, Hume's definition can be tweaked, but another counterexample is found, and so on and on until we reach the pinnacle of analytic attempts at the problem – John Mackie's (1965) INUS condition of causality (a cause is an "insufficient but necessary component of an unnecessary but sufficient condition" of the effect), which is a binary order of magnitude more complex than its predecessors.

Since all of these analyses assume at core that there is nothing more to causality than universality, all of them have failed in the end to exclude accidental or spurious relationships. As a symptom of such failure, they fail in some cases to distinguish between causes and effects – that is, what is obviously an *effect* may well satisfy the conditions these analyses put upon causes (e.g., the shadow of the Empire State Building "causes" the height of the building on some such analyses, in an example of Bromberger, 1970).

A second difficulty with all of these analyses is that they assume determinism. It may be that Laplacean determinism is just true: that given a fully specified state of the world at some time and all the laws true of the world, the subsequent states of the world are simply logically implied. But even if true, this is not true *a priori*. For example, a natural interpretation of quantum mechanics (QM) is that it supports an indeterministic universe; whereas many problems have been raised for QM, very few have had the temerity – in the face of its very substantial experimental support – to argue that QM should be rejected because determinism is true. We can also use a simple Humean argument to see that whether or not determinism is true, it is not *a priori* true. It is trivial to describe a world which is *not* deterministic: let a world W have the (completely specified) state A, followed by state B, followed by state A, followed by state C. This is not an incoherent sequence; and it is also not deterministic, since the same state A fails to determine its immediate successor state. We do not know, prior to investigating, whether *our* world is W or not. But that is exactly the point! If we do not know *a priori* whether the world is deterministic, our analyses of causality have no business deciding the matter without reference to the results of science. All of the Humean analyses of causality, including Mackie's, have simply *assumed* that determinism is true, and therefore they are all of them philosophically inadequate.

In recent times, initially in the work of Reichenbach (1956), Good (1961–62) and Suppes (1970), and subsequently in the writings of Salmon (1980) and Cartwright (1983), an explicitly indeterministic view of causality, *probabilistic causality*, has been advocated and developed. The probabilistic conception of causality has an immediate advantage over any deterministic conception: even if determinism turns out to be true, a probabilistic conception may still apply, since extreme probabilities are probabilities; however, if indeterminism is true, no deterministic analysis of causality is likely to make sense of the world.

The probabilistic view, particularly in light of the indeterministic laws of quantum mechanics discovered in the first decade of this century, has had severe difficulty getting a hearing. This may be in part due to the fact that science progresses in many cases precisely by replacing indeterministic relations with something closer to universal relations. Thus, in the middle ages people had no clue what distinguished those houses where the Black Death struck from those where it did not; similarly, we can now predict genetically caused brain disorders, such as Huntington's disease, much more exactly than before. In general, science progresses by reducing uncertainty about the world and thus by pushing the probabilities in stochastic laws towards one or zero.

However, to move from this observation – important as it may be for methodology – to the metaphysical conclusion that *all* scientific laws are in principle deterministic requires, as Salmon puts it (1980, p. 137), a *deterministic leap of faith which is as pointless as it is unjustified.* The deterministic faith is pointless because: First, as we've seen, it is an *a posteriori* affair whether this world is deterministic or not; hence we shall in any case have to provide an analysis of causality that is consistent with indeterminism. Second, as a practical matter it is highly implausible that our sciences will develop to the point of being able to select out and assess *all* causally relevant variables in *every* causal system of interest. How many variables are relevant to the flight of a golf ball? Of course, there's impact momentum, spin, air pressure, wind, humidity, the golfing surface, etc. But if you consider that every air particle disturbed in turn influences the flight of the ball, you will see that a *precise* accounting is quite hopeless. Michael Scriven's favorite example is also worth noting: syphilis is known to cause paresis, in the perfectly normal sense that first having an untreated case of syphilis is the only known way of subsequently succumbing to the brain disorder paresis. On the other hand, only about 20 per cent of untreated cases go on to paresis, among men who live 20 or more years after the syphilis infection (Merck, 1992). The determinist will say: Yes, but there is some unknown ingredient, some additional medical condition, which distinguishes the 20 per cent from the rest and medical science will some day uncover it. But this is a deterministic act of faith indeed! What if there simply is no such further condition? What if it is simply a brute fact that the syphilis-paresis connection is stochastic, precisely as some quantum state relations appear to be? Would we then be forced to conclude that paresis is uncaused? I prefer to say that any analysis which does force such a conclusion fails to do justice to our pre-analytic notion of causation.

The deterministic faith, in claiming the reducibility of *all* possible sciences to a deterministic bedrock, is, as Salmon also noted, entirely unjustified: what would justify such a sweeping faith would have to be an equally sweeping range of supportive evidence. If deterministic analyses were successful wherever they had yet been attempted, then we would have some *weak* inductive evidence on its behalf. But so far from that we find many cases where deterministic analyses have been abandoned in favor of stochastic analyses: statistical mechanics, population genetics, chaotic systems, econometrics, medicine, engineering, and – not least – quantum physics. The idea that these many failures of deterministic analyses ought to be construed as support for determinism is, mildly put, a curious notion.

My plan for this article then is to begin by reviewing the discussion of probabilistic causality, somewhat in the style of Salmon (1980), but bringing the story current.[4] This is an interesting story in its own right, again one of postulated definitions of causality and of counterexamples to the same. It is not always clear whether the participants are attempting to eliminate causal language in favor of probabilistic language, but either way there appears to be determinable progress towards a better accounting of causality. In the end I proffer my own probabilistic analysis of causality which I believe useful and interesting, not because it eliminates the concept of causality,[5] but because it illuminates methodological issues about science and about automating scientific methods in artificial intelligence.

2. PRELIMINARIES

In order to present this story in a readable way we need to adopt certain conventions about notation and clarify the intended way of understanding certain crucial phrases such as 'probability' and 'statistical relevance'.

2.1 *Probability*

First, I very briefly set out my position (without defence) on the interpretation of probability. I concur with Rudolf Carnap (1962), and many others before and after, that there are (at least) *two* concepts of probability that are relevant to concerns about inductive inference: epistemic (or subjective, or Bayesian) probabilities and physical probabilities. If causal structure is probabilistic and if it describes a world that is largely independent of the epistemic agent, then it can hardly be that agent's changing subjective probabilities that figure in the causal structure. What Bayesian probabilities describe are the learning processes that (idealized) agents undergo; what physical probabilities describe are the systems about which they learn. I agree with David Lewis's (1980) Principal Principle that what connects them in a rational agent is the willingness to overwrite subjective by physical probabilities when the latter become known. Finally, I believe that physical probabilities are best understood via the propensity interpretation of probability, due to Popper (1959); that is, physical probabilities at ground just describe the causal tendencies of systems to undergo change. (For a defence of these views, see Korb, 1992.) In the sequel, where probabilities are mentioned, physical probabilities will be meant, unless a different interpretation is specifically required.

2.2 *Notation*

What we are primarily interested in here is making sense of such locutions as '*A* causes *B*', where *A* and *B* designate events or, perhaps, *kinds* of events. I shall not be much worried about just what events may be; they are things like floods, eating soup, supernovae, etc. A more exact accounting is beyond my metaphysical abilities. A more practical concern is keeping *token events* distinct from *event types*: The confusion of the two in fact figures in the discussion below. A *token event* is an individual occurrence: for example, my eating some soup *now*. An *event type* is a class of such occurrences, such as all instances of my eating soup, or all instances of *anyone* eating soup. It is proper to report that a token event *caused* (in a specific place and time) another token event, and it is proper to report that an event type *causes* (e.g., everywhere and everywhen) a different event type – but, as one can see, subject and verb need matching.

In order to keep the distinction clear, I adopt the convention of using calligraphic letters, such as $\mathcal{A}$, $\mathcal{B}$, . . . , $\mathcal{Z}$ to identify event types, factors, or variables occurring in general causal laws; I may abbreviate by calling these 'events', 'causes' and 'effects', but so long as they are in calligraphic typeface, they must be *types*. An instantiation of an event type $\mathcal{A}$ will be designated by *A* and their relation by $A \in \mathcal{A}$. If not explicitly varied the letters $\mathcal{C}$ and $\mathcal{E}$ denote a cause event type and an effect event type respectively. And I shall use the following further notation:

$\mathcal{C} \rightarrow \mathcal{E}$ indicates that $\mathcal{C}$ causes $\mathcal{E}$.
$\mathcal{C} \rightarrow \neg\mathcal{E}$ indicates that $\mathcal{C}$ prevents $\mathcal{E}$ (or, equivalently, $\mathcal{C} \not\rightarrow \mathcal{E}$)

I shall also say that C is a positive causal factor for E in the first case, or a negative causal factor in the second, or simply a causal factor in either case.

2.3 *Statistical relevance*

Marginal probabilities, such as $P(A)$, describe a "background" probability of a type of event (or, its prior probability). The conditional probability of one event given another is written $P(A|B)$ and is understood as the probability of an event of type A occurring when an event of type B has occurred.[6] Given these tools we can define the *statistical relevance* of B to A:

$$SR(A,B) = P(A|B) - P(B)$$

From this definition if A and B are positively correlated, negatively correlated or independent, then their statistical relevance will be positive, negative or zero respectively. It should be noted that statistical relevance is symmetrical. In the case of positive statistical relevance B raises the probability of A and B lowers the probability of A in the case of negative statistical relevance. How exactly statistical relevance bears upon causal relevance, and vice versa, is of course the main subject matter of this paper.

2.4 *Objective homogeneity*

The concepts of marginal and conditional probability, statistical relevance, etc. apply equally to event tokens, in direct analogy to event types; e.g., $SR(A,B)$ identifies the value $P(B|A) - P(B)$ for two individual events. Just what these probabilities of individual events are will depend upon what their proper reference classes (or "chance setups") might be. I shall require that there be some *objectively homogeneous* reference class for any such event – i.e., that there is an event type containing the event in question such that no statistically relevant subtype can be identified prior to that event.[7] If, on the contrary, the probability supported by a given reference class can be altered by looking at an available subclass, then the initial probability reflects our ignorance about which available subclass the event is in, rather than reflecting the underlying causal structure of the event. To take a classic case of physical probabilities consider dice: we would ordinarily take the probability of a die landing with a 6 up to be one-sixth, taking the class of all potential rolls of the given die as the right reference class. But we would change our minds about this were we to discover that when the die starts out with the 6 up and we roll it in some particular fashion then the 6 comes up, say, one-fifth of the time – for then we have found a statistically relevant way to partition the initial reference class *prior* to the event in question. Hence, our initial reference class did not determine the physical probability of the event, but merely identified the limits of our previous knowledge.

I understand causality as giving rise to physical probabilities and, therefore, involving objectively homogeneous reference classes. It is a different question whether we have in hand the right reference class for a problem. If we do not, then what we believe is the causal structure of the system will differ from the true causal structure, which is a divergence that empirical testing can reveal.[8]

Symbolically, what I am requiring for objective homogeneity is that for any event

A there is an $\mathcal{A}$ such that $A \in \mathcal{A}$ and for any $\mathcal{A}' \subset \mathcal{A}$ (where $A \in \mathcal{A}'$ and this fact is epistemically accessible prior to A) $P(\mathcal{A}') = P(\mathcal{A})$; and I take $P(\mathrm{A}) = P(\mathcal{A})$. For discussions of objective homogeneity, single-case propensities and related issues see Popper (1959), Giere (1973), and Salmon (1984).

2.5 *Conditional independence*

Two dependent events $\mathcal{A}$ and $\mathcal{B}$ are conditionally independent with respect to a third event $\mathcal{C}$ iff the following equality is true: $P(\mathcal{B}|\mathcal{A}.\mathcal{C}) = P(\mathcal{B}|\mathcal{C})$. That is, given knowledge that $\mathcal{C}$ the addition of knowledge that $\mathcal{A}$ does not affect the probability of $\mathcal{B}$. This is also a symmetrical relationship and furthermore is equivalent to the following decomposition: $P(\mathcal{A}.\mathcal{B}|\mathcal{C}) = P(\mathcal{A}|\mathcal{C})\ P(\mathcal{B}|\mathcal{C})$. In the philosophical literature, following Reichenbach, this relationship is often called *screening off*: for example, if you learn of a low barometer reading this will suggest a probability of a rain storm, but if you also learn (infallibly, somehow) what the air pressure actually is, this will screen off the barometer reading – the barometer reading is providing no additional information, or rather less information, since the barometer may be inaccurate.

With these conceptual tools in hand we are ready to start examining strands in the history of the concept of probabilistic causality.

3. PROBABILISTIC CAUSALITY I

3.1 *Reichenbach*

In *The Direction of Time* (1956) Hans Reichenbach undertook to produce a causal theory of time – that is, his goal was to define time in terms of causal relations, on the plausible grounds that time just is a measure of change in the world and all changes are the result of causal processes. One of his main goals therefore, unlike those of others to come, was to extract temporal asymmetry from causal asymmetry; in other words, rather than *assume* or *prescribe* that a cause precedes its effect, Reichenbach attempted to define temporal precedence by reference to causal structure.

Underlying Reichenbach's work, and my own, is the following matter of faith, which I am happy to set against the faith of determinists: that enduring correlations between event types are not magical. There are plenty of correlations to be found in the world that are accidental – the pronouncements of Wall Street gurus come to mind: when a "guru" has been right about the direction of the last ten market moves, this is likely better explained by the number of gurus around vying for attention than by any great insight of the person at hand. Such correlations are not enduring and need no causal explanation. Other correlations are not only enduring but appear to indicate some direct causal connection: for example, there is a correlation between evidence of major asteroidal impacts and evidence of major extinctions on the earth. But the most problematic correlations are those which appear to have no direct connection. These latter have frequently been misused by statisticians intent on arguing that causal relations cannot be inferred from correlational relations, and are commonly called cases of "spurious correlation" or "spurious causality." The faith I am preaching here is that there is nothing spurious about either the correlation (that should be obvious!) or the causality (which should be less so). This faith is enshrined in Reichenbach's *Principle of the Common Cause* (1956, p. 157): "If an improbable coincidence has occurred, there must exist a common cause." (It is a bit odd that Reichenbach ignored here the possibility of explanatory *chains* of causation, but presumably he was most interested in what is most problematic.)

For example, suppose we find that there is a correlation between ice cream consumption in Rome and baseball scores in the United States. There are a large number of possibilities, but only a few which are likely, given a passing acquaintance with the world we are in: one of these is that the association is accidental. This can be tested by increasing the sample size. Suppose we are satisfied that the effect is not accidental. Since it is not likely that there is a causal chain of events leading from one of these to the other, we are left with either a common cause, such as warm (summer) weather in the northern hemisphere, or magic. The faith in question asserts that magic does not exist. In other words, since we have accepted that the association is enduring, it must be explained in terms of the events in question being embedded in some causal structure – either a chain between the two (which we've ruled out in this case) or a common ancestral event (which here would obviously be the case). Either way, of course, there is nothing spurious about the causality relating the two events; the only valid point anti-causal statisticians can make is that where there is (enduring) correlation there may not be a *simple-minded, direct* causal connection between them. But then no one advocating causal inference from correlational structure has been advocating simple mindedness.[9]

The faith I promote here is simply that which underwrites all scientific research: that associations between events are to be explained causally in some manner or other. This faith may in the end let us down, because the world, or some part of it, may in the end be inexplicable. But that point of view, of course, is one that we will arrive at only by being driven to it by a recalcitrant world, not by a statistician's metaphysics.

In order to understand the probabilistic nature of common causes Reichenbach introduced the concept of a *conjunctive fork*:

(D2) $\mathcal{C}$ is a *conjunctive fork* with respect to $\mathcal{A}$ and $\mathcal{B}$ iff

1. $P(\mathcal{A}.\mathcal{B}|\mathcal{C}) = P(\mathcal{A}|\mathcal{C})P(\mathcal{B}|\mathcal{C})$
2. $P(\mathcal{A}.\mathcal{B}|\neg\mathcal{C}) = P(\mathcal{A}|\neg\mathcal{C})P(\mathcal{B}|\neg\mathcal{C})$
3. $P(\mathcal{A}|\mathcal{C}) > P(\mathcal{A}|\neg\mathcal{C})$
4. $P(\mathcal{B}|\mathcal{C}) > P(\mathcal{B}|\neg\mathcal{C})$

Reichenbach's idea was that these relations fully capture the probabilistic correlate of a common cause; that is, that conjunctive forks are the "statistical model" of common causation. We can see that the conditions for conjunctive forks are plausibly applied to common causes. Conditions (1) and (2) simply require that two effects be conditionally independent given the state of their common cause, as we saw in the case of the barometer, for example. Conditions (3) and (4) express the ordinary understanding of causes, that they are positively related to their effects, rendered in the language of probabilism as positive statistical relevance.

With the concept of *causal betweenness* Reichenbach extended his probabilistic analysis to causal chains:

(D3) An event B is *causally between* events A and C iff

1. $P(C|B) > P(C|A) > P(C)$
2. $P(A|B) > P(A|C) > P(A)$
3. $P(C|A.B) = P(C|B)$

Conditions (1) and (2) again incorporate the plausible notion of positive statistical relevance for causally related events. If there is a sequence of event types which cause each other in turn, and if raising probabilities is one way in which causal connections are manifested, then the middle variable should be probabilistically more closely related to each end variable than the ends of the chain are to each other. Condition (3) is a special case of the *Markov property*: that where two events are causally related only through a mediating event or set of events then they are conditionally independent given the state of the mediating event(s). For the Markov property to make sense the mediating variables must be exhaustive, there must be no hidden connection between the end events. For example, a dirty kitchen may lead to food poisoning by way of ingesting the food on the menu; eating the menu items (with some specified level of bacterial contamination) will render the health of the customer and the state of the kitchen conditionally independent (via the Markov property), but only so long as there is no way for customers to *otherwise* come in contact with the kitchen, e.g., through off-menu items of food.

A more precise account of the Markov property may be provided via Judea Pearl's concept of *d-separation* (Pearl, 1988). In a directed acyclic graphical (dag) model of causal structure a set of variables $\mathbf{Z}$ d-separates two variables X and Y just in case for each undirected path between X and Y the set $\mathbf{Z}$ either (i) contains a common cause of X and Y (occurring in that path), or (ii) contains a mediating variable in a causal chain connecting them, or else (iii) *fails* to contain a common effect of X and Y (or any descendent of such a common effect). D-separation records the conditional independence properties of common causes and causal chains that Reichenbach captured in his definitions above in (i) and (ii) but also records the conditional *dependency* induced by common effects. An example of the latter phenomenon would be a case where the probability of tuberculosis and influenza are marginally independent of each other, however if we know of a patient (only) that he or she has a severe cough, we get a dependency: learning the patient has TB would then reduce the probability of the patient having influenza, and vice versa. Reichenbach anticipated d-separation; indeed, it was precisely the asymmetry between common causes inducing independence and common effects inducing dependence that he relied upon to introduce the time asymmetry of causation. The point of immediate relevance to our discussion is that Pearl's concept of d-separation supplies a criterion of probabilistic independencies among the variables, but only so long as the causal model in fact corresponds to the real causal structure appropriately – that is, all of the independencies implied by the model must be real independencies. When this holds true, we shall say that the causal model has the Markov property (Pearl says that the model is an "independency map", or I-map, of the true causal structure).

The Markov property requires then that a causal model be in some sense complete. Assuming that our causal models have the Markov property allows for very substantial simplifications. For a simple example, in a graphical representation of a probability distribution in three variables connected as a chain we can compute the full joint probability of $P(A.B.C)$ as $P(A|B)P(B|C)P(C)$ instead of as $P(A|B.C)P(B|C)P(C)$. Although the savings in time and space complexity are small in this example (substituting $P(A|B)$ for $P(A|B.C)$), they grow in potential as the complexity of the model grows and such savings are the primary source of the explosion of interest in graphical methods of probabilistic reasoning in recent years (see Pearl, 1988, and Neopolitan, 1990).

Wesley Salmon, Reichenbach's student, has launched highly influential attacks on both of these key elements of Reichenbach's analysis, that causal relations induce positive statistical relevance and that a complete causal account of a system will have the Markov property. I believe both of Salmon's criticisms are wrong, but before getting into that argument I first present Suppes' probabilistic analysis of causality, since it is very closely related to that of Reichenbach.

3.2 *Suppes*

Patrick Suppes (1970) was more directly concerned than Reichenbach with reducing the concept of causality to probabilistic relations between events. Since it was not his goal to analyze time, he was able to employ temporal precedence directly, as had Hume, in the definition of what he called *prima facie causality* (1970, p. 12):

(D4) C is a *prima facie cause* of E iff

1. C precedes E
2. $P(C) > 0$
3. $SR(C,E) > 0$

Note that I have been forced out of the type notation because event types do not *occur* and so do not occur prior to one another. Only token events can be *prima facie* caused or *prima facie* causes.[10]

An event C satisfying these conditions may be causal *prima facie* only because it is correlated with E via a common ancestral event – i.e., C and E may merely be common effects. A *genuine cause* therefore will be a *prima facie* cause which is not screened off in that fashion (1970, p. 25):

(D5) C is a *spurious cause* of E iff C is a prima facie cause of E and there exists a partition of events $\{F_i\}$ at some time prior to C such that for each event F_i:

1. $P(C.F_i) > 0$
2. $P(E|C.F_i) = P(E|F_i)$
3. $P(E|C.F_i) \geq P(E|C)$

The introduction of a partition of prior events is Suppes' attempt to identify that the screening off is being done by an event *type* rather than a specific token event (thus, for example, the full state of the weather screens off the barometer from rainfall *whatever* that state may be). Genuine causes then are *prima facie* causes which are not spurious – which are not screened off by any prior factor and which are either directly or indirectly positively relevant to their effects.

3.3 *Negative relevance*

The first of Salmon's criticisms of Reichenbach's and Suppes' analyses that I consider (from Salmon, 1980) is the rather surprising notion that the cause of an event can be *negatively* related to its effect statistically. Now this is not the banal observation that *prevention* is a kind of causal process, but rather the more radical assertion that what *ought* to have prevented the subsequent event instead caused it to happen – and so that Suppes' condition (D4.3) and Reichenbach's (D3.1), both requiring positive statistical relevance of causes, are confused. To make the case that negative statistical relevance of positive causes is possible Salmon draws upon an example of Deborah Rosen (a student of Suppes): let *A* be Jones' teeing off at a par three hole, *B* be that Jones' shot hits a tree next to the green and *C* be the spectacular result that the ball bounces in for a hole in one. Salmon notes that $P(C|B) \ll P(C|A)$ – the probability of a hole in one given that the ball has hit the tree is far less than the probability of a hole in one given only that the person has teed off. But then we have the probability of an effect, the hole in one, given the intermediate factor of striking the tree, being less than the probability of the effect given the initial cause, teeing off. This violates causal betweenness – condition (D3.1) – and more than that, it violates the positive statistical relevance of *prima facie* causes since striking the tree was clearly the proximate cause of the ball falling in. Since the teeing off, the striking of the tree and the falling in the hole clearly form a single causal chain, Salmon reasons, any definition of causality which rules out any portion of that chain is defective. But since striking the tree is also clearly negatively relevant to the outcome, positive causal factors may be negative statistical factors.[11]

Attempting to save positive statistical relevance, one kind of response (adopted by Rosen herself, 1978) is to play upon the ambiguity of the event types implicit in the story. Thus, if we were to know *precisely* where on the tree the ball struck, at what angle and with what spin, etc., then perhaps the ball hitting the tree *in precisely that way* would no longer be negatively relevant to the ball falling in. I agree with Salmon's rejection of this line of defence: this substitutes for the deterministic faith a *pseudo-deterministic faith* – that some more precise rendering of the details of the story will always yield positive statistical relevance – which is distinguishable from the determinist's faith only to within an arbitrarily small error term. One can always hope that positive relevance will be thus regained, but there is little evidence that the world is so accommodating.

A second response has been that if we tighten up the initial reference class – the class of teeings-off – then, within the right reference class, *B* will turn out to be positively relevant. Thus, for example, if we restrict the reference class to cases where Jones has sliced off the tee and the ball is headed within some small angular distance of the tree, then the probability of a hole in one must be extremely small. It is quite plausible that within *this* reference class striking the tree will turn out to be positively related to holes in one. So much is true, but it fails to save positive relevance. It is a simple matter to rewrite the story so that negative statistical relevance is re-established, without in any way impairing the causal story. All that is needed is to erect a shaped wall behind the tree which *guarantees* a hole in one (as in miniature golfing) should the ball *miss* the tree. In that case striking the tree, given that the ball flew off in the direction of the tree, may: dramatically reduce the probability of a hole in one; cause a hole in one.

Even were the story of negative relevance irreparable, this second defence of positive relevance would be missing the point. As I emphasized previously, the probabilities of events are specified relative to a reference class, so that in effect any physical probability statement is implicitly indexed by a reference class which is picked out by some problem context. Furthermore, the reference classes involved must be homogeneous, since by definition an inhomogeneous reference class reflects at best our ignorance rather than something physically undetermined. For example, the probability of a normal coin flip turning up heads may be one-half. But the probability of heads given that I either choose to use a normal coin or choose a double-headed coin is undefined through lack of a homogeneous reference class.[12] Now it is arguable – indeed, I shall argue explicitly later – that causal statements are likewise true or false only relative to a problem context, including a reference class. If that is true, then the consequences of moves to change the reference class can tell us nothing about the truth of causal claims relative to the original reference class. Hence, the claim – even were it true – that striking the tree is a positive statistical factor for a hole in one in *some context or other* would fail to refute Salmon's claim that some positive causal factor is negatively statistically relevant to its effect *in one and the same context*. Salmon's claim is nevertheless wrong.

Before explaining why, let us look at one of the difficulties Salmon raises for Reichenbach's analysis of common causes, namely an example due to Crasnow (reported in Salmon, 1984). Suppose Ms. Brown normally goes to work on the 8.30 bus and arrives to an empty office and makes herself a cup of coffee. On some occasions she catches the 7.30 bus instead and arrives in her office to find the coffee already made by her secretary and a visitor waiting for her. This is a perfectly ordinary, unobjectionable state of affairs, and may well be true of some Brown. However, if we recast the story in Reichenbachian terms we get a most extraordinary story: since it is only when she takes the 7.30 bus that there is coffee ready and a person waiting, the catching of the 7.30 bus forms a conjunctive fork with the other two events. In other words, $\mathcal{C}$ is a conjunctive fork with respect to $\mathcal{A}$ and $\mathcal{B}$ according to (D2):

$\mathcal{A}$: The coffee is ready on arrival.
$\mathcal{B}$: Someone is waiting for Brown.
$\mathcal{C}$: Brown takes the 7.30 bus.

So, according to Reichenbach's analysis of common causes, Brown's taking the 7.30 bus causes the coffee to be made and a visitor to show up shortly thereafter. But it is obvious that the true common cause has been left out of the story: someone has made an early appointment the previous day (say) and Brown's actions, her secretary's actions and that someone's actions in the morning all follow from that earlier event. If in some fairy story we were asked to take the Reichenbachian account seriously, what would be demanded of us would be the belief that boarding a bus could somehow independently control the actions of distant people who are not directly aware of the boarding of the bus – through "mind waves" or some such magic (relative to our world).

Despite this apparently devastating objection it is an easy matter to fix Reichenbach's definition of conjunctive forks. First we may note that Suppes' analysis of causality already deals with the example appropriately: that is, in our elaboration of the Crasnow story there is a prior partition of events – namely, the states of the variable describing appointments logged by the secretary – which screens off the bus boarding behavior from coffee making and visitor presence; therefore, the bus boarding behavior will be recognized as a spurious cause rather than a common cause. But then all we need do to Reichenbach's definition is to enhance it with just this requirement from Suppes, that the common cause not be spurious; i.e., we add to (D2)

(D2.5) For every $A \in \mathcal{A}$, $B \in \mathcal{B}$, $C \in \mathcal{C}$ satisfying conditions (D2.1)–(D2.4) there is no $\mathcal{F}$ such that for any possible $F \in \mathcal{F}$ occurring prior to A,B, and C,F screens off C both from A and from B.

This is not the kind of condition Reichenbach could endorse, since it invokes temporal precedence, but that is no impediment to our adopting it.

A substantial difficulty with conjunctive forks remains, however. As is suggested by Reichenbach's Common Cause Principle, one of the primary reasons why conjunctive forks are interesting is that they explain why common causes correlate their effects; that is, from the definition of conjunctive forks it follows that the effects of a common cause are positively statistically relevant to one another. Yet we have lately seen that there are plausible, apparent counterexamples to positive statistical relevance in causal chains. We can also produce an apparent counterexample to the positive correlation of effects of a common cause. Suppose Mr. Smith learns that he is HIV positive. Suppose also that Smith has an unusual psychology, so that his response to this news is to set a really unlikely goal for himself to achieve, namely winning the New York marathon. Now winning the New York marathon is in any case an unlikely event, but however unlikely it is surely even less likely given that one has been HIV positive for eight years. Nonetheless, this is precisely what happens to Smith: after eight years of hard training, he wins the New York marathon. Even though eight years of being HIV positive and winning the New York marathon are clearly negatively statistically related, in this case becoming infected with HIV caused both of them to occur and so, by definition, is their common cause. Hence, Reichenbach's concept of conjunctive forks cannot provide a fully adequate criterion of common causation.

I assume that Salmon would be agreeable to this line of reasoning. Salmon (1984) holds that what has gone wrong with Reichenbach's and Suppes' analyses of probabilistic causality is that they are essentially attempts at *reducing* the concept of causality to probability (and time). What the fairy-tale interpretation of the Crasnow example suggests is that we need to be concerned not just with probability relations but also with the *mechanism* which induces those probabilistic relations. What Salmon insists upon is that in order to talk about A causing B we must be assuming that there is a causal process running through spacetime which *connects* A and B. If there is such a causal process, then it doesn't really matter what probability A confers upon B,A must in any case be a cause of B.[13] Although I agree that there must be a connecting causal process between any cause and its effect, this process is at most only half the story. There are huge numbers of causally irrelevant "causal processes" going on all the time. Thus, it is pretty clear that sunlight causes skin cancer. But since sunlight impinges on just about every event on the surface of the earth, on Salmon's account it seems that sunlight causes just about *everything* we experience: not just cancer, but recovery from cancer as well; not just global warming, but ice ages and asteroidal impacts. Indeed, the sun and stars are more influential apparently than in the wildest dreams of the astrologers![14]

Salmon is no doubt correct that Reichenbach and Suppes both suffer for attempting to substitute probabilistic for causal concepts and for ignoring the metaphysics of causal processes. But Salmon is guilty of an equally large error, that of ignoring the distinction between event *types* and event *tokens*. Each of the examples demonstrating the causal relevance of statistically negatively relevant events were examples of token causality. Hitting a tree may *in particular instances* cause golf balls to drop into holes, but even Salmon would have to gag at the thought that hitting trees generally causes people to get holes in one – or, more idiomatically, that the way to improve your golf score is to hit trees (even if some appear to be trying to follow such advice). Undoubtedly, causality does not simply reduce to a few probabilistic relations between events; but it is just as undoubted that *type* causality cannot be understood without recourse to probabilistic relations between events and, in particular, that it is just nonsense to say of two event types that one causes the other while lowering its probability.[15] Any event *type*, putatively causal, which lowers the probability of another event type is either going to be screened off by a common cause or it will be recognized as a *negative* causal factor – i.e., as a preventative factor.

It is not clear to me what the requirements are of token causal attributions. As we've seen, it is not enough to find chains of spacetime regions between cause and effect tokens which are continuous or contiguous. I must simply confess unclarity about the metaphysics of token causation; therefore, for the remainder of this paper I shall confine myself to dealing with type causation and its probabilistic and experimental aspects.[16]

3.4 *Interactive forks*

A second major objection Salmon raises against Reichenbach's and Suppes' analyses is to the screening off condition for conjunctive forks and for spurious causality. That is, Salmon argues that there are common cause situations where the ancestral event *fails* to screen off one effect from the other, even though there is no causal chain between the effects themselves. Such situations he calls *interactive forks*. If there are interactive forks, then systems which embody them cannot be modeled using Bayesian networks or causal models, since the interactive fork guarantees failure of the Markov property (the model cannot be an independency map, in Pearl's terminology). If our models of the world imply many non-existent independencies, then they will not be of much use to us, since the failure to model dependencies may distort predictions arbitrarily; indeed, it is just such unwarranted assumptions of independence which dogged AI implementations of reasoning under uncertainty prior to Bayesian net methods, from early medical applications (e.g., Dombel *et al.*, 1972) to Duda's so-called Bayesian PROSPECTOR (1978) to Buchanan and Shortliffe's "certainty factors" (neatly torn apart by Heckerman, 1986). In other words, if Salmon is right, then either we must find that interactive forks are few and far between or, it seems, we must abandon our goal of automating causal inference.

To support his claim, Salmon invokes the Compton scattering effect. Consider an electron with energy E_1 and a photon with energy E_2. If the photon collides with the electron, then there is a 10 per cent chance that the electron will emerge with energy $E'_1 \neq E_1$ and the photon with energy $E'_2 \neq E_2$, but there is a 90 per cent chance they will remain unchanged; let us call this the collision law. The law of conservation of energy requires $E_1 + E_2 = E'_1 + E'_2$ so that *if* the electron emerges with energy E'_1 then the photon *must* emerge with energy E'_2. Let C be the event of the photon colliding with the electron, A be the event of the electron emerging with energy E'_1 and B the event of the photon emerging with energy E'_2. We would expect C as a common cause of A and B to form a conjunctive fork with A and B, however this does not turn out to be the case:

$P(B|C.A) = 1$ by the conservation law; but
$P(B|C) = 0.1$ by the collision law

alternatively:

$P(A.B|C) = 0.1$ by the collision law; but
$P(A|C)P(B|C) = 0.01$

Hence, the collision does not screen off the energy levels of the electron and photon from each other. Acknowledging the possibility of such cases, we may define *interactive fork* in the same way as a *conjunctive fork*, but replacing

(D2.1) $P(A.B|C) = P(A|C)P(B|C)$

by

(D2.1′) $P(A.B|C) > P(A|C)P(B|C)$

while adding a proviso that there is no direct or indirect causal chain between A and B, since any such case would not be anomalous.[17]

The Einstein–Podolsky–Rosen (EPR) paradox is perhaps the most (in)famous example of an interactive fork (Einstein, Podolsky and Rosen, 1935). In their thought experiment a composite physical system is composed of two subsystems, I and II. The subsystems are allowed to interact in such a way that a physical quantity A pertaining to I and a physical quantity B pertaining to II are coordinated by a quantum mechanical wave function; components I and II are subsequently separated so that they *cannot* interact – i.e., they are space-wise separated – and are not otherwise disturbed. When physical quantity A of I is then measured, the coordinated value B of II can be predicted with certainty; and vice versa. However, there is no information available about the common cause (the interaction event) which suffices to screen off the state of one subsystem from the other (which is a consequence of Bell's theorem, Bell, 1966).

It appears that at the quantum level there just are such strange things as interactive forks. If that were the extent of the problem, then we could be reasonably happy restricting our automated causal methods to dealing with the macro world, and granting what we already knew, that the quantum world is strange indeed. However, Salmon argues that there exist similar phenomena at the macro level.

In his macro example Salmon asks us to consider a pool player who is attempting to sink the 8 ball in one of the far corner pockets. The cue ball and 8 ball are so positioned, however, that if the 8 ball drops then it is almost certain that the cue ball will fall in the opposite pocket. It is also assumed that the player has a 50 per cent chance of dropping the 8 ball, given that it is hit. Let C be the collision of the cue ball with the 8 ball, A the holing of the 8 ball, and B the holing of the cue ball. Then the probability of sinking the 8 ball given that it's struck is: $P(B|C) = 0.5$. However, the probability that it sinks given also that the cue ball drops in is: $P(B|C.A) = 1$. Hence, we have an interactive, rather than conjunctive, fork.

This kind of example, unlike Compton scattering, can be countered by demanding a more precise description of the common cause. That is, if the angle of collision, momentum and spin are described, rather than the mere fact that there is or is not a collision, then we will have a much more exact notion whether or not the 8 ball will go in – indeed, we can fairly easily obtain enough information so that knowledge of the fate of the cue ball becomes irrelevant. Nor can we, in answering thusly, be accused of that pseudo-deterministic faith that never sees an obstacle in replacing knowledge of the less certain with knowledge of the more: in this case it is quite clear that the additional information required for a conjunctive fork is available. If the uncertainties governing pool balls were so great that knowledge of collision angles and the rest could not tell us enough about the trajectory of the 8 ball, then it would be impossible in the first place to position the cue and 8 balls so that one will drop if and only if the other drops. That is, the ability to set up such trick conditions depends upon a significant practical understanding of spin, momentum etc. and strongly suggests the possibility of developing such knowledge of the methods of manipulating these conditions explicitly;[18] therefore, it appears to be a precondition of making sense of Salmon's example that it be amenable to treatment as a conjunctive fork.

Another proposed example of interactive forks is due to Davis (1988). Suppose we have a flaky switch on a TV set, so that when it is turned on sometimes the TV comes on and sometimes not. Then we will have an interactive fork in the following variables:

A: The screen comes on.
B: The sound comes on.
C: The TV is switched on.

Because, clearly, $P(B|C) < P(B|C.A)$, since whenever the screen *does* come on so too does the sound. Spirtes, et al. (1993), however, point out that this story is simply radically incomplete, leaving out a crucial missing variable:

D: The circuit closes.

Graphically, the TV system should be modeled as in Figure 1, where the flakiness is represented by a stochastic relation between C and D. Hence, we have D forming a conjunctive fork with B and A, rather than an interactive fork.

It might also be thought of my responses to the macro examples that it is improper for me to require the addition of a new circuit variable or the substantial refinement of Salmon's collision variable. After all, I noted previously that probabilities and causal relations are relative to a problem context and here I am evading a problem – the existence of interactive forks – simply by switching problem contexts. Of this I am guilty as charged. If Salmon's point is merely that when events are described at a

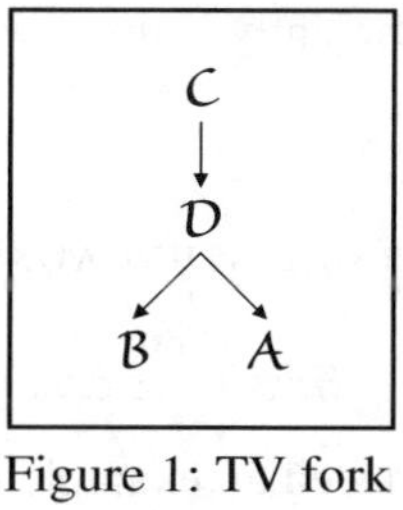

Figure 1: TV fork

sufficiently crude level, then interactive forks will *apparently* be found, then the point is valid. But what is also true is that causal models lacking key variables are misleading and that ill-specified event types do not in fact have physical probabilities associated with them. In Salmon's case, the crude class of collisions can surely be partitioned on the basis of knowledge available prior to the collision (e.g., observed spin of the cue ball) in ways that are statistically relevant to the dropping of the 8 ball. If we are truly talking of physical probabilities, then we are not talking of carving up nature at a level so crude as to ignore the physical probabilities, rather we are talking of carving it up "at the joints." In general it is clear that if we fail to find the right level of specification for our causal models, then we are likely to fail to have a model with the Markov property, and so the predictive and explanatory reasoning based upon the model will likely go amiss.

Apparently, the examples of interactive forks are either examples of the operation of quantum mechanical laws which, as was pointed out, admit of no refinements via further specification of causal variables (i.e., Bell's theorem) or they are examples of incompletely (inhomogeneously) specified causal models. Still, one must wonder why our automated systems should not be able to learn or discover quantum mechanical theories as we have done. The difficulty I raised for the automated learning systems is just that interactive forks violate the Markov property, and any system using a causal model lacking the Markov property may fail to understand the true relation between distinct variables, with its predictions being off by arbitrary amounts. However, the Markov property for any interactive fork can be recovered simply by adding a causal relation between the two effects of the common cause. This would be an implausible maneuver with any of the macro examples offered, since we appear to know in advance that such causal relations fail to exist. But in such cases we apparently can insist on a refinement of variables – i.e., that the variables in the example fail to capture homogeneity. In quantum mechanical cases the claim that such causal relations between effect variables are impossible, while supported by many, is less obvious; indeed, it seems that Reichenbach's Common Cause Principle, coupled with a denial of hidden variables and the experimental support for a correlation between variables measuring subsystems in EPR-type arrangements (Aspect *et al.*, 1982), demands non-local causation.

In short, there is no clear impediment to taking the Markov property as a foundational assumption of methodology. When it fails, we can use this as an indication of the incompleteness of our models, rather than the hopelessness of further scientific investigation. In particular, when the Markov property fails, we may conclude that either our model contains insufficiently refined variables or it is missing essential causal links.

4. PROBABILISTIC CAUSALITY II

4.1 *Hesslow's example*

Germund Hesslow (1976) introduced an example of a causal structure which again allows for negative statistical relevance between a cause and its effect, but which also illustrates some other difficult features. There is some evidence that taking contraceptive pills (C) can cause thrombosis (T), a kind of blood clotting. Now, pregnancy (P) can also cause thrombosis, so what Hesslow asked us to consider is the possibility that even though the pill causes thrombosis directly it may be that pregnancy is more efficacious, so that overall taking the contraceptive pill is negatively statistically relevant to thrombosis. That is, it may be the case that:

$$P(T|C) < P(T)$$

because $C \rightarrow \neg P$ and $P \rightarrow T$ even though $C \rightarrow T$. Cartwright (1989) describes this as a case where the causal factor has "dual capacities": both to produce the effect and, by another route, to inhibit it.

One thing this example suggests is that we must be very careful in identifying the populations (reference classes) relative to which a factor is being claimed to be causally efficacious. In fact, I agree with Sober (1984) that it just makes no sense to talk about causal factors without a commitment, implicit or explicit, to its being a causal factor for a specific effect relative to a population *and* its environment (that is, relative to a homogeneous reference class).

What Hesslow has asked us to consider is that across the entire human (adult? female? industrialized?) population, given its other current reproductive and health practices, the contraceptive pill may reduce the incidence of thrombosis via a reduction of pregnancies. Clearly, when we make the reference class as broad as this we will not have a homogeneous reference class – in particular, the statistical relevance of the pill to thrombosis will vary between sterile and fertile women, for example. What we presumably have here is the need to relativize the causal claim to specific homogeneous subpopulations. Amongst fertile women likely to become pregnant otherwise, the pill may be preventative of thrombosis; amongst women who are otherwise unlikely to become pregnant, for whatever reason, the pill may be a cause of thrombosis. If this is a proper description of Hesslow's case, as I believe, then we do not have a counterexample to the concept of probabilistic causality: as always, in the *right* reference classes, probability and causality flow in the same direction.

This view is by no means widely accepted. Many who have paid attention to subpopulations while writing on probabilistic causality have claimed that *contextual unanimity* is an essential feature; that is, they claim that the statistical relevance between cause and effect found in any one subpopulation, whether positive or negative, must have the same sign in all subpopulations. On the contrary I find nothing necessary in "contextual unanimity": the world may simply not be as neat and orderly as some would desire. Some may continue to insist that the pill causes thrombosis *simpliciter*: there is a specific chemical mechanism that leads from ingestion to blood clotting (a Salmonian kind of response); but, of course, there is just as much a mechanism involved in inhibiting pregnancies. Given the lack of an all-encompassing homogeneous reference class I do not see how sense can be made of talk about causality *simpliciter*; all that we can draw upon that applies to the class of all women is ignorance.[19]

4.2 *Tooley*

In rather stark contrast to my view, Michael Tooley argues that the existence of cases such as Hesslow's demonstrates that a probabilistic interpretation of causality is impossible. In particular, Tooley asks us to consider the possibility that two diseases which cause death may interact so that one of them mitigates the other (in Sosa and Tooley, 1993; cf. Tooley, 1987). Thus, it may be that for two diseases $\mathcal{F}$ and $\mathcal{G}$ and death over a period of time $\mathcal{D}$:

$$P(\mathcal{D}|\mathcal{F}.\neg\mathcal{G}) = .1 > P(\mathcal{D}|\neg\mathcal{F}.\neg\mathcal{G}) \approx 0$$

reflecting the causal efficacy of $\mathcal{F}$, while

$$P(\mathcal{D}|\mathcal{F}.\mathcal{G}) = .1 < P(\mathcal{D}|\neg\mathcal{F}.\mathcal{G}) = .8$$

because $\mathcal{F}$ interferes with the normal progression of disease $\mathcal{G}$. In this case, as Sosa and Tooley note (1993, p. 20), if we suppose for some population that half will be infected with disease $\mathcal{G}$, then the probability of death given the known cause of death $\mathcal{F}$ is lower than the marginal probability of death, and so we have negative statistical relevance for a factor (event type) which *ex hypothesi* is a positive causal factor. Therefore, positive relevance cannot be a precondition for causality.

Whereas Tooley is quite right that positive relevance across an inhomogeneous population, as in the case above, is no precondition for causality, it seems clear that the difficulties with his example stem from a presupposition that the unanimity requirement will be adhered to.[20] But dropping the unanimity requirement makes immediate sense of the case probabilistically: the disease $\mathcal{F}$ *both* causes and prevents death, under different circumstances. Clearly, were you to discover that you were infected with disease $\mathcal{G}$ you would *expose yourself* to the *disease* $\mathcal{F}$ if at all possible, regardless of its description as a *disease*. On the other hand, it would be foolish to expose yourself to $\mathcal{F}$ otherwise, since in a population unexposed to $\mathcal{G}$ it is a probabilistic cause of death. There are real cases of factors which operate in the way which Tooley describes. Chemotherapy in the absence of cancer would be fool-hardy as it suppresses the immune system and increases the chances of death; but in the presence of some cancers it reduces the chances of death. In the latter case, we in fact describe chemotherapy as a negative causal factor for death, even though in any particular instance (token causality) it may lead directly to a patient's death.[21]

In short, failure of the unanimity requirement does not cut against probabilistic causality for the simple reason that the unanimity requirement itself is wrong-headed. What is right-headed is the relativization of causal claims to homogeneous reference classes, subpopulations, and/or environments. Otherwise, in the absence of homogeneity, causal claims about women in general or diseases across all of humanity, etc.

must be viewed either as implicit claims about a dominant or salient subpopulation or as rubbish.

4.3 *Simpson's paradox and Cartwright's CC*

Nancy Cartwright (1983) was one of the first to consider the role of subpopulations explicitly, apparently in response to her reading of how medical experiments are conducted. Her treatment was also developed in response to "Simpson's paradox" (Simpson, 1951; originally treated in Yule, 1904). Simpson's paradox shows that in general any sample correlation (or a lack of correlation) between two variables that shows up in a population may disappear or be reversed across all subpopulations. For example, we might have a treatment for a disease which we would like to test. Let T be the treatment, D be death due to the disease and F and M be female and male respectively. Then an experiment might generate the results of Table 1. Examining subpopulations we see that the treatment is effective:

$$P(\neg D|T.F) = 1.0 > P(\neg D|\neg T.F) = .89$$
$$P(\neg D|T.M) = .25 > P(\neg D|\neg T.F) = .17$$

Table 1: Simpson's Paradox

	T		$\neg T$	
	F	M	F	M
$\neg D$	3	24	84	1
D	0	73	10	5
	1.0	.25	.89	.17

However, if we simply compare survival rates between the control and experimental groups, combining subgroups as experimenters ordinarily do, we get a different story:

$$P(\neg D|T) = .27 < P(\neg D|\neg T) = .85$$

So apparently we should avoid the treatment like the plague!

As Cartwright notes, Simpson's paradox shows that "any association – $Prob(A|B) = Prob(A)$; $Prob(A|B) > Prob(A)$; $Prob(A|B) < Prob(A)$ – between two variables which holds in a given populations can be reversed in the sub-populations by finding a third variable which is correlated with both" (1983, p. 24). From this point of view, Hesslow's case is likewise an example of Simpson's paradox: the positive (direct) association between the contraceptive pill and thrombosis can be reinforced, neutralized or reversed through a joint association with pregnancy.

It should be clear enough that an account of causality that relativizes it to homogeneous subpopulations will have no difficulties with Simpson's paradox: since F and M are statistically relevant to the outcome, the control and experimental groups are not homogeneous. It is also true that standard methods have no difficulty dismissing Simpson's paradox in experimental settings since sex (or whatever other factor) will either be matched explicitly across control and experimental groups (if it is known or suspected to be causally relevant) or be matched implicitly (stochastically) in a randomized design.

The moral Cartwright wants to draw from Simpson's paradox (1983, p. 25) is that averaging causal effects over divergent populations fails to get at the underlying causal structure and may be positively misleading. Of course, we have already agreed to this. But it is interesting to observe that Simpson's paradox alone does not force the issue, since both Cartwright's CC condition for causality and randomized experimental design differ on the issue of averaging, while both cope with the paradox, because both break the association with the third factor through (explicit or stochastic) matching.

Cartwright's *CC condition* asserts:

(D3) C causes E iff $\forall_i\ P(E|C.S_i) > P(E|\neg C.S_i)$

where the S_i are subpopulations which are determined by holding fixed *all* (and only) causal factors except for C itself and except for any intermediate causal factors between C and E. Thus, if there were only the causal factors $F_1, \ldots, F_n$ other than C, and if they were all binary, then there would be 2^n subpopulations to consider determined by the state descriptions from $F_1 \wedge \ldots \wedge F_n$ through to $\neg F_1 \wedge \ldots \wedge \neg F_n$.

Although this condition demonstrates awareness of experimental design and its concern to match relevant factors, it is not identical to the definition of causality which is implicitly operative in accepting the results of randomized experimental studies as indicating causal connections. I suggest instead that such a definition would be the following *condition RE*:

(D4) C causes E iff $\sum_i P(E\ C.S_i)P(S_i) > \sum_i P(E|\neg C.S_i)P(S_i)$

That is, randomized experiments, while stochastically matching subpopulations, look at the result within each subpopulation and weight its contribution according to size of the subpopulation to get a weighted average measure of causal efficacy.

Ronald Giere (1979) introduced a criterion of causality which shares all the important features of RE – indeed, he introduced it in the context of explicating experimental methods. This has been called the "all or none" counterfactual approach:

> C is a positive (negative) causal factor for E in population P iff there *would be* a higher (lower) frequency of E if every individual in P had C than there would be if no individual in P had C.

It is to be understood that, other than the counterfactual introduction (withdrawal) of C to (from) every member of P, the two hypothetical populations are exactly the same in every respect.[22] Giere's claim is that randomized experimental design, as exemplified in RE, is an attempt to approximate this idealized (counterfactual) test of causality. In this I believe he is right, but it turns out to be no advantage.

RE, or Giere's criterion, clearly can give a verdict different from that of Cartwright's CC since a failure of unanimity is consistent with a verdict of causality. Cartwright says of this that averaging measures of causality may be useful for policy making, but operate at too gross a level to capture the underlying causal structure (1989, pp. 103f); her idea is that governmental policy, say the regulation of carcinogens, should be based in part upon the size of the subpopulations directly affected, which RE accommodates and CC does not. I think this gesture to RE is too generous, however: the neutral Hesslow case (i.e., where the dual capacities (conditional probabilities) are tweaked until $P(T|C) = P(T|\neg C)$) shows that, say, a policy decision to allow a carcinogen to be *freely* used because of a null average effect could be wildly mistaken. There simply is no substitute for understanding the probabilistic relations at the *causal* level.

By this claim I do not mean to deny that randomized experimental design has a place in science; on the contrary, its invention was surely one of the most significant methodological achievements in the history of science. Whenever RE reports a causal relation (barring unfortunate sampling fluctuations or the result being swamped by prior probabilities), we indeed must have a causal relationship. The caveats I wish to record are: RE may *miss* causal relationships due to balanced dual capacities; and, when reporting a causal relationship, RE may be reporting a gross-level of relationship, one which supervenes on an underlying structure which is radically underdetermined by what RE does report. Hence, a positive report from an experimental study can only be the *beginning* of an investigation into the structure of the relevant causal process.[23] This point will be no mystery to scientists themselves, for it is already made intuitively by pointing out that the proof of the existence of a causal relationship may well come before anyone understands the causal *mechanism* which realizes that relationship; for example, Alfred Wegener (despite his biased American detractors) proved the existence of continental drift long before the mechanism of convection currents in the earth's core was established (namely, in 1912 versus 1966).

4.4 *Eells and Sober*

Cartwright's CC condition contained a vexatious difficulty: namely, how to avoid having intermediate causal factors screen off a cause C from its effect E. As I've presented it above there appears to be no difficulty here: I simply ruled out intermediate factors from the matching process used to identify subpopulations. However, Cartwright believes that identifying which factors are intermediate between C and E by examining the *time* when the various factors are instantiated is flawed, and this leaves the intermediate factors not fully identified. Cartwright identifies them as factors which are caused by C; however, as she acknowledges (1983, p. 30), such factors may in particular cases have been caused by something *other* than C and so should have been held fixed.

Eells and Sober (1983) argue contrariwise that examining the time of instantiation is exactly how causal factors to be matched should be identified. In particular, they defend the following (ES) condition for causality which, in common with ordinary experimental design, fixes all causal factors *as of the time* C is applied:

(D5) C causes E iff $\forall_i [P(E|C.S_i) > P(E|\neg C.S_i)]$

where the S_i are determined by fixing all and only causal factors (other than C) at the start of the experiment only.

Intuitively, Eells and Sober seem to be right that the intermediate causes that we want to leave unfixed are just those which arise *after* the application of C. This will surely leave alone the factors intermediate between C and E; furthermore, it handles cases where these factors have been caused by something other than C, since *those* causes (or else their successor events) will have been fixed at the time when the experiment is initiated, and so they will have been "paid their due." Nevertheless, Cartwright has developed an ingenious example to show that this intuition, and the experimental method on which it is founded, is wrong.

4.5 *Cartwright's CC* and the Star algorithm*

Cartwright employs the dual capacities in Hesslow's case to find a weakness in the idea of fixing factors in an experiment at the time when the putative causal factor is applied – as is, in fact, the universal practice in science (when factors are fixed at all). In order to illustrate the difficulty with greatest economy, let us suppose that thrombosis in Hesslow's case is caused directly by some chemical condition $\mathcal{H}$ induced by taking the contraceptive $\mathcal{C}$ and also caused directly by pregnancy $\mathcal{P}$; furthermore, let us suppose that these are all binary propositional variables related specifically via the causal model:

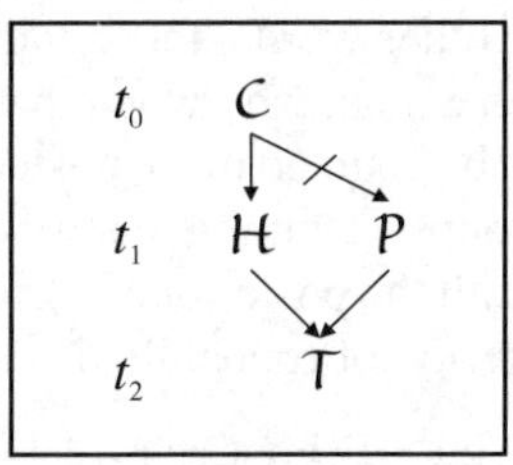

Figure 2: Hesslow's Case

Then the causal effect upon the state of the system at t_0 on $\mathcal{T}$ is fully determined by its state at t_1 (via the Markov property) and, in particular, the probability of thrombosis at t_2 is determined by which of four posterior subpopulations an individual is in at t_1: $\beta_0(\neg\mathcal{H},\neg\mathcal{P}), \beta_1(\neg\mathcal{H},\mathcal{P}), \beta_2(\mathcal{H},\neg\mathcal{P}), \beta_3(\mathcal{H},\mathcal{P})$. If we suppose that the presence of $\mathcal{C}$ has any effect on $\mathcal{T}$, then by the Markov property it must have this effect by redistributing membership in these subpopulations. And we can make some plausible story here: it seems that $\mathcal{C}$ reducing the incidence of pregnancy means that some members of subpopulations 1 and 3 will be relocated to one of the other subpopulations;[24] and its causing thrombosis must mean that members of subpopulations 0 and 1 are redistributed by it to the other subpopulations. Let us call the subpopulations occurring in the absence of $\mathcal{C}$ β_i (as above) and those occurring in the presence of $\mathcal{C}$ B_i (and indexed by the same ordering of properties as above).

According to the Hesslow story, we can be pretty clear that the result of applying $\mathcal{C}$ will be as Cartwright describes it:

> The group that has both effects already [i.e., $\mathcal{H}$ and $\neg\mathcal{P}$] must necessarily grow bigger; and the group with neither effect will in the end be smaller; what happens in the other two depends on whether the tendency of the contraceptives to induce [$\mathcal{H}$] is stronger or weaker than its tendency to inhibit pregnancy. The net result for thrombosis of all these changes is unpredictable without the numbers. . . . Anything can happen to the overall probability. (Cartwright, 1989, p. 101)

But if anything can happen to the overall probability, then a method (or criterion) of causal discovery that compares the efficacy of C across the subpopulations S_i matched at the beginning of the experiment, such as ES, and ignores the missing numbers concerning the intermediate subpopulations, must be incapable of providing us with the right causal story. What is required, according to Cartwright, is that we compare the efficacy of the putative cause across subpopulations that *would have arisen* in the absence of the causal factor. We want to know how effective C is in inducing T, not amongst those, for example, who start out not pregnant and without H, but amongst those who *would be* in that state at t_1 *were* they denied the contraceptive. Hence, one target of comparison will be subpopulations at t_1 in the control group; the other target will be subpopulations at t_1 in the experimental group – however, those subpopulations must be rearranged, or "collimated," to account for cases where it is the action of C itself which has moved individuals from one subpopulation to another, so as not to lose account of its effectiveness via such rearrangements. Since the latter is precisely *how* C influences the outcome T or $\neg T$ (at least in the hypothetical Hesslow case under discussion), if we do not collimate these posterior groups we will guarantee a failure to find any effect due to C. In other words, those individuals who start out at t_0 not pregnant and without H (i.e., in S_0) but who would have become pregnant and ended up in β_3 were it not for C should be counted in the posterior subpopulation 3 even when they are in the experimental group. Assuming that we can identify such individuals in the experimental group, let us call the posterior experimental subpopulations γ_i. Then Cartwright's reformulated criterion of probabilistic causality is *Condition CC**:

(D6) C causes E iff $\forall_i\ P(E|C.\gamma_i) > P(E|\neg C.\gamma_i)$

Cartwright does not directly concern herself with *how* one might collimate experimental subpopulations. But there is a fairly straightforward procedure if we assume that the control subpopulations at the start of the experiment are well matched with the experimental subpopulations and are sufficiently large to provide good estimates of the relevant probabilities of transition from one subpopulation to another during the life of the experiment. I illustrate what I call the *Star Algorithm* (in recognition of Cartwright's naming convention) with cases (using values which appear to preserve the main features of Hesslow's example). Doing so readily reveals that, despite Cartwright's express criticisms of Eells and Sober, under a perfectly natural interpretation the ES and CC* conditions for causality are equivalent.

4.5.1 Case 1: $S_i = \beta_i$

This is a very simple case in which there is *no* movement between subpopulations except for that caused by the application of C. In this case it is trivial that ES and CC* are equivalent.

Since by assumption the only direct effect on T is via the individual's state at t_1, we have $P(T|\beta_i) = P(T|B_i)$. In particular, suppose

$$T_B = T_\beta$$

$$= \begin{pmatrix} \beta_0 & \beta_1 & \beta_2 & \beta_3 \\ .01 & .2 & .1 & .3 \\ .99 & .8 & .9 & .7 \end{pmatrix}$$

where the top row is $P(T|\beta_i)$ and the bottom row is $P(\neg T|\beta_i)$. C works by shifting some individuals from an initial S_i to some B_j. In particular, suppose this works via the transition matrix

$$X_C = \begin{array}{c} \begin{matrix} S_0 & S_1 & S_2 & S_3 \end{matrix} \\ \begin{pmatrix} .8 & .72 & 0 & 0 \\ 0 & .09 & 0 & 0 \\ .2 & .18 & 1 & .9 \\ 0 & .01 & 0 & .1 \end{pmatrix} \end{array}$$

where the rows correspond to B_0 through B_3.

This provides what we need to calculate the ES criterion, which was just: $P(E|C.S_i) > P(E|\neg C.S_i)$. The left hand side of the inequality (LHS) in the case at hand is

$$\begin{aligned} P(T|C.S_i) &= P_C(T|S_i) \\ &= \sum_j P_C(T|B_j)P_C(B_j|S_i) \\ &= T_B \times X_C \\ &= \begin{matrix} S_0 & S_1 & S_2 & S_3 \end{matrix} \\ &= \begin{pmatrix} .028 & .0462 & .1 & .12 \\ .972 & .9538 & .9 & .88 \end{pmatrix} \end{aligned}$$

The right hand side (RHS) is

$$\begin{aligned} P(T|\neg C.S_i) &= P(T|\neg C.\beta_i) && \text{by Case 1 assumption} \\ &= P(T|\beta_i) && \text{by the Markov property} \\ &= T_\beta \\ & \quad\; \begin{matrix} \beta_0 & \beta_1 & \beta_2 & \beta_3 \end{matrix} \\ &= \begin{pmatrix} .01 & .2 & .1 & .3 \\ .99 & .8 & .9 & .7 \end{pmatrix} \end{aligned}$$

Following ES, we may conclude that C causes T within subpopulation S_0, that it prevents T within subpopulations S_1 and S_3 (but to different degrees of effectiveness) and that it is otherwise causally irrelevant. (Of course, this is all true by construction of the example.)

Now I prove that CC* yields the exact same conclusion, by the exact same calculations. CC* asks us to examine the inequalities: $P(E|C.\gamma_i) > P(E|\neg C.\gamma_i)$. LHS is $P(T|C.\gamma_i)$. γ_i and β_i are supposed to have the same probabilistic properties by construction (I defer the construction of γ_i to the next case, where it is problematic). Therefore, LHS = $P(T|C.\beta_i)$; by the Case 1 assumption, this is just $P(T|C.S_i)$. RHS is $P(E|\neg C.\gamma_i)$. By definition, in the control group $\gamma_i = \beta_i$. Hence, RHS = $P(E|\neg C.\beta_i) = P(E|\neg C.S_i)$. Applying the Markov property, we find these to be identical to the LHS and RHS of the ES condition.

4.5.2 Case 2: $S_i \neq \beta_i$

This case is complicated by the assumption that individuals "wander" from subpopulation to subpopulation without the influence of the causal factor under test.

This wandering requires a transition matrix from the S_i to the β_i for the control population; suppose it is:

$$X_{\neg C} = \begin{array}{c} \begin{matrix} S_0 & S_1 & S_2 & S_3 \end{matrix} \\ \begin{pmatrix} .9 & .01 & 0 & 0 \\ .1 & .99 & 0 & 0 \\ 0 & 0 & .9 & .01 \\ 0 & 0 & .1 & .99 \end{pmatrix} \end{array}$$

This provides sufficient information to compute the ES criterion, $P(E|C.S_i) > P(E|\neg C.S_i)$. The LHS is unchanged from Case 1. The RHS is

$$\begin{aligned} P(T|\neg C.S_i) &= T_\beta \times X_{\neg C} \\ &= \begin{array}{c} \begin{matrix} S_0 & S_1 & S_2 & S_3 \end{matrix} \\ \begin{pmatrix} .029 & .1981 & .12 & .298 \\ .971 & .8019 & .88 & .702 \end{pmatrix} \end{array} \end{aligned}$$

This gives us the result that C prevents T among subpopulations S_0 and S_2 (with property $\neg P$) and otherwise causes T, although with different rates in all cases. This *appears* to be intuitively the right result, since we expect C to prevent T, wherever it does so, by keeping women who are not pregnant from becoming pregnant. However, following Cartwright's argument above, these findings are likely to be misleading for they do not properly reflect the causal role of C (they also don't reflect the proper interpretation of H and P, as we shall see).

For CC* things are now different. RHS is just $P(E|\neg C.\gamma_i) = P(T|\neg C.\beta_i) = P(T|\beta_i)$. LHS is

$$\begin{aligned} P(E|C.\gamma_i) &= P(T|C.\gamma_i) \\ &= \sum_j P_C(T|B_j)P_C(B_j|\gamma_i) \\ &= T_B \times X_{CC*} \end{aligned}$$

where X_{CC*} is a new transition matrix under C reporting the values $P_C(B_j|\gamma_i)$, which need to be computed, via the "Star Algorithm." Consider for example $P_C(B_1|\gamma_0)$. This is the transition probability to B_1 given C from the inverse image of β_0 under $X_{\neg C}$. From $X_{\neg C}$ we see that 98.9 per cent of β_0 comes from S_0 and 1.1 per cent from S_1.[25] To find the transition vector for γ_0 we compute the (normalized) weighted sum of columns of X_C, with the weights representing the contribution of each S_i to β_0, which is equal to the product of the β_0 row of $X_{\neg C}$ with the sizes of the S_i subpopulations. In this case we get

$$X_{CC*} = \begin{array}{c} \begin{matrix} \gamma_0 & \gamma_1 & \gamma_2 & \gamma_3 \end{matrix} \\ \begin{pmatrix} .7991 & .7273 & .0000 & .0000 \\ .0010 & .0817 & .0000 & .0000 \\ .1998 & .1818 & .9989 & .9092 \\ .0001 & .0091 & .0011 & .0908 \end{pmatrix} \end{array}$$

Completing the computation of LHS we have

$$P(T|C.\gamma_i) = \begin{array}{c} \begin{matrix} \gamma_0 & \gamma_1 & \gamma_2 & \gamma_3 \end{matrix} \\ \begin{pmatrix} .0282 & .0445 & .1002 & .1182 \\ .9718 & .9554 & .8998 & .8818 \end{pmatrix} \end{array}$$

Comparing this with T_B we find that by the CC* criterion $\mathcal{C}$ causes $\mathcal{T}$ in subpopulations with property $\neg\mathcal{P}$, namely γ_0 and γ_2 (although the latter, with property $\mathcal{H}$, is very nearly unchanged) and $\mathcal{C}$ prevents thrombosis in the subpopulations with property $\mathcal{P}$. This is exactly the opposite conclusion from that of the ES criterion! That criterion, recall, appeared to have the intuitively correct result that $\mathcal{C}$ prevents $\mathcal{T}$ when it does by keeping women who are not pregnant from becoming pregnant. It hardly seems likely that an agent that causes thrombosis directly will *reduce* the incidence among pregnant women, when pregnancy *also* causes thrombosis!

But that is the wrong reading of the outcome of CC*. The right reading takes into account the counterfactuality of the γ subpopulations. CC* compares the outcome of, for example, the subpopulation of women who actually *have* properties $\neg\mathcal{H}$ and $\neg\mathcal{P}$ only in the control group; it compares this subpopulation with the subpopulation in the experimental group of women who *would have* ended up in class B_0 (with $\neg\mathcal{H}$ and $\neg\mathcal{P}$) *were it not for* the action of $\mathcal{C}$. Amongst this latter group (γ_0) the numbers in the example reflect the fact that none of them will become pregnant (admittedly, the example is idealized), but some of them will acquire property $\mathcal{H}$ through the administration of $\mathcal{C}$. Naturally enough, CC* finds that among women who would not have become pregnant in any case $\mathcal{C}$ causes thrombosis, whereas among women who would have become pregnant $\mathcal{C}$ prevents it. This is exactly what we would expect from the Hesslow story. Thus, on the right – counterfactual – reading of properties $\mathcal{H}$ and $\mathcal{P}$ the ES criterion gets the story exactly backwards.

CC* – with the Star Algorithm – identifies the *relevant* counterfactual subpopulations for comparison. Experimental methods which fix subpopulations based upon matching measured properties at the initiation of the experiment may find almost any relationship between the causal factor under test and the effect, when the factor under test has the dual capacities both to induce and to prevent the effect via some third causal factor. This kind of case is in some sense an elaboration of Simpson's paradox: by holding fixed (segregating) the subpopulations identified at the beginning of the experiment, we *mix* the subpopulations that matter for identifying causality. And the remedy is again in some sense an elaboration of the remedy for Simpson's case: we normalize the subpopulations that are to be compared with one another – although in this case normalization means more than just getting the weights right, it requires also getting the constitution of the subpopulations right. This is a significant change to the current practice in the analysis of experiments, one which is well worth commending.

4.5.3 Unification of ES and CC*

As I suggested above, despite the failure of the ES criterion in the Hesslow case, there remains an important and relevant sense in which the ES criterion of causality is fully correct. The ES criterion obliges us to fix *all* of the causal factors for $\mathcal{E}$ other than $\mathcal{C}$ at the time of applying $\mathcal{C}$ to the experimental group. According to Cartwright (1989, p. 102), the difficulties for the ES criterion arise from the fact that some causes have dual capacities, for in those cases the failure to collimate subpopulations results in the criterion mixing those dual capacities, leading to any possible aggregate outcome. However, Cartwright misdescribes the problem with ES here: the problem is that on the naive reading of ES we have been using the criterion does not guarantee that the subpopulations being examined are objectively homogeneous. Since this ES procedure does not collimate subpopulations, there are indeed factors which are epistemically accessible prior to the occurrence of E which could be used to partition the S_i into subsets that are statistically relevant to the outcome – namely, we could divide these subpopulations into subsubpopulations of those destined to end up in β_0, etc. The Star Algorithm does nothing more nor less than apply this information *retroactively* to the t_0 subpopulations, so as to render them objectively homogeneous.

If the subpopulations *started out* objectively homogeneous, then, by definition, there could be no wandering between subpopulations not under the influence of C; and we've already seen that the result is that ES and CC* are provably equivalent (in Case 1 above). Hence, when we apply the Star Algorithm to generate an artificial objective homogeneity, we again must have equivalence: i.e., CC* and ES* – ES amended to employ the Star Algorithm – are equivalent. Of course, this is perfectly obvious by examination of (D5) and (D6), since they then read identically. But the point is that the Star Algorithm is most naturally understood simply as the *means* whereby *all* the causally relevant properties are fixed "in advance" of applying the factor under test. "All relevant properties" does not mean: all those which we know how to measure at t_0; "all properties" means *ALL* properties, including those we do not know how to measure at t_0, in particular the counterfactual properties "of whether [women] would have [H] and P if C did not operate" (Cartwright, 1989, p. 102). But, in fact, this is just Cartwright herself portraying how CC* identifies subpopulations matched for counterfactual properties at the start of the experiment. It is only by adopting an uncharacteristically Humean construal of the properties available to Eells and Sober that Cartwright has been able to persuade herself that the ES criterion and CC* are actually in opposition to each other.[26]

5. INTERACTION AND UNANIMITY

5.1 *Unanimity and CC**

Both the ES criterion and CC* require unanimity across the relevant subpopulations. On the other hand, during the discussion above I have suggested that ES finds that C causes E in this subpopulation and CC* finds that C causes E in that subpopulation, etc. But taken literally both ES and CC* imply that the C and E discussed above are not causally related! It will not suffice to say that it is the *combination* of C with some interfering factor (P in Hesslow's case) which is the cause.

Consider the example of Figure 3 from Eells (1986, p. 59; note that the four subpopulations are understood to have equal sizes). Eells introduces this with the (unlikely) interpretations: S is smoking; H is heart attack; and X is exercise. So, in the presence of exercise, smoking reduces the risk of heart attack and otherwise increases it. The first point is that here it seems clear that the combined property $S.X$ prevents H and that $\neg S.X$ causes H, but since there is no univocal opposite to either $S.\neg X$ or $\neg S.\neg X$ the causal status of the latter two is indeterminate. And yet we know perfectly well what causal role all of the combinations play.

As Eells points out, CC* is in even more trouble in this example.[27] First, supposing

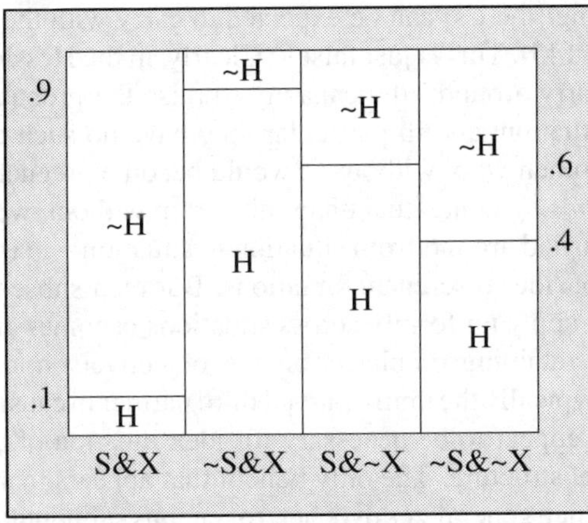

Figure 3: Eells's interaction example

S to be a causal factor, X cannot also be a causal factor for H since unanimity fails; but then in testing S we are not allowed to hold X fixed,[28] with the result that $P(H|S) = .35 < P(H|\neg S) = .65$, implying according to CC* that S is simply a preventative of H. If we begin by supposing instead that X *is* a causal factor, then S cannot also be a causal factor; but furthermore, we will find that X is *not* a causal factor, since $P(H|X) = .5 = P(H|\neg X) = .5$. But we know that neither verdict is adequate: both S and X are causal factors for H through their causal interaction with each other.

Cartwright, in her 1989 book, does not explicitly resolve this problem. However a natural resolution suggests itself. She does explicitly agree with Eells and Sober (and me) that it is appropriate to consider type causality as a three place relation between cause and effect *within some population* or, more exactly, within an objectively homogeneous reference class. She calls claims about such "very local" causal relations 'causal laws' (1989, p. 144). However, given that we have made this shift to explicit (or implicit) reference to a homogeneous population when making a causal claim, it follows trivially that – if there is to be a probabilistic criterion of causality at all – contextual unanimity must hold, for there is only one context under consideration. To handle Eells's case, all we need to do is require that all (and perhaps only) factors that enter into causal laws be held fixed for CC* (these are what I would call 'causal factors'), rather than all (and perhaps only) *causes*, as determined by some broader unanimity requirement. We might label this homogeneity view *contextual identity* to contrast it with the broader contextual unanimity accounts.

Cartwright nevertheless continues to advocate the merits of a broader unanimity requirement. In particular, she advocates that we de-relativize causal claims, so that instead of referring to specific, homogeneous subpopulations they refer to larger groupings of such subpopulations. She is surely right that when doing so we shall avoid a certain amount of trouble if we also restrict the groupings to subpopulations which are unanimous in supporting positive (or negative) statistical relevance between cause and effect. But Cartwright suggests that by requiring contextual unanimity and de-relativizing we capture something fundamental about causality, namely the concept of *causal capacity*:[29] "That capacity [to raise (lower) the probability of E by virtue of being *C*] is something [the *C*s] can be expected to carry with them from situation to situation" (1989, p. 145). This is just false.[30] Clearly, in the Hesslow case, contraceptive pills do *not* "carry around" the capacity to raise the probability of thrombosis from situation to situation, and in particular they have no such capacity among the subpopulation of women who, without *C*, would become pregnant.

Of course, by *imposing* contextual unanimity from without, we can guarantee that "the capacity" is carried around from situation to situation – inasmuch as we are not prepared even to consider discrepant situations. But then subsequently finding that "causal capacities" apply uniformly across situations becomes an unimpressive discovery. By instead retaining emphasis on the objectively homogeneous subpopulations, which are typically the unmentioned third partner in causal relations, we have all the features that appear to be necessary for identification of causal laws and the attributions of causal structure. The only benefit that appears in store for developing a causal language which generalizes over heterogeneous subpopulations is the explication of ordinary language uses of causal locutions – and such ordinary language usage seems likely to be sufficiently complex to evade such a simple maneuver as imposing contextual unanimity. In any case, the main point is that there is a significant downside to adopting the unanimity maneuver: it obscures the difference between causal laws where the action of some *C*, while being in the same direction, differs greatly in its causal effectiveness across distinct homogeneous subpopulations (and, in such cases, we hardly seem to be confronted with an opportunity for generalizing talk about *individual* causal capacities).

5.2 *Dupré on unanimity*

John Dupré spends much of his chapter on probabilistic causality in his recent *The Disorder of Things* (1993) denouncing the contextual unanimity thesis. So much would, perhaps, be fine, except that remarkably Dupré conflates contextual unanimity with the contextual identity requirement of objectively homogeneous reference classes (cf. pp. 198–201), which as I have argued is a precondition for having physical probabilities and, hence, any sensible probabilistic account of causality. His central criticism is that homogeneity ("unanimity") accounts require access to reference classes that are largely unknown. This is a charge that we can accept with equanimity, since it clearly is part of the business of science to uncover classes showing greater degrees of homogeneity, as I noted previously. It is hardly incumbent upon an account of causality to wait upon the completion of science, and the contrary belief confuses the ontology of causation, the attempt to characterize what sorts of relations science is attempting to discover, with its epistemology, the attempt to characterize how it is that we discover such things.

In an extended discussion Dupré points out that, for example, we have little idea what the relevant factors are for a baseball player hitting a home run on any particular occasion. Of course, we can use a reference class of the particular hitter's frequency of hitting home runs over the past three years; but surely also the amount of sleep the player had in the last two days would be relevant, or again the amount and kind of food consumed, etc. If we examine the total number of such factors that might be relevant, particularly when they take unusual values, we shall have some huge number of factors. Even if these factors were merely binary valued, attaining a homogeneous partition would require two to the number of factors involved, and we should very quickly run out of baseball players, or at bats, or even particles in the universe to make up the partition cells!

This is all true. So we cannot hope to find objectively homogeneous classes for baseball in any strict sense. The science of baseball is most unlikely to ever achieve a set of probabilistic causal laws which are baseball's bedrock, which could never be improved upon. But that is no impediment to sabermetricians' (from "Society of American Baseball Research") discovery of more and more useful probabilistic relations between hitting conditions and hitting home runs (indeed, it is instead a precondition upon such a capacity to improve), which is a proper analogy to scientists improving upon the probabilistic laws in their various domains. To make sense of this we require the ideal concept, of objective homogeneity, towards which these efforts strive.

Dupré calls this sort of response hiding behind a "metaphysical figleaf" and suggests that we not adopt any metaphysical doctrine "for which we neither have, nor possibly could have, empirical evidence of applicability" (p. 201). The former is as may be, but the claim of methodological inscrutability confuses a definition as an ideal and goal with the satisfaction of the definition as a methodological prerequisite. *Pace* Dupré it is a substantial advantage of the probabilistic causality view that it directly *explains* why scientists are not satisfied with the bare verdict of a randomized experiment – say that C is a positive causal factor for E – and seek to discover its mechanism of action.

5.3 *Eells on interaction*

Ellery Eells, while saying similar things to what I have had to say about unanimity, does not take the final step of simply restricting our causal attributions to those cases involving objectively homogeneous reference classes. He insists that there is importance (or at least validity) to ascribing a univocal causal role to a factor C across different subpopulations even when the probability of the effect varies through those subpopulations, so long as these effects are unanimously positive (or negative). And this leads him to the following definition of causal interaction, which applies only to cases where unanimity fails; he defines *causal interaction* to be:

> the (not always symmetrical) relation on which factor C interacts with a factor F, with respect to E as the effect, if and only if C raises the probability of E in the presence of F and lowers the probability of E in the absence of F, or vice versa. (Eells, 1986, pp. 54–5)

Eells's account of the smoking and exercise example is then that these two interact and so are causally related to heart attack within the general population, but not positively, negatively nor neutrally. Instead, to specify a precise causal relationship we need to look at specific subpopulations. In particular, within the subpopulation of exercisers smoking is preventative of heart attack; or, within the subpopulation of smokers, exercise is preventative of heart attack, and so forth.

This is also my own analysis of that case. However, dropping the idea of unanimity naturally leads us to a different account of causal interaction, one which recognizes the possibility that causal factors may interact by *reinforcing* one another as well as by one reversing the impact of the other on the effect in question. To take a more plausible example than Eells's, it may well be that smoking causes emphysema and that air pollution causes emphysema, and further that smoking combined with air pollution induces emphysema, but not as an additive function of their separate effects, instead in an interactive way producing a much higher probability of emphysema.[31]

Non-interactive causes then, are those whose joint probabalistic impact on the effect (within some reference class) is a function of the probabilities of the effect conditioned upon the two causal factors separately, but not collectively. Interactive causes violate this restriction in requiring that the joint conditional probability *not* decompose into a function of the individual conditional probabilities, whether because the joint conditional probability either exceeds or falls short of some simple combination of the individual probabilities.[32]

Interaction, contra Eells and all of the "unanimity" theorists, may reinforce a causal tendency, undermine it, or do both simultaneously in different background contexts.

5.4 *Frequency interaction*

Frequency interaction occurs when the causal factor C interacts not with some quite different factor F but with other instances of C itself, within a population. Eells discusses a nice example (1986, pp. 60–3): raising one's hand in a lecture (C) typically serves to get one noticed by the lecturer (E); however, if *everyone* else in the lecture hall has raised his or her hand, then raising your hand will *reduce* your chances of being noticed. Such cases are interesting, but not as interesting as Eells believes. In particular, Eells claims that this case defeats Giere's "all or none" criterion of causality, and others have taken similar cases to do the same (Sober, 1982) or to represent serious challenges to Giere's account (Collier, 1983; Giere 1984). The argument goes: in order to test the thesis that raising one's hand is a causal factor for getting attention, according to Giere, we must examine the probability of being noticed in two hypothetical populations, one with everyone having raised a hand and the other where no one has raised a hand; however, in that case "Giere's theory will tell us that raising a hand is (simply) causally *irrelevant* to getting noticed, for (plausibly) there would be the same frequency of noticed students if they were all to raise a hand as there would be if none were to raise a hand" (Eells, 1986, p. 61).

This argument, while clever, is confused. The problem is perhaps one of taking the word 'population' too literally. What Giere's criterion (or RE) requires is that C be applied uniformly to some experimental group and denied similarly to some control group. But the members of the group need not make up a population in any biological sense, of course: we may test whether chemicals dissolve in some solution just as well as whether they cause cancer in rats. But then the question putatively at issue in the lecture hall is whether raising *one's* hand will attract attention, not whether raising *everyone's* hand (if that were a possible action) would attract attention to you. The population of concern for such a claim is emphatically *not* the population of people attending the lecture, but instead the population (reference class!) of possible time-slices of *lectures* which are relevantly similar to the current lecture. By (hypothetically) applying the factor under test, raising *one's* hand, to (all, and then none, of) the population of interest – relevantly similar lectures – one could find out whether raising one's hand is a positive causal factor for being noticed. The answer, as we all know in advance, will depend upon the frequency of hand-raising in those relevantly similar lecture time-slices.

Interpreting the test and control groups appropriately puts Giere's criterion, and randomized designs, at no disadvantage to contextual unanimity accounts. The disadvantage to Giere's account is instead that which we canvassed earlier, that randomized designs mix together causal factors to find an average effect within some non-homogeneous reference class. Although avoiding such averaging across classes representing dual capacities, Eells's contextual unanimity account suffers similarly in mixing non-homogeneous reference classes that happen to pull in the same direction.

The problem with contextual unanimity is not merely that it is wrong but that it draws attention away from what is important – the basic causal structure of the phenomena under study – toward what is unimportant, vague similarities in causal structure across diverse reference classes. In short, my view on contextual unanimity is that it is probably at best an ill-conceived attempt to do some ordinary language analysis; what we need to give an account of learning causal structure is already in hand.

6. CAUSAL MODELING

I shall now turn to the automation of the induction of causal theories. Interest in the problems posed in the attempt to automate causal induction has exploded in recent years within artificial intelligence. This is primarily due to advances in the past decade in representing causal theories via Bayesian networks (Pearl, 1988; Neapolitan, 1990). These have allowed moderately complex situations involving uncertainty to be modeled normatively correctly,[33] as in medical diagnosis. This in turn has led to the realization that there is need for machine learning methods to supplement "knowledge engineering" (recording what human experts say) in probabilistic expert systems, since expert systems generally encounter the "knowledge bottleneck": human experts often don't exist for a domain or don't know how to describe what they know.

In approaching the automated induction of causal structure a wide range of models might be considered. Bayesian networks in particular can represent any probability distribution, and so any kind of probabilistic causal relation, including non-linear interactions. However, rather than deal with the full range of possible causal models, I will limit my discussion here to the automated induction of linear causal models. The purpose is to gain a useful simplification in mathematical and statistical problems involved in induction. This same purpose has led most of statistical research into causal models in the social sciences to emphasize linear models and methods suited to them. In adopting this simplification I am following the path of Clark Glymour and his group at Carnegie Mellon University (Glymour et al., 1987, and Spirtes et al., 1993), more of which later, and of Nancy Cartwright (1989).[34]

Because of the lack of interaction of parent variables in linear causal models it is possible to represent the strength of the causal relation between a parent (exogenous) variable and child (endogenous) variable in a single coefficient, independent of the strengths of relation between other exogenous variables and the endogenous variable at issue. Sewall Wright (1934) already demonstrated that in this case there is a straightforward method of generating a set of simultaneous equations relating the coefficients required by a model and the correlations which are implied by the model. In particular, Wright demonstrated that

$$\rho_{ij} = \sum_k [\prod_{x_m \to x_l \in \Phi_k} p_{lm}]$$

where p_{lm} is the coefficient[35] of the arc $x_m \to x_l$ and that arc is on the *path* Φ_k between variables i and j. In other words, the correlation between two variables is equal to the sum of the products of linear coefficients along each path between them. The definition of a path is related to Pearl's concept of d-separation:

> Φ_k is a path between x_i and x_j iff it is a sequence of variables such that each variable is connected to its predecessor and to its successor (if any) by an arc and
> 1. the sequence contains no variable twice;
> 2. it does not "go backwards" (against the direction of an arc) after having once gone forward;
> 3. it does not contain more than one bidirectional arc.

The last condition is imposed because Wright was not restricting himself to fully specified causal models: some relations (subnetworks) were replaced directly by bidirectional arcs labeled with correlations (rather than linear coefficients). The second condition implies that no correlation will be found between two variables which are only connected through a common effect (or common descendant): having proceeded forwards from one parent, one cannot return to the other. Common cause structures are handled appropriately, since a path may *initially* move backwards from some effect to an ancestor and subsequently proceed forwards to a correlated effect.

Given a set of equations determined by Wright's rules, and assuming that the equations are solvable (which depends upon the model's structure), then we can directly compute the path coefficients required by a structure given a correlation matrix for its variables. Typically, what we have available are not the population (true) correlations, but instead sample correlations based upon some finite sample from the real population. What we can do then is estimate the path coefficients from the sample correlations by various standard statistical methods (such as iterative minimization of squared error in predicting dependent variables).

In any case, the point is that given a causal model's structure in advance there are relatively straightforward means of finishing the picture by adding in the numerical parameters using sample-based estimates via Wright's rules (or equivalent methods, of which there are more than a few). The algorithm for causal discovery suggested by this picture of causal modeling is then:

1. Find some sample correlations that are puzzling.
2. Invent a potentially explanatory causal structure.
3. Use the sample data to estimate (parameterize) the model.
4. Use the completed model to generate predictions.
5. Check the predictions. If the outcome is satisfactory, stop; else, loop at step 2.

If we allow a human user to identify what's puzzling about some sample, then we can pretty well imagine how to implement the algorithm in computer code – all except for the mysterious second step, since "invent" is not a primitive operation in standard computer architectures. Finding a search technique to substitute for human invention is the main problem for automated causal induction. We could attempt brute force search through the space of directed acyclic graphs, except for two problems: the space of such models is large, being exponential in the number of variables;[36] the space of such models is even larger than that, since the sampling may have missed important variables which will need to be "invented" in order to account for the data. There are, however, some useful properties, particularly conditional independencies implied by causal structure (assuming the Markov property, or I-mapness, which Salmon was doubting), which can directly assist the search.

In the domain of linear causal models, the role of the d-separation criterion for conditional dependence and conditional independence is played by partial correlation. Partial correlations report the linear association (correlation) between two variables when the associating effect between them of a different set of different variables has been "partialled out". When that set is a singleton, the partial correlation between x_1 and x_2 with x_3 partialled out is defined as:

$$\rho_{x_1 x_2 \cdot x_3} = \frac{\rho_{x_1 x_2} - \rho_{x_1 x_3}\rho_{x_2 x_3}}{\sqrt{1-\rho^2_{x_1 x_3}}\sqrt{1-\rho^2_{x_2 x_3}}}$$

Higher order partial correlations for larger sets may be defined recursively.

Partial correlation in linear models has properties in direct parallel to d-separation. The partial correlation between two variables connected only via a common ancestor, when that ancestor's influence is held constant (partialled out), is zero, just as in such a case the variables are d-separated by their ancestor. The marginal correlation of two parents connected only through a common effect is zero, but becomes non-zero when the effect variable is partialled out, corresponding to an induced dependence. This leads fairly naturally to the idea adopted by Glymour's research group for learning causal models: pretending for the moment that there will be no sampling error, the presence or absence of partial correlations in sample data can answer questions for us about d-separability in the causal model that underlies those data; therefore, we can use correlation information not just to fit a model to the data by finding optimal coefficients for it, but more importantly by using it as a test for d-separability, and so as a means for discovering the existence (or otherwise) of the causal connections between variables.

7. STATISTICAL EQUIVALENCE AND TETRAD II

An interesting consequence of the conditional independence properties of causal models with the Markov property (I-maps) is that such causal models can be grouped into classes of statistically equivalent models. This means that given the structure of any two of the models within an equivalence class (i.e., looking at the models *without* numerical parameters) they can both be parameterized equally well using a given sample. In other words, they will both account equally well for the data in terms of a likelihood measure, for example (Chickering, 1995).[37] The demarcation of the equivalence classes can be determined by a simple graphical criterion, as Verma and Pearl prove (1990; reported in Chickering's terminology):

Theorem 1 Two causal structures are statistically equivalent if and only if they have identical skeletons and v-structures.

The *skeleton* of a model is just its graphical structure, disregarding the orientation of its arcs. *V-structures* refer to triples of nodes such that in the subgraph in the three nodes one of them is the common effect of the other two and the two parents are non-adjacent (undirected v-structures will then be three nodes connected in a chain with the end nodes not connected). For example, Figures 4(a) and (b) are statistically equivalent (there are no v-structures), while (c) and (d) are not, since in (c) node 4 is the center of a v-structure whereas in (d) node 3 is and not vice versa. This can be seen most directly by observing that d-separation implies that holding fixed nodes 1 and 4 induces a conditional independency between 2 and 3 in Figure 4(d) but not in 4(c).

It is worth noting in passing that such properties are dependent strictly upon the structure of the causal model and not upon its parameters. Any causal structure can be parameterized so as to imply the same sample distribution as any other (in the same variables), for example by labeling all arcs with the coefficient 0. More interestingly, Simpson's paradox suggests classes of structures which can be parameterized to be statistically equivalent with non-zero coefficients: in (a linear version of) Hesslow's case, for example, if the coefficient of the arc connecting the contraceptive pill to thrombosis has the same magnitude but opposite sign to the product of those on the path through pregnancy, then there will be no marginal correlation between the contraceptive pill and thrombosis; as a result, Hesslow's model will be equivalent to a v-structure with arcs pointing inward to the pregnancy variable. Therefore, given Hesslow's model as the source of sample data, it will be impossible using the sample data alone to discover the original model, if the simpler v-structure is preferred a priori. Such limitations are intrinsic to non-experimental studies, and may be overcome by extending the scope of automated discovery by using prior information and also by employing experimental interventions.

Regardless of such difficulties, it is clear that non-equivalencies abound: there are a great variety of models which differ from each other in the conditional independencies they imply, and these implications are not dependent upon their parameterizations. This opens the way for automated inductive methods to take advantage of the implied independencies, using sample information supporting the existence or failure of conditional independence to support one causal structure over another. This

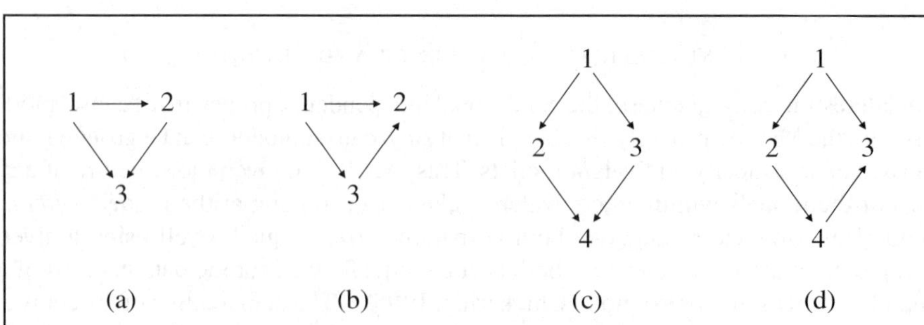

Figure 4: Equivalent and non–equivalent structures

furthermore makes plain the pointlessness of the contextual unanimity account of probabilistic causality: all of this inductive detective work depends upon conditional independence, a property which may obtain between variables all of which are uniformly positively (negatively) related to one another just as much as between variables which are not. If we were constrained to ignore independencies which arose among all positively (negatively) related variables, attempts to recover causal structure would be severely handicapped, if not utterly paralyzed.

Glymour and company make direct use of the concept of statistical equivalence in two principles for discovering causal structure in their TETRAD II program (Spirtes et al., 1990). The first principle is:[38]

> **Principle I**: Given a set of measured variables **M** and X, $Y \in$ **M**, X and Y are directly causally connected (i.e., there should be an arc between them) iff for every **S** $\subset$ **M** such that X, $Y \notin$ **S** X and Y are not conditionally independent given **S**.

This is the means proposed by Spirtes et al. for determining whether or not direct causal relations exist between two variables (disregarding latent or hidden variables, which might introduce intervening causal connections). Given reliable access to knowledge of conditional independencies, following Principle I appears to be the means to construct a minimal undirected I-map. In the context of statistical inference for linear models, the substitute for a test for conditional independence is just to test for whether the corresponding partial correlation is zero. Since Spirtes et al. are working within the framework of orthodox statistics (Glymour is a noted anti-Bayesian; see his 1980), TETRAD II employs a significance test for the partial correlations being zero for each subset .

Having by such means discovered a skeleton graph corresponding to the data, the next step is to recover as much information as possible about the orientation of arcs. Verma and Pearl's theorem on equivalence says that, from the data alone, we cannot hope to recover anything more than the directed v-structures (and anything implied by them together with the failure of other undirected v-structures to be directed v-structures). Spirtes et al. employ a second principle which applies information that can be extracted from the v-structures represented in the data:

> **Principle II**: If variables X, Y and Z form an undirected v-structure with Y the middle variable, then form a directed v-structure iff X and Z are conditionally dependent given any **S** $\subset$ **M** such that X,$Z \notin$ **S** and $Y \in$ **S**.

Assuming again the availability of reliable information on the existence of conditional independencies and dependencies, this principle will allow the discovery of the class of statistically equivalent models in the form of a hybrid graph with both undirected and directed arcs, called a "pattern" by Verma and Pearl (1990). And the test for Principle II is implemented in TETRAD II via significance tests for partial correlations given sets containing the middle variable of all undirected v-structures produced by Principle I.

The methods employed by TETRAD II to recover fully measured models (those without latent, or hidden, variables) appear to be intuitively justified by the results on statistical equivalence. Furthermore, it is clear from their published results (and some of my own, see Wallace et al., 1996, and Dai et al., 1997) that TETRAD II indeed successfully discovers linear causal models reasonably well in simple cases. There are two problems with TETRAD II, however. First, the use of orthodox significance tests for partial correlations means TETRAD II is susceptible to all the ills of ordinary significance tests, in particular those due to the failure to take into account prior probabilities for the existence of a causal relation. This may be considered a minor point, especially since in the first instance we may be interested in developing methods of induction without background knowledge being available, as a precondition for the automation of any kind of interesting induction.[39] However, there are in fact Bayesian methods of automating causal induction which go part way towards addressing such concerns, which I will outline below.

The second difficulty is the exponential requirement that all subsets of **M** excluding X and Y be examined for introducing a zero partial correlation between the latter (in Principle I, Principle II having a similar requirement). Presumably, this implies a limit on the size of causal models with which TETRAD II can deal (although the Bayesian method described below has similar limits). TETRAD II does do a bit better than a straightforward implementation of Principle I would, however. It iteratively examines partial correlations from low order to high order (in its "PC" algorithm; Spirtes et al., 1994), partialling out only nodes which are adjacent to the two nodes in question. Since all nodes are initially treated as adjacent, if the tests succeed in only removing arcs during prior iterations which do not obtain in the true model, then this procedure will eventually examine the parent set of the two nodes, which will be sufficient to capture all of the conditional independencies in the original model between those two nodes (cf. Pearl's theorem 5, loc. cit.). Since arcs are removed in this fashion, in the average case the number and size of higher order partial correlations that need to be examined will be much reduced. Nevertheless, the reliance on significance tests for high order partial correlations suggests that TETRAD II will be unlikely to recover the structure of large models without very large samples available: since a high order partial correlation estimate depends upon estimates of the marginal correlations for each pair of variables involved, the uncertainties associated with each estimate will accumulate, which results in high standard errors (variance) for high order partial correlation estimates and therefore in the need for very large samples to get significant results.[40] In fact, when testing a model with 27 variables and 54 direct causal connections forming a large number of v-structures and with all coefficients of magnitude 0.4 or larger, MML (using a Monte Carlo sampling technique) recovered exactly the original model structure, whereas TETRAD II was unable to orient 35 of the 52 arcs it discovered, despite the fact that almost all arc directions in the model were required to retain statistical equivalence (Wallace and Korb, 1997).

The requirement to examine the partialling out of all subsets excluding X and Y also implies a conceptual difficulty in the individual tests for vanishing partial correlations: whereas it is quite correct to hold parents and/or ancestors of variables constant in order to determine if there remains a correlation between two variables that needs to be explained by a direct causal connection, it is not correct to hold their common effects constant, as Principle I does since it examines *all* subsets of excluding the variables in question. As a result, in applying Principle I to large models as the order of the partial correlation is increased not only is the influence of common causes on the correlation between two variables reduced, as is desirable, but also the induced dependencies between variables by holding common effects constant will increase, if they have multiple common effects. The induced dependencies via common effects may either reinforce or weaken the dependencies between two variables and so may lead to spurious links being introduced or real links being ignored. The result will be the induction of causal structures which are inferior as models for the sample data.[41]

The concept of probabilistic causality (ES*) suggests a solution for this problem:[42] only one partial correlation need be tested in principle and that is the one that removes the influence of all the direct parents of either of the variables in question. Variables which cannot be causal for the effects in question, because they occur after these variables take their values or because they are otherwise known not to impinge upon them, should not be partialled out. The difficulty with this is that in circumstances of primitive induction (no background knowledge) we do not know the temporal order of variables nor which are the parent variables, so we cannot directly implement this suggestion. However, we can implement it indirectly.

8. BAYESIAN CAUSAL INDUCTION

A Bayesian method of automating inductive processes, called Minimum Message Length (MML) inference (e.g., Georgeff and Wallace, 1984; Wallace and Patrick 1993; Quinlan and Rivest, 1989, use a related method, Minimum Description Length (MDL), for automated classification), has been steadily gaining recognition within the machine learning community over the last fifteen years. Put simply, the idea is to find an efficient encoding method (efficient in the information-theoretic sense of minimizing entropy) for both theories and the data relevant to assessing the theories. Given some data we can use the encoding technique to search for that theory which results in the minimum joint encoding for both theory and data. MML explicitly plays the well known trade-off between further elaborating a theory so as to more precisely account for the data (resulting in a greater likelihood value) and simplifying the theory so as to maximize its prior probability; this is both because theory and data are part of the message whose length is to be minimized and because more complex theories may afford a savings in that part of the message recording the data (e.g., if a theory strictly implies a certain property of the data, then it will not be explicitly encoded at all). From Shannon's equation (Shannon and Weaver, 1959) relating the information content (message length) of the message communicating hypothesis and data to probability (and applying Bayes' theorem) we can maximize the posterior probability of our hypothesis h (ignoring the constant term $-log_2P(e)$) by minimizing the message length: $I(h\&e) = -log_2P(h) - log_2P(e|h)$.

If the encoding method specifically takes a full joint prior probability into account, this is an explicitly Bayesian method of determining posteriors.[43] In my joint work with Wallace and Dai (Wallace et al., 1996; Dai et al., 1997), we have thus far assumed no specific prior knowledge and have employed encoding techniques which reward theories (give them smaller encodings) for having fewer variables and for moderate degrees of connectivity.[44] Wallace et al. (1996) describes a method of efficiently encoding a causal model, including parameters, and a joint sample in the model's variables. Using the MML measure to evaluate causal models, we employ a standard greedy search to the space of measured models attempting to find that model for which the MML measure is minimized. In other words, given an initial model (which we construct based upon strong marginal correlations between variables) we apply operators to add, delete or reverse arcs between variables. After each change we prefer whichever of the new or the last model gives the better MML performance. We continue with this until all single changes fail to produce an improvement.[45] The result is a program, MML–CI (for MML Causal Inducer), which equals or betters TETRAD II in discovering the causal model which generated the sample (or a statistically equivalent model) in a variety of cases.[46]

The search technique that we employ is heuristic, rather than exhaustive. An exhaustive search would be prohibitively computationally expensive. To the extent that the search is effective it will guide the program away from models useless in explaining sample data, so we can hope that important alternative models will not be missed (of course, the evidence of MML–CI finding good models supports such optimism). Furthermore, additional kinds of search are being examined (such as Gibbs sampling and genetic algorithms; Neil and Korb, 1996, applies genetic algorithms using MML to an analogous graphical search problem).

On the assumption that the search is examining all models that might relevantly be compared in their ability to account for the sample data, the problems raised for TETRAD IIdo not arise for MML–CI. The first problem mentioned was that prior information supporting or undermining a specific causal relation cannot be accommodated in partial correlation significance tests. In MML, however, there is no intrinsic difficulty in encoding arcs in a manner biased by the prior probability of their obtaining.[47] The second difficulty I raised was that Principle I holds fixed common effect variables as well as common causes. This appears unavoidable without prior knowledge of what causal structure is being looked for, or at least the knowledge of the temporal ordering of variables. MML–CI copes with this, however, assuming that the search technique reaches the right causal models for examination. In particular, where Principles I and II lead to a causal structure which is not statistically equivalent to the original model, as in the footnoted example above, MML–CI in comparing that model with the original (assuming it makes that comparison) will find that their likelihoods differ. This may not lead to the adoption of the correct model if the difference in prior probabilities is greater than the likelihood difference; however, as sample sizes increase the correct model (or its equivalence class) will come to be preferred. This cannot occur with TETRAD II since any bias introduced by partialling out common effects is systematic (assuming that the possible sets of variables for partialling out are examined in the same order).

MML provides, if not exactly a Bayesian posterior evaluation of models, then something very like it; indeed, it provides an evaluation directly formally related to Bayesian posteriors. The result is a method which extracts all relevant information from the sample data in assessing models, and, in particular, which does not ignore relevant information about conditional dependencies and independencies subsequent to the failure of a significance test, as does TETRAD II.

In short, we believe that we have made a promising start to the automation of causal discovery, grounding this process on the same Bayesian principles that have proven to be successful in qualitative analyses of scientific method (see Howson and Urbach,

1993), but in this case employing quantitative Bayesian methods. The limitations which we have adopted so far in both the hypothesis space (linearity, fully measured variables, etc.) and the search methods must be freely acknowledged, but we have no reason to believe they may not be removed piecemeal as progress is made.

9. THE DEFINITION OF PROBABILISTIC CAUSALITY

Finally, I close with the "definition" of probabilistic causality.[48] *Condition CM*:

(D7) C causes E iff $p_{EC} \neq 0$ in the true causal model incorporating the two variables.

I shall pass on the subject of what is a *true* causal model, although I think for a causal model to describe a physical system completely the system would have to be closed, so whatever the concept is it must be an ideal rather than a condition which we can expect to satisfy in ordinary life. The inductive methods I have been describing here rely upon the systems under study being closed and, in particular, there not being any relevant variables that have not been measured in the joint sample. Glymour's group, and others, have been applying various methods to the discovery of latent (hidden) variables as well; and factor analysis has been doing similar work for social scientists since the beginning of the century. So our inductive methods are surely extensible in like manner. However, the goal towards which all of these methods strive is just the discovery of a complete model accurately describing a physical system of interest – which is presumably not strictly obtainable. Nevertheless, something like the above idealized account of causality must be correct if any such program of learning causal structure by learning models from sample information has any merit whatsoever.

There are two respects (at least) in which this condition differs from ES*. Causal modeling is primarily concerned with variables that take continuous, rather than discrete, values. More importantly, linear causal models cannot represent non-linear causal interactions, which are in fact both common and important (as in the case of emphysema, smoking and air pollution). As automated modeling methods are extended to non-linear cases, which is beginning to happen,[49] the implied criterion for a probabilistic causal relation – that a parameter value for the relation between C and E is required in the true model – will be extended in scope as well, presumably in a way that subsumes both CM and ES*.

Monash University

NOTES

1 I thank Wesley Salmon, Gurol Irzik, Michael Dickson, Chris Wallace, James Forbes and Honghua Dai for discussions helpful in the writing of this paper. I also thank Howard Sankey for organizing the symposium on laws and causation at the 1996 Australasian Association for the History, Philosophy and Social Studies of Science, where this paper was presented. Finally, I especially thank Linda Wessels for her seminar on the subject.

2 This is itself undoubtedly false, contra Russell: any number of scientific papers discussing, for example, biological function or cosmogony would simply be unwritable without the use of the concept of causation.

3 That story is analogous to, if longer than, the story in AI of attempts by logicists to produce rules for

defeasible logics that capture something like human defeasible inferences while avoiding nonsense inferences; see Korb, 1995.

[4] I make no attempt at an *exhaustive* review. For example, I pass over Good's treatment of causality (1961–62) because, although it is quite interesting, it has strengths and weakness very different from those of the analyses that feed more directly into my account of causality.

[5] In any case, I agree with Nancy Cartwright that both causality and physical probabilities (i.e., *propensities*) are likely part of some grand circle of concepts that cannot be eliminated in accord with empiricist principles.

[6] I shall also occasionally write $P(A|B)$ as $P_B(A)$ on the ground that conditionalized probability functions are themselves probability functions.

[7] I do not, however, require that reference classes be the *broadest* available objectively homogeneous reference class – i.e., that they be the largest such classes. That is a requirement that Salmon (1967) introduced and can be readily motivated in the context of understanding scientific explanations, which do not allow the introduction of properties that irrelevantly reduce the size of reference classes. For problems of statistical inference, such as whether or not a causal law obtains, the introduction of irrelevancies is itself irrelevant (*pace* Cartwright, 1983, pp. 36–9).

[8] Naturally, if determinism is a fact, then we will *never* have the right reference classes in hand, unless we have those leading to probabilities of 1 or 0.

[9] Sober (1987) objects to the common cause principle that we routinely find correlations between variables that are unrelated causally, for example, bread prices and the Venetian water level. Indeed, the correlation between time-dependent variables is frequently cited as evidence against the viability of causal inference. Spirtes, et al. (1993, p. 63) claim that it is a sufficient response to treat the changing relationship between variables as a succession of causal structures. Since we know on independent grounds that mixing causal structures will lead to incorrect inferences about causal relationships (because we need objective homogeneity; see Spirtes, et al., 1993, pp. 59–60), we can attribute the spurious relation between price and sea levels to such mixing. But we might wonder whether we in fact have hold of the real causal structure if it is metamorphosing in our grasp: shouldn't the *underlying* causal structure of a physical system be time invariant? If we suppose so, we can nevertheless immediately see that Sober's objection provides no counterexample to Reichenbach's principle: as the very phrase 'time-dependent' suggests, there is indeed a common causal factor operative, and one which is as measurable as any, namely time. If we ignore time as a variable, then, just as when we ignore any important causal factor, we are likely to infer direct causal connections that turn out not to exist.

[10] In this I am being consistent with Suppes, but also note that a Suppesian definition of type causality could be given by applying temporal constraints to the instances of the relevant types.

[11] Another case that has generated considerable and convoluted discussions (e.g., Eells, 1991; Menzies, 1989) is one of Cartwright (1983, p. 28): suppose a certain defoliant kills plants with the same probability that striking trees fails to lead to holes in one; we spray a plant with the defoliant; it nevertheless lives on. It has been widely claimed that this and the golfing example are asymmetrical. Whereas striking the tree *really does* (somehow) cause the hole in one, spraying defoliant really does *not* cause the plant to survive, even if the probabilistic structures are the same – we are much happier saying, for example, that the plant lives *despite* the defoliant than we are saying that the hole in one occurred *despite* the ball's striking the tree. But I think it worth pointing out that the linguistic facts are likely due to real differences in probabilistic structure (revealed counterfactually: What would have happened if the ball missed the tree? If the plant were not sprayed?). The claimed asymmetry can just be denied: the striking of the tree forms a crucial part of the causal trajectory of the hole-in-one process; but equally the spraying, absorption and (say) neutralization of the defoliant is a crucial part of the causal trajectory of the life of the plant.

[12] Jerzy Neyman (1950, pp. 53–4), although not directly concerned with homogeneity, emphasized the importance of identifying the appropriate reference class for a problem, which he called "Fundamental Probability Sets," and provided an amusing example where, by ignoring reference classes, he was able to generate a "probability" of 1.3.

[13] Salmon does not appear to put the matter so baldly as this. However, this is surely the natural interpretation of much of what he does say. For example, in summing up his treatment of the Rosen golfing example it is simply the fact that there is a continuous causal process involving the tree limb that matters, and not any probabilistic relation (Salmon, 1980).

[14] In a recent exchange with Hitchcock (1995), who has also made this point, Salmon accepts the need for

a combined account of causal explanation, employing both causal processes and statistical relevance (in Salmon, unpublished).

15 Evidently Salmon also has noticed that positive statistical relevance is necessary, for at the end of his argument that it is necessary to allow for negative relevance he does a remarkable backflip, endorsing what he calls 'internal positive relevance' (1980, p. 68). Suppose we have a game involving two ways of winning, for example in craps you can win by rolling seven or eleven or you can win the "hard way" by rolling, say, a three and then making your point (rolling three again before rolling seven or eleven). Making a second three is less likely than rolling a seven or eleven and losing, so whenever you do this you have won via a negatively relevant intermediate causal factor, namely getting the initial three. Salmon's thought is that being in this state is nevertheless positive to winning relative to *something* – namely, walking away from the craps table. This waffle is supposed to account for our intuition that positive relevance is *somehow* involved in causality. Most causal processes however do not provide any opportunity to "opt out"; clearly the quantum mechanical cases do not allow for the electron and photon, for example, to pack up their bags and visit another world – nor is it plausible to view the golf ball's striking the tree as positively relevant to a hole in one relative to, say, the golf ball's exploding. But also in game-playing one wants to respond that even should there be no opportunity to opt out it remains possible that one should win the "hard way."

16 None of this is intended to say that the relations between token causal processes and probability cannot be got right. Ellery Eells attempts to do just that in the "probability trajectory" approach to token causality (see Eells, 1991, chapter 6; Dowe, 1996).

17 Salmon instead defines an interactive fork as an intersection of two causal processes such that "they undergo correlated modifications that persist after the intersection" (1984, p. 170) and then goes on to suggest an *analogy* with conjunctive forks, modifying the probabilistic conditions as above. This amounts to the same thing, however, since the subsequent correlation between the processes that interacted is either fully explained by the common cause (the initial interaction) or it is not. If it is fully explained, we have a conjunctive fork, so the remainder, interactive forks, are *required* to satisfy the modified probabilistic condition.

In Salmon's treatment, he is concerned to emphasize that causal relations, such as interactive forks, cannot be fully analyzed in probabilistic terms (see especially Salmon, 1990). While sharing his overall viewpoint, I do not share his concern: even were causal concepts analyzable into physical probabilities, this would be no victory for reductionism, since *physical probability* is not likely to be intelligible without the causality concept.

18 Note that this is an outstanding difference between pool and golfing off trees: there are no trick golf shots off tree branches.

19 My remarks about contextual unanimity, here and subsequently, apply equally to those versions which use Pareto-style comparison, requiring a causal factor to be positively relevant in at least one subpopulation and non-negative elsewhere (see Skyrms, 1980).

20 Tooley's summing up of this kind of case makes his presupposition of inhomogeneity quite clear (1987, p. 234): "The upshot of this is that if causation entails positive statistical relevance, then [it] follows that whether or not events of one type cause events of other types will depend, in some cases, upon the relative probabilities of still other types of events." The "still other types" of events in question are those which establish homogeneous subpopulations on my account.

21 Cow pox and smallpox appear to be examples as well, although there is no advantage to exposure to cow pox after the fact.

22 It is a bit odd that Giere, a noted propensity theorist, employs *frequencies* in his criterion, rather than probabilities. But Giere was after a practical criterion of causality that would be responsive to the measured frequencies of the effect in experimental settings.

23 John Dupré, on the other hand, asserts dramatically, and thoroughly wrongly, that probabilistic causality accounts are constrained to regard randomized experiments as not providing "any information at all about causality" (1993, p. 202) because of their averaging of causal effects.

24 If this is truly the right interpretation, though, C is not (just) a contraceptive pill but (also) an abortion pill.

25 Assuming that the S_i are the same size; the adjustment otherwise is trivial.

26 Note that my account reunites what Dupré would split assunder (1993, p. 195): "In fact there are two causally significant modes of variation that should be distinguished. The first . . . consists of the different causally relevant contexts in which the cause may occur. In addition, the cause may affect the probability

of its effect in more than one way, most obviously through the production of intermediate effects with differing or even opposite influences on the probability of the final effect. Different routes may link the cause to the effect." But the homogeneous reference class – homogeneous also in the counterfactual properties of which intermediate state would be attained without the influence of the causal agent – subsumes both ways of influencing the probability of the effect relative to the cause.

27 His remarks concern only CC directly, since Cartwright had not presented CC* in 1986; however, they apply just as readily to her later criterion.

28 This because Cartwright does not countenance holding fixed non-causal, supposedly irrelevant, factors in testing, a proviso which is not well motivated (see Eells and Sober, 1983, for a discussion). Nevertheless, simply removing that proviso will not work either. The argument against it, after all, is just that holding other facts fixed *is* (or ought to be) irrelevant to the causal verdict. But in this case it manifestly is not irrelevant. Therefore, the right conclusion seems to be that the unanimity requirement for identifying causal factors is suspect, at least within the context of CC*.

29 Although elsewhere Cartwright appears to mean by *capacity* roughly what I mean by *propensity*.

30 As is the even stronger and more peculiar assertion of hers (p. 146) that "where capacities are at work . . . one can infer from one causal law directly to another, without ever having to do more tests." This does not recommend itself as an addition to experimental scientific method, particularly as the evidence available that a capacity (in this sense) *is* at work appears simply to be that we have unanimity in statistical relevance across subpopulations.

31 Eells comes close to acknowledging the need for a more general account of causal interaction (cf. 1986, footnote 5, p. 55); however, he does not pursue his own hint and evidently does not realize that this leads to a rejection of the importance of the concept of contextual unanimity.

32 Some preliminary work on the discovery of causal models with both interactive and non-interactive causal factors may be found in Neil *et al.* (1999).

33 This is by contrast with such earlier attempts to model reasoning under uncertainty as MYCIN's certainty factors, which deviated dramatically from accuracy when strong probabilistic independence assumptions failed (see Heckerman, 1986).

34 It should be noted that there are a variety of methods of converting non-linear into linear models, and so extending the range of the techniques developed in the social sciences. For example, taking logs of some non-linear functions will convert them into linear form; again, Langley et al. (1987) describe heuristics for discovering quantitative relations which have similar effect. These techniques nevertheless do not apply to nearly as wide a range of relations as can be represented in Bayesian networks.

35 I am following Wright in writing the index for the causal node second. Note that Wright's rule assumes that variables have been standardized.

36 The number of directed acyclic graphs is $3^{n(n-1)/2}$ minus the number of cycles possible in n variables.

37 Chickering, following Verma and Pearl (1990), defines statistical equivalence to mean that the causal structure of one model can be parameterized to represent any probability distribution that the other can represent. In terms of linear models the corresponding concept is that one model can be parameterized to represent any correlation matrix that the other can represent.

38 This principle may be compared to Pearl's (1988) theorem 5, p. 98.

39 It is also true that TETRAD II incorporates prior information in the form of allowed or disallowed temporal relations between variables. But that is a very limited approach to incorporating prior information.

40 My thanks to Chris Wallace for pointing this out.

41 Suppose, for example, that in Figure 4(c) we add a link from node 2 to 3 and parameterize with the following path coefficients: $p_{21} = .2$; $p_{31} = .5$; $p_{32} = .1$; $p_{42} = .373$; $p_{43} = .373$. Since $\varrho_{23.4} = 0$ it is possible for TETRAD II to deny the link between nodes 2 and 3 (depending on the order in which it examines partial correlations). Assuming then it adopts the skeleton of Figure 4(c) (dropping the link between 2 and 3), all the arcs will have to be oriented upwards (assuming no sampling error on correlation values), following Principle II. Examining the original and the recovered models using Wright's rules and the correlation values implied by the path coefficients above we will also find that parameter estimates for the recovered model are inconsistent.

Note that there is a possible defence in pointing out that such situations arise only given specific population parameters – in principle, the correlations induced by partialling out descendants must exactly match those induced by ancestors and direct causal links before a causal link will be incorrectly deleted. And such configurations of parameter values have measure zero. (Of course, there is no parallel defence to the spurious introduction of causal links due to induced dependencies when partialling out descendants.) It does not follow, however, that these configurations are unimportant or that TETRAD's method is trouble-free: although in the limit, when we have infinite amounts of data, the probability of encountering this problem will be zero, in the real world our inductive algorithms must attempt to learn causal structure with finite and noisy data. (We examine some of the implications of finite data for parameter estimation in causal discovery in Wallace and Korb, 1997.)

[42] To be sure, this solution is already apparent from Pearl's theorem 5, loc. cit.

[43] And if such a full prior is available, there are established means of generating encodings which accurately reflect them (cf. Cover and Thomas, 1991).

[44] The more heavily connected a dag is, the more parameters need to be encoded and also the fewer linear extensions it has (topological sorts respecting its dag structure; our encoding method rewards models which have larger numbers of linear extensions).

[45] Of course, we can improve on the search in a variety of ways, such as looking ahead to the result of multiple changes to the model, which will help the search out of a local minimum.

[46] Those being seven cases from the social science literature in Wallace et al. (1996) and six artificial models in Dai et al. (1997), which in particular shows MML–CI outperforming TETRAD II at small sample sizes.

[47] However, we have not in fact done that yet; the arc encodings we have used so far assume a constant probability of 1/2 that any pair of nodes is connected by a direct link.

[48] Cf. Irzik and Meyer (1987).

[49] E.g., Heckerman and Geiger, 1995; Wang, 1997. TETRAD II can also learn discrete-valued networks, applying a χ^2 test instead of the partial correlation test (Spirtes et al., 1993, pp. 128ff.).

REFERENCES

Aspect, A., P. Grangier and G. Roger (1982), 'Experimental Realization of Einstein-Podolsky-Rosen-Bohm *Gedankenexperiment*: A New Violation of Bell's Inequalities,' *Physical Review Letters* **49**, 91–4

Bell, J.S. (1966), 'On the Problem of Hidden Variables in Quantum Mechanics,' *Reviews of Modern Physics* **38**, 447–52

Bromberger, S. (1970), 'Why Questions,' in B. Brody (ed.), *Readings in the Philosophy of Science*, Prentice-Hall

Carnap, R. (1928), *Der logische Aufbau der Welt*, Berlin

Carnap, R. (1936/37), 'Testability and Meaning,' *Philosophy of Science* **3**, 419–71, and **4**, 1–40

Carnap, R. (1962), *The Logical Foundations of Probability*, second edition, Chicago: University of Chicago

Cartwright, N. (1983), 'Causal Laws and Effective Strategies,' in N. Cartwright *How the Laws of Physics Lie*, Oxford: Clarendon Press, 21–43

Cartwright, N. (1989), *Nature's Capacities and their Measurement*, Oxford: Clarendon Press

Chickering, D. (1995), 'A Transformational Characterization of Equivalent Bayesian Network Structures,' *Proceedings of the 11th Conference on Uncertainty in Artificial Intelligence*, Morgan Kaufmann, 87–98

Collier, J. (1983), 'Frequency-Dependent Causation: A Defense of Giere,' *Philosophy of Science* **50**, 618–25

Cover, T. and J. Thomas (1991), *Elements of Information Theory*, New York: John Wiley and Sons

Dai, H., K. Korb, Wallace, C.S. and Wu, X. (1997) A Study of Causal Discovery with Weak Links and Small Samples, *Fifteenth International Joint Conference on Artificial Intelligence*, Nagoya, Japan, 23–9 August, 1997

Davis, W. (1988), 'Probabilistic Theories of Causation,' in J. Fetzer (ed.), *Probability and Causality*, Dordrecht: Dordrecht Reidel

de Dombel, F.T., D.J. Leaper, J.R. Staniland, A.P. McCann and J.C. Horricks (1972), 'Computer-Aided Diagnosis of Acute Abdominal Pain,' *British Medical Journal* **2**

Dowe, P. (1996), 'Chance Lowering Causes: Old Problems for New Versions of the Probabilistic Theory of Causation,' in D. Dowe, K. Korb and J. Oliver (eds) *Information, Statistics and Induction in Science*, Singapore: World Scientific, 226–36

Duda, R., P.E. Hart, P. Barrett, J.G. Gaschnig, K. Konolige, R. Reboh and J. Slocum (1978), *Development of the PROSPECTOR Consultation System for Mineral Exploration: Final Report*, Menlo Park: SRI International

Dupré, J. (1993), *The Disorder of Things: Metaphysical Foundations of the Disunity of Science*, Harvard University Press
Eells, E. (1986), 'Probabilistic Causal Interaction,' *Philosophy of Science* **53**, 52–64
Eells, E. (1991), *Probabilistic Causality*, Cambridge: Cambridge University
Eells, E. and E. Sober (1983), 'Probabilistic Causality and the Question of Transitivity,' *Philosophy of Science* **50**, 35–57
Einstein, A., B. Podolsky and N. Rosen (1935), 'Can Quantum-Mechanical Description of Physical Reality Be Considered Complete?' *Physical Review* **47**, 777–80
Georgeff, M. and C. Wallace (1984), 'A General Selection Criterion for Inductive Inference,' *Advances in Artificial Intelligence*, Amsterdam: North Holland, 219–28
Giere, R.N. (1973), 'Objective Single Case Probabilities and the Foundations of Statistics,' in P. Suppes *et al.* (eds.) *Logic, Methodology and Philosophy of Science IV*, Amsterdam: North Holland, 467–83
Giere, R.N. (1979), *Understanding Scientific Reasoning*, New York: Holt, Rinehart and Winston
Giere, R.N. (1984), 'Causal Models with Frequency Dependence,' *Journal of Philosophy* **81**, 384–91
Glymour, C. (1980), *Theory and Evidence*, Princeton
Glymour, C., R. Scheines, P. Spirtes and K. Kelly (1987), *Discovering Causal Structure*, New York: Academic Press
Good, I.J. (1961–62), 'A Causal Calculus I and II,' *British Journal for the Philosophy of Science* **11**, 305–18, and **12**, 43–51 ("Corrigenda" volume **13**, 88)
Heckerman, D. (1986), 'Probabilistic Interpretations for MYCIN's Certainty Factors,' in L.N. Kanal and J.F. Lemmer (eds.), *Uncertainty in Artificial Intelligence*, Amsterdam: North Holland
Heckerman, D. and D. Geiger (1995), 'Likelihoods and Priors for Learning Bayesian Networks,' *Neural Information Processing Systems 95 Workshop on Learning in Bayesian Networks and Other Graphical Models*
Hesslow, G. (1976), 'Two Notes on the Probabilistic Approach to Causality,' *Philosophy of Science* **43**, 290–2
Hitchcock, C. (1995), 'Discussion: Salmon on Explanatory Relevance,' *Philosophy of Science* **62**, 304–20
Howson, C. and P. Urbach (1993), *Scientific Reasoning: the Bayesian Approach*, second edition, La Salle, Ill.: Open Court
Hume, D. (1739/1962), *A Treatise of Human Nature*, Cleveland: World Publishing
Irzik, G. and E. Meyer (1987), 'Causal Modelling: New Directions for Statistical Explanation,' *Philosophy of Science* **54**, 495–514
Korb, K. (1992), *A Pragmatic Bayesian Platform for Automating Scientific Induction*, PhD dissertation, Indiana University
Korb, K. (1995), 'Inductive Learning and Defeasible Inference,' *Journal of Experimental and Theoretical Artificial Intelligence* **7**, 291–324
Langley, P., H. Simon, G. Bradshaw, J. Zytkow (1987), *Scientific Discovery*, Cambridge, Mass.: MIT Press
Lewis, D. (1980), 'A Subjectivist's Guide to Objective Chance,' in R. Jeffrey (ed.), *Studies in Inductive Logic and Probability, vol. II*, University of California Press, 263–93
Mackie, J. (1965), 'Causes and Conditions,' *American Philosophical Quarterly* **2**, 245–64
Mellor, D. (1995), *The Facts of Causation*, London: Routledge
Menzies, P. (1989), 'Probabilistic Causation and Causal Processes: A Critique of Lewis,' *Philosophy of Science* **56**, 642–63
Merck Research Laboratories (1992), *The Merck Manual of Diagnosis and Therapy*, Rahway, NJ: Merck Research Laboratories
Neapolitan, R. (1990), *Probabilistic Reasoning in Expert Systems*, New York: Wiley
Neil, J. and K. Korb (1996), 'The MML Evolution of Classification Graphs,' in D. Dowe, K. Korb, J. Oliver (eds.), *Information, Statistics and Induction in Science*, Singapore: World Scientific, 78–89
Neil, J., C.S. Wallace and K.B. korb (1999), 'Learning Bayesian Networks with Restricted Causal Interactions', *uncertainty in Artificial Intelligence*, San Mateo: Morgan Kaufman, pp. 486–493
Neyman, J. (1950), *First Course in Probability and Statistics*, New York: Holt
Pearl, J. (1988), *Probabilistic Reasoning in Intelligent Systems*, Morgan Kaufmann
Popper, K. (1959), 'The Propensity Interpretation of Probability,' *The British Journal for the Philosophy of Science* **10**, 25–42

Quinlan, J.R. and R.L. Rivest (1989), 'Inferring Decision Trees Using the Minimum Description Length Principle,' *Information and Computation* **80**, 227–48

Reichenbach, H. (1956), *The Direction of Time*, University of California Press

Rosen, D. (1978), 'In Defense of a Probabilistic Theory of Causality,' *Philosophy of Science* **45**, 604–13

Russell, B. (1912), 'On the Notion of Cause,' reprinted in B. Russell *Mysticism and Logic*, London: Allen and Unwin, 132–51

Salmon, W. (1967), *The Foundations of Scientific Inference*, Pittsburgh: University of Pittsburgh

Salmon, W. (1971), *Statistical Explanation and Statistical Relevance*, Pittsburgh: University of Pittsburgh

Salmon, W. (1980), 'Probabilistic Causality,' *Pacific Philosophical Quarterly* **61**, 50–74

Salmon, W. (1984), *Scientific Explanation and the Causal Structure of the World*, Princeton: Princeton University

Salmon, W. (1990), 'Causal Propensities: Statistical Causality vs. Aleatory Causality,' *Topoi* **9**, 95–100

Salmon, W. (unpublished), 'Causality and Explanation: A Reply to Two Critiques'

Shannon, C.E. and W. Weaver (1959), *The mathematical theory of communication*, Urbana: University of Illinois

Simpson, E.H. (1951), 'The Interpretation of Interaction in Contingency Tables,' *Journal of the Royal Statistical Society, Series B* **13**, 238–41

Skyrms, B. (1980), *Causal Necessity*, New haven: Yale University Press

Sober, E. (1982), 'Frequency-Dependent Causation,' *Journal of Philosophy* **79**, 247–53

Sober, E. (1984), 'Two Concepts of Cause,' *PSA 1984*, volume 2, 405–24

Sober, E. (1987), 'Parsimony, Likelihood, and the Principle of the Common Cause,' *Philosophy of Science* **54**, 465–9

Sosa, E. and M. Tooley (1993), 'Introduction.' In E. Sosa and M. Tooley (eds.), *Causation*, Oxford: Oxford University Press

Spirtes, P., C. Glymour and R. Scheines (1990), 'Causality from Probability,' in J.E. Tiles, G.T. McKee and G.C. Dean (eds.), *Evolving Knowledge in Natural Science and Artificial Intelligence*, London: Pitman

Spirtes, P., C. Glymour, R. Scheines (1993), *Causation, Prediction and Search*, Berlin: Springer Verlag

Spirtes, P., C. Glymour, R. Scheines, C. Meek (1994), *TETRAD II: Tools for Causal Modeling*, Lawrence Erlbaum

Suppes, P. (1970), *A Probabilistic Theory of Causality*, Amsterdam: North Holland

Tooley, M. (1987), *Causation: A Realist Approach*, Oxford: Clarendon Press

Verma, T. and J. Pearl (1990), 'Equivalence and Synthesis of Causal Models,' *Proceedings of the 6th Conference on Uncertainty in Artificial Intelligence*, Morgan Kaufmann, 220–7

Wallace, C. and K. Korb (1997), 'A Study of Causal Discovery by MML Sampling,' forthcoming in M. Slater (ed.), *Causal Models and Intelligent Data Analysis*, Springer Verlag

Wallace, C., K. Korb and H. Dai (1996), 'Causal Discovery via MML,' *Machine Learning: Proceedings of the 13th International Conference*, Morgan Kaufmann, 516–24

Wallace, C. and J. Patrick (1993), 'Coding Decision Trees,' *Machine Learning* **11**, 7–22

Wang, D. (1997), *Gibbs Sampling for Learning a Bayesian Network via MML*, Masters Thesis, School of Computer Science and Engineering, University of New South Wales, Australia

Wright, S. (1934), 'The Method of Path Coefficients,' *Annals of Mathematical Statistics* **5**, 161–215

Yule, G. (1904), *An Introduction to the Theory of Statistics*, London: Griffin

PETER MENZIES

INTRINSIC VERSUS EXTRINSIC CONCEPTIONS OF CAUSATION*

I. INTRODUCTION

Hume begins his famous discussion of causation in the *Enquiry* with these words. "There are no ideas, which occur in metaphysics, more obscure and uncertain, than those of power, force, energy, or necessary connexion, of which it is every moment necessary to treat in all our disquisitions" (VII, pp. 61–2). It is well-known how he goes on to subject these ideas to a 'sceptical doubt', arguing that they are incoherent because they do not have their origins in any kind of sensory impression. Hume's own 'sceptical solution' to this doubt is also well-known: he argues that the only sensorily verifiable definition of causation must be drawn from things "extraneous and foreign", in particular, from the relations of temporal priority, spatial contiguity and regularity. He argues that the conception of the causal relation as consisting in a necessary connexion is due to our projecting onto the world a "felt determination of the mind" to pass from cause to effect.

It is an open question whether or not this conventional interpretation of Hume is accurate. (See Galen Strawson (1989), John Wright (1983) for an alternative interpretation of Hume as a sceptical realist; and Simon Blackburn's (1990) critique of this interpretation.) My interest in this interpretation lies, not in the question of its historical accuracy, but in the fact of its pervasive influence: it is an interpretation that has shaped generations of philosophers in thinking about what counts as a Humean or an anti-Humean theory of causation. I am especially interested in the competing ontological conceptions of singular causation – the different conceptions of the truthmakers for singular causal judgements – appealed to by this interpretation. On the naive conception of causation opposed by Hume, the truthmaker for a singular causal judgment is an intrinsic relation – a relation of power, energy, or necessary connexion – holding as a local matter of fact; whereas on the Humean replacement conception the truthmaker is a complex extrinsic matter determined partly by spatiotemporal relations and partly by global patterns of occurrences in the form of regularities. The conventional interpretation has it that Hume's great philosophical achievement, if one agrees with him, is to have shown that causation is not a local, intrinsic relation, but actually an extrinsic relation depending on widespread patterns of occurrences.

It is these contrasting ontological views of causation – the naive conception of causation as an intrinsic relation and the Humean conception of it as an extrinsic relation – that I want to investigate further in this paper. I wish to clarify the ontological conceptions of causation advanced by these competing views: to spell out what it is

H. Sankey (ed.), Causation and Laws of Nature, 313–329.

to think that causation is an intrinsic relation or that it is an extrinsic relation. Although contemporary philosophers are divided into two rival camps advocating the two conceptions of causation – under the heading singularist and non-singularist theories of causation – they have not successfully articulated precisely what is at stake between these contrasting views. If we are to make sure progress in deciding between these conceptions, we need to get a firmer grip on what the difference between them amounts to. That is the issue I tackle in this paper.

In Section 2 of this paper I consider the way in which contemporary philosophers have tried to distinguish rival conceptions of causation in terms of a distinction between singularist and non-singularist theories of causation. I show how this distinction is best understood as a distinction between conceptions of causation as an intrinsic relation and as an extrinsic relation. In Section 3 I offer an explanation of the difference between intrinsic and extrinsic relations in terms of an élite class of perfectly natural properties and relations. In Section 4 I clarify this explanation by responding to an objection to the effect that it carries an excessive burden of metaphysical commitments. In Section 5 I relate the distinction between conceptions of causation as an intrinsic relation and as an extrinsic relation with an increasingly-popular way philosophers have distinguished between theories of causation in terms of a thesis called Humean supervenience. I argue that the proposed distinction between conceptions of causation drawn in terms of intrinsicality/extrinsicality crosscuts the distinction drawn in terms of Humean supervenience. I make some brief concluding remarks in Section 6.

2. SINGULARIST VERSUS NON-SINGULARIST THEORIES OF CAUSATION

In recent years a number of philosophers of causation have advanced an approach to causation that is supposed to be fundamentally opposed to the Humean approach. On this so-called *singularist approach*, the truthmaker for a singular causal claim is supposed to be a local relation holding in single instances – a relation that does not depend on the existence of widespread patterns of occurrences. (Exponents of the singularist approach include G.E. Anscombe (1983), D.M. Armstrong (1983; this volume), John Bigelow and Robert Pargetter (1990), C.J. Ducasse (1927), David Fair (1979), Evan Fales (1990), Michael Tooley (1990; 1993).) It is best to explain the singularist approach by way of its motivating intuition.

Suppose I drop a piece of sodium into a beaker of acid, which event causes an explosion to take place. The intuition emphasised by singularists is that the causal relation holding between these events depends entirely on the local character of the events and the process which links them; and does not depend on anything else happening in the world. For example, suppose that another person is waiting in the wings, ready to drop a piece of sodium into the beaker of acid if I do not. Does the presence of this alternative cause, which would come into play if I do not drop the sodium, make any difference to whether the causal relation exists between my dropping the sodium and the explosion? The singularists argue that it does not: the presence of an alternative cause is neither here nor there to the causal relation that exists between the actual cause and effect. The causal relation does not depend on any other events occurring in the neighbourhood: the causal relation is intrinsic, in some sense, to the relata and the process connecting them.[1]

This is common ground among singularists. However, singularists disagree about the properties possessed by the intrinsic relation that counts as the causal relation. One school of singularists (Anscombe, Armstrong, Ducasse, and Fales) think that the singular causal relations are sometimes observable. They argue that, contrary to Hume, there are sometimes sense-impressions corresponding to the idea of causation; and indeed this fact is part of the evidence for the intrinsic character of causation. An opposing school of singularists (Bigelow and Pargetter, Fair, Tooley) believe that the intrinsic relation that counts as the causal relation is a theoretical or unobservable relation. Again, some singularists (Anscombe, Armstrong, and Tooley) believe that there is no *a priori* reason for thinking that the causal relation must automatically fall under a law. These *anomic singularists*, as I shall call them, claim that it is conceptually open that a causal relation can hold in the absence of any covering law.[2] Some anomic singularists see this claim as being supported by the motivating intuition of singularism: if the causal relation is an intrinsic relation that can hold independently of whatever events occur in its neighbourhood, it should also be thought of as a relation that can hold independently of whatever laws obtain. In contrast to this view, the *nomic singularists* (Bigelow and Pargetter, Ducasse, Fair) say that the causal relation, consisting in a certain kind of intrinsic relation, cannot be so independent that it can hold in the absence of a covering law. The intuition that guides singularism does not, in their view, go so far as to undermine the *a priori* principle requiring a causal relation to fall under a law.

Notwithstanding this disagreement, singularists of both persuasions agree on a central ontological thesis: namely, when a singular causal claim is true, it is made true by the holding of a certain kind of intrinsic relation. Singularists dispute among themselves about the properties possessed by this relation, in particular whether it is observable and whether it is necessarily law-governed. But the uncontested point for all of them is that the truthmaker – the ontological ground – of a singular causal statement is a local, intrinsic tie between events.

Despite the intuitive appeal of the idea that causation involves a local tie between events, many contemporary philosophers still regard this idea with suspicion, no doubt motivated by the belief that Hume's critique of the idea conclusively demonstrated its untenability. Hume's 'sceptical doubt' was explicitly directed at the idea that causation consists in a necessary connection, or a relation of power, force, or energy between events. Generalising slightly, however, we can interpret his critique as an attempt to undermine any theory stating that causation consists in a local, intrinsic tie connecting events. In place of this conception of causation, Hume proposed an alternative conception which he explained in slightly different ways. In the *Enquiry*, he proposed the following definition among others: "Suitably to this experience, therefore, we may define a cause to be *an object, followed by another, and where all the objects similar to the first are followed by objects similar to the second*" (VII, p. 76). This definition of causation, as well as the others he offered, have generally been regarded as unsatisfactory. Any number of attempts have been made to repair Hume's theory while preserving its essential insights. (See J.S. Mill (1879), R.B. Braithwaite (1965), C.G. Hempel (1965), J.L. Mackie (1974), Paul Horwich (1989).) These successor theories to Hume's theory have been called 'regularity theories' because of the central place they give to regularities in the determination of causal relations. In supposing that regularities are part of the truthmakers for causal claims, these theories take causation to be an extrinsic relation.

This distinctive feature of the ontology of regularity theories is made obvious in their very statement. Still, the ontology of a theory is not always evident from its surface form. A case in point is David Lewis's (1973) theory of causation, which analyses causal claims in terms of counterfactuals. In the version of the theory that assumes determinism, the claim that one event *c* caused another event *e* is analysed as the claim that there is a chain of counterfactual dependences running from *c* to *e*, where one event counterfactually depends on another just in case if the second event had not occurred the first event would not have occurred. It is not so clear from this bare statement of the theory how it actually makes causation an extrinsic phenomenon. This becomes clearer, however, when we ask after the truthmakers for the counterfactuals that analyse causal claims. On Lewis's possible worlds theory of counterfactuals, the truth conditions for counterfactuals are stated in terms of similarity relations between the actual world and other possible worlds; but the actual truthmakers for the statements of the similarity relations are facts about the actual world, in particular facts about the history of occurrences of the actual world and facts about the laws of nature of the actual world. (See Lewis (1979).) But Lewis also advances a regularity theory of laws according to which the the truthmakers for statements of law are global patterns of regularity. (See his (1986b), pp. 121–31.) When these two parts of Lewis's overall theory – his similarity theory of counterfactuals and his regularity theory of laws – are taken in conjunction, it becomes evident that the truthmakers for causal claims, for Lewis, ultimately come down to widespread patterns of particular occurrences in the actual world.

From the ontological point of view that focuses on the truthmakers of causal statements, Lewis's counterfactual theory of causation is most similar to explicit regularity theories, whatever their differences in other respects. From the ontological perspective, both kinds of theory are opposed to singularist theories in claiming that causal claims are made true, not by intrinsic, local ties between events, but widespread patterns of occurrences. They are aptly grouped together as *non-singularist theories*.

In my view, the dispute between singularist and non-singularist theories of causation is one of the most interesting in the philosophy of causation. At the heart of this dispute is the fundamental ontological issue of whether causation is intrinsic or extrinsic in character. It is this crucial issue that yields the most significant ontological division between philosophical theories of causation. On the one hand, we have singularist theories that take causation to be intrinsic; and on the other side, we have non-singularist theories that take it to be extrinsic. Just counting heads, one would have to say that the non-singularist position represents the orthodoxy. Far and away the majority of theories assign causation an extrinsic character, mainly through the central role they assign to regularities as truthmakers for singular causal relations. My own view is that this orthodoxy is mistaken: the intuition that causation is an intrinsic relation between events, the intuition that Hume sought to dismiss as incoherent, forms the absolute crux of our concept of causation. This view must, however, await elaboration and defence in another place.

So far our discussion of singularist and non-singularist theories has appealed to the notions of intrinsic and extrinsic in an informal way, relying on an intuitive grasp of the distinction. It is time to clarify these notions and to see if we can arrive at a precise characterisation of the distinction. With this clarification achieved, what I take to be the fundamental ontological question about causation will emerge more clearly.

3. THE INTRINSIC/EXTRINSIC DISTINCTION

What does it mean to say that causation is intrinsic or that it is extrinsic in character? Because the ontological character of the causal relation itself has rarely been the focus of philosophical attention, very few philosophers have addressed these questions with any degree of explicitness. Nevertheless, there is a substantial body of philosophical literature which addresses the prior question of what makes properties and relations intrinsic or extrinsic in general. (I.L. Humberstone (1996); Jaegwon Kim (1982); Rae Langton and David Lewis (1998); David Lewis (1983a; 1986a).) We can appeal to this more general discussion to answer the specific question about causation.

In my view, the most promising account of the intrinsicality and extrinsicality of relations such as the causal relation is one which appeals to a special conception of properties and relations. By way of explanation of this special conception, let us consider two very different conceptions of properties and relations: a conception of them as abundant in number and a conception of them as sparse in number. (I am indebted to Lewis (1983a; 1986a) for the following way of distinguishing the two conceptions.) Under the first conception, there is a huge number of properties and relations, more than there are predicates of any possible human language. Indeed, there are as many properties as there are sets, because for any set, there is the property of belonging to that set. These abundant properties and relations can be very extrinsic, very gruesome, and very disjunctive in nature. They group things in ways which do not heed the real qualitative joints in nature, so that sharing these properties and relations does not count for true similarity: two things which are perfect duplicates in some intuitive sense may differ with respect to many of these abundant properties and relations. There are different ways of representing properties and relations on this conception, but one perspicuous way is to treat a property as the set of its possible instances; a dyadic relation as the set of pairs which are its possible instances; and, more generally, an n-adic relation as the set of n-'tuples which are its possible instances.

Under the second conception, properties and relations are sparse in number. Indeed, there are only as many as is required to characterise the way things are in reality comprehensively and without redundancy: an inventory of the sparse properties and relations is a non-linguistic counterpart of a primitive vocabulary for a language capable of describing reality exhaustively. It is certainly not the case that there is a property or relation of this kind corresponding to every predicate of a human language: human languages contain many predicates which do not match up with any of the sparse properties or relations and there may be sparse properties and relations which are not expressed by any predicate of any language. Moreover, the sparse properties and relations are highly specific and not at all gruesome or disjunctive in nature. They carve nature at its joints, so that sharing of them makes for true qualitative similarity: objects which have the same sparse properties are in some sense perfect duplicates.

It is the sparse properties and relations which are relevant to the characterisation of the ontological character of causation. Armstrong (1978) calls these sparse properties and relations *universals* and offers a theory according to which they are immanent in the objects which instantiate them – immanent in the sense that they are non-spatiotemporal parts of these objects. Lewis (1983a; 1986a) calls them *natural properties* and *natural relations*; and considers a number of different theories of them between which he is officially neutral. On one theory,the natural properties and relations are an élite subset of the abundant properties and relations: they are classes of possibilia like the abundant properties and relations, but they are distinguished from them by virtue of their being natural in some primitive, unanalysable way. On the other theory, natural properties and relations are understood in terms of a primitive relation of objective resemblance holding among objects: a natural property is a set of objects all of whose members resemble each other and fail to resemble the non-members of the set. In the subsequent discussion, I shall adopt Lewis's terminology of natural properties and relations, though this should not to be taken as indicating a preference for either of the theories he considers over Armstrong's theory. It is not necessary for my purposes to decide between these competing theories about the ultimate character of natural properties and relations; and so I shall remain neutral between them.

So far the distinction between natural properties and relations and the unnatural ones has been presented as an absolute distinction. But in fact those philosophers who advocate the distinction usually allow that it is a distinction that admits of degrees. For example, Lewis claims that a few properties and relations are *perfectly natural*, while others are less natural in being somewhat disjunctive or extrinsic in character; these can be obtained by not-too-complicated chains of definability from the perfectly natural properties and relations. Thus, the colours are less natural than fundamental physical properties such as mass and charge; but the colours in turn are more natural than grue and bleen. Without ignoring the fact that the distinction admits of degrees, our immediate concern is with the perfectly natural properties and relations. For these, we are told by Armstrong and Lewis, form the basis on which the complete qualitative character of everything else supervenes: given the complete distribution of the perfectly natural properties and relations in a world, the distribution of all the less natural properties and relations is fixed.

Physics provides us, Armstrong and Lewis say, with an inventory of the perfectly natural properties and relations that are instantiated in the actual world. Included in this inventory are the fundamental physical properties – the properties of particles such as their masses, their charges, their spins, and their colours – and the physical relation of spatiotemporal distance: these represent the basic features in terms of which physics attempts to give a complete description of the actual world. However, physics, even in an ideally completed form, does not provide an exhaustive inventory of all the perfectly natural properties and relations. For there are worlds where physics is different and deals with a different set of perfectly natural properties and relations. Furthermore, there are non-physicalist worlds in which some of the fundamental properties and relations of things are not physical at all. In these worlds, the distribution of the actual fundamental physical properties and relations does not provide a complete qualitative description of the way things are: that requires the introduction of perfectly natural properties and relations which are *alien to the actual world* in the sense of not being instantiated in it. The notion of alien properties and relations should not be misunderstood. It does not mean that the perfectly natural properties and relations vary from world to world. That is not the case: the perfectly natural properties and relations form a fixed set which are common to all the worlds. What varies from world to world is which perfectly natural properties and relations are instantiated: in the actual world, one set of perfectly natural properties and relations are instantiated – all physical properties and relations if physicalism is correct; and in another world, a different set of perfectly natural properties and relations – perhaps including many alien, non-physical properties and relations – are instantiated.

The point of introducing natural properties and relations is to explain the intrinsicality and extrinsicality of properties and relations. The explanation I shall give here closely follows an explanation given by Lewis (1983a, pp. 355–6; 1986a, pp. 61–2). Let us say that two objects *a* and *a′ are duplicates* if and only if they have exactly the same perfectly natural properties.[3] Then one can say that a property is *intrinsic* if and only if, for any two duplicates, either both or neither have the property. In other words, an intrinsic property is one which can never differ between duplicates. Correspondingly, a property is *extrinsic* if and only if there is some pair of duplicates one of which has the property and the other of which lacks it. While it follows from this definition that all perfectly natural properties are intrinsic, it does not follow that all intrinsic properties are perfectly natural. For, as Lewis notes, a disjunction of intrinsic properties is itself intrinsic since it can never differ between duplicates, but it is not a perfectly natural property because of its disjunctiveness. The distinction between perfectly natural and unnatural properties yields, but does not coincide with, the distinction between intrinsic and extrinsic properties.

The explanation of the intrinsicality of relations is slightly more complicated because it turns out that relations can be intrinsic in two different ways. The first way in which a relation can be intrinsic is if it supervenes on the intrinsic properties of its relata: if the pair $\langle a,b \rangle$ stand in this kind of intrinsic relation but the pair $\langle a',b' \rangle$ does not, then there must be a difference in intrinsic properties between *a* and *a′* or between *b* and *b′*. More precisely, a relation is *intrinsic to its relata*, to use Lewis's terminology, if and only if, whenever *a* and *a′* are duplicates and *b* and *b′* are duplicates, then both or neither of the pairs $\langle a,b \rangle$ and $\langle a',b' \rangle$ stand in the relation. Relations intrinsic to their relata correspond to the traditional internal relations. For example, the internal relation of congruence of shape is intrinsic to its relata in this way.

There is, however, another way in which a relation can be intrinsic. Consider, for instance, the relation of spatiotemporal distance. It does not supervene on the intrinsic properties of its relata, for it may happen that *a* and *a′* are duplicates and *b* and *b′* are duplicates, but the pair $\langle a,b \rangle$ stands in a different spatiotemporal relationship to the pair $\langle a',b' \rangle$. Yet it is true that the relation of spatiotemporal distance is intrinsic in a different sense. Let us say that $\langle a,b \rangle$ and $\langle a',b' \rangle$ are *duplicate pairs* if and only if *a* and *a′* have the same perfectly natural properties, and so do *b* and *b′*, and also the perfectly natural relations between *a* and *b* are exactly those between *a′* and *b′*. Then, again adopting Lewis's terminology, we can say a dyadic relation is *intrinsic to its pairs* if and only if, whenever $\langle a,b \rangle$ and $\langle a',b' \rangle$ are duplicate pairs, both or neither of the pairs stands in the relation. The relation of spatiotemporal distance is intrinsic to its pairs even though it is not intrinsic to its relata, because it is true that whenever *a* and *a′* have the same natural properties, *b* and *b′* have the same natural properties, and the natural relations between *a* and *b* are exactly those between *a′* and *b′*, then the pairs $\langle a,b \rangle$ and $\langle a',b' \rangle$ must stand in the same relation of spatiotemporal distance. The traditional *external relations* correspond to relations that are intrinsic to their pairs, but not intrinsic to their relata.

There is a third category of relations that will be important for our discussion. These are relations that are neither intrinsic to their relata nor intrinsic to their pairs. Consider, for example, the relation of belonging to the same owner. It does not fall into either of the two categories discussed since it is not fixed by the perfectly natural properties of its relata, nor by these taken in conjunction with the perfectly natural relations between them. For whether two things belong to the same owner depends on the existence of other things besides the two objects: it depends on the existence of the owner and everything else which is necessary for the institution of ownership. Let us call such relations which are neither intrinsic to their relata nor intrinsic to their pairs *extrinsic relations*.

The key intuition about causation that singularists emphasise is that it is a relation that is intrinsic to its pairs but not to its relata; or in more traditional terminology, it is an external relation. It is not a relation intrinsic to its relata since it is not fixed by the perfectly natural properties of the events that are its relata: if the events c and e stand in the causal relation but c' and e' do not, then it is not necessarily the case that either c and c', or e and e' must differ in their intrinsic properties. Nonetheless, causation is a relation that is intrinsic to its pairs since it is fixed by the perfectly natural properties of the events which are its relata, taken in conjunction with the perfectly natural relations they stand in: if the events c and e stand in the causal relation but c' and e' do not, then either c and c' differ in their perfectly natural properties, or e and e' do, or the pairs $\langle c,e \rangle$ and $\langle c',e' \rangle$ stand in different perfectly natural relations. If causation is a relation intrinsic to its pairs, then it is not an extrinsic relation like that of having the same owner: it is not a relation that depends on extraneous events besides the events that are its relata and whatever events may be involved in the perfectly natural relations connecting them.

As I see things, the fundamental ontological divide in the philosophy of causation is between those theories that take causation to be an extrinsic relation and those theories which take it to be an intrinsic (to its pairs) relation. (Now that I have distinguished the two ways in which a relation can be intrinsic and singled out for attention the second way – the way in which a relation is intrinsic to its pairs – I shall drop the qualification 'to its pairs': henceforth all talk of causation as an intrinsic relation should be understood as shorthand for talk of it as a relation intrinsic to its pairs.) The distinctive mark of Humean theories of causation is that they take causal relations to be extrinsic relations between events, extrinsic precisely in the sense that they are supposed to be fixed by features of reality extraneous to the causally related events, in particular by widespread regularities holding throughout the world. In contrast, the distinctive mark of anti-Humean, singularist theories of causation is that they take causal relations to be intrinsic relations, intrinsic precisely in the sense that they are determined by the causally related events, taken in isolation from everything else happening in the world, in particular by the perfectly natural properties of the events and the perfectly natural relations holding between them.

4. A CLARIFICATION

The merits of the above account of the intrinsic/extrinsic distinction as applied to the causal relation are considerable. Nonetheless, one feature of the account may invite objection; and in this section I wish to clarify it.

It may be objected that the account is unsatisfactory to the extent that it relies on the 'metaphysical' doctrine of natural and unnatural properties and relations. It may be granted that the distinction between intrinsic and extrinsic relations has some intuitive basis, but objected that the proposed explanation of this distinction in terms of the doctrine of natural and unnatural properties and relations carries an excessive burden of metaphysical commitments.

To be sure, the proposed account of the distinction between intrinsic and extrinsic relations does carry a commitment to a serious metaphysical division between kinds of properties and relations. However, in justification of this commitment, it must be said that any systematic metaphysics will need to invoke some such distinction at some point. Lewis argues (1983a; 1986a) – compellingly in my view – that the invocation of an élite class of properties and relations which are sparse in number, which carve nature at its joints, and the sharing of which makes for true similarity, helps to solve a whole range of otherwise unsolvable metaphysical problems. By appealing to this special class of properties and relations, he shows it is possible to provide a convincing explanation of the distinction between intrinsic and extrinsic properties and relations, to formulate a number of important supervenience theses, to give a plausible characterisation of the otherwise elusive nature of physicalism, and to solve some pressing problems about the content of language and thought. I find his arguments thoroughly persuasive. In my view, it is a non-negotiable requirement of any systematic metaphysics that it draw some distinction between natural properties and relations, on the one hand, and unnatural ones, on the other hand. With that said, however, I wish to remain neutral on the issue of exactly which theory of natural properties and relations is correct.

Still, it might be wondered if there is some more metaphysically neutral way of explaining the intrinsic/extrinsic distinction. For, even apart from its appeal to the primitive natural/unnatural distinction, the proposed account involves some substantial metaphysical commitments. For example, it assumes that there is an élite class of perfectly natural properties and relations; it assumes that the members of this élite class are intuitively intrinsic; and finally it assumes that the perfectly natural properties and relations in this élite class will serve as the basis on which the complete qualitative character of everything there is, and everything there could be, supervenes.

As it happens, there is a way of explaining the intrinsic/extrinsic distinction that does not buy into these substantial commitments. The explanation is one developed by Rae Langton and David Lewis (1998), starting from an idea of Jaegwon Kim (1982). Langton and Lewis define an ordered pair $\langle a, b\rangle$ as *accompanied* if and only if it coexists with some contingent object wholly distinct from both *a* and *b*. Otherwise they define the pair as *lonely*. The idea they explore is that a relation is intrinsic if and only if it is independent of accompaniment or loneliness. This is to say that the following cases are possible: a lonely ordered pair can stand in the relation, a lonely ordered pair can fail to stand in the relation, an accompanied ordered pair can stand in the relation, an accompanied ordered pair can fail to stand in the relation. The intuition this is meant to capture is that if an intrinsic relation holds or fails to hold, it does not do so in virtue of being accompanied or lonely.

Still, this does not capture the entire content of the idea of a relation's being intrinsic, as a disjunctive relation too can be independent of accompaniment or loneliness. In order to define, and so rule out, the disjunctive relations, Langton and Lewis say, it is necessary to appeal to some distinction between natural and unnatural relations. But it need not be a metaphysically robust distinction: it can be some 'vegetarian' substitute that seeks to draw a line around the natural relations in terms of the role that they play in our thinking or theorising. (For one such theory of properties see Barry Taylor (1993).) Given some or other notion of natural relations, they define the disjunctive relations as those relations that can be expressed by a disjunction of (conjunctions of) natural relations; but that are not themselves natural relations. (Or if naturalness admits of degrees, they are much less natural than the disjuncts in terms of which they can be expressed.)

With the disjunctive relations defined in this manner, Langton and Lewis rule them out as intrinsic in the following way. The *basic intrinsic relations*, they say, are those relations that are (1) independent of accompaniment or loneliness; (2) not disjunctive relations; and (3) not negations of disjunctive relations. They define two ordered pairs as *duplicates* if and only if they stand in exactly the same basic intrinsic relations. A relation is *intrinsic*, then, if and only if it can never differ between duplicate pairs.

The metaphysical commitments of this explanation are indeed fewer than that of the explanation offered in the last section. The Langton and Lewis explanation does not need to insist that there is an élite class of perfectly natural properties and relations; or that all the members of the élite class will be intrinsic; or that the élite class will serve as a basis for a complete qualitative description of everything there is or could be. To be sure, their explanation does buy into a distinction between natural and unnatural relations, but, as they say, it need not be understood in a full-blooded metaphysical way. There is good reason anyway for thinking that it is impossible to elucidate the intrinsic/extrinsic distinction without appealing at some point to substantial (not purely logical) notions such as natural properties and relations. (See also Sider (1996).)

If this is correct, why have I chosen to work in terms of the explanation of the last section instead of the Langton and Lewis explanation? I have done so because the explanation of the last section is simpler and easier to grasp. Formulating philosophical doctrines about causation in terms of this explanation of the intrinsic/extrinsic distinction makes it easier to see their implications. In the next section I shall compare the rival philosophical views about the intrinsicality or extrinsicality of causation with a view that states that causation conforms to a Humean supervenience thesis. It would not be so easy to see the relationships between these views if the competing views about the intrinsicality or extrinsicality of causation were formulated in terms of the Langton and Lewis explanation.

In any case the chief advantage of the Langton and Lewis explanation is that it is more parsimonious in its metaphysical commitments. It is not difficult, however, to modify the explanation of the last section so that it makes as few metaphysical commitments. Suppose, for instance, that there are no perfectly natural properties and relations, only an infinitely descending sequence of more and more natural ones. The explanation of the last section can be modified to take account of this possibility. Compatibly with the existence of this infinitely descending sequence, one can talk of certain natural properties and relations, but ones that are natural-enough for the given purposes of the discussion. In this case, talk of the intrinsicality of relations must be understood as relative to a contextually-determined level of naturalness. To say that the causal relation is intrinsic would be to say that it is a relation determined jointly by properties of its relata and the relations holding between its relata that are natural-enough for the purposes at hand. Again, if it is thought desirable, one can relativise the explanation of the last section so that the distinction between natural and unnatural is relative to a theory. (In the manner of Taylor (1993), for example.) On this way of looking at matters, to say that the causal relation is intrinsic would be to say that it is a relation that supervenes on certain properties of its relata and certain relations holding between its relata, properties and relations that are deemed natural by the theory in question.

If these modifications make the explanation of the intrinsicality/extrinsicality of causation more acceptable, then the explanation should be understood as so modified. It will not affect any of the subsequent discussion to read these modifications into the explanation. I myself prefer the original formulation of the explanation, not just because of its appealing simplicity, but because of the congeniality of its metaphysical assumptions.

5. HUMEAN SUPERVENIENCE

I have described the Humean position on causation as rejecting the view that causation is an intrinsic relation in favour of the view that it is an extrinsic relation. But there is an alternative characterisation of the Humean and anti-Humean positions that has become increasingly popular in recent years. This alternative characterisation formulates the difference between the positions, not in terms of a thesis about the nature of the causal relation, but in terms of a supervenience thesis. It is claimed that what is distinctive of a Humean theory of causation is that it implies that causation conforms to the thesis of Humean supervenience; and that what is distinctive of an anti-Humean theory is that it entails that causation does not conform to this supervenience thesis. In this section, I shall compare this alternative characterisation of the Humean and the anti-Humean positions with the characterisation I have offered above.

What is the thesis of Humean supervenience? This thesis was first formulated by David Lewis (1986b), who argued that much of his work, particularly his work on laws, counterfactuals, and causation, could be seen as falling into place as part of a prolonged defence of the thesis. Lewis introduced the thesis in the following terms:

> Humean supervenience is named in honour of the great denier of necessary connections. It is the doctrine that all there is to the world is a vast mosaic of local matters of particular fact, just one little thing and then another. (But it is no part of the thesis that these local matters are mental.) We have geometry: a system of external relations of spatiotemporal distance between points. Maybe points of spacetime itself, maybe point-sized bits of matter or aether or fields, maybe both. And at those points we have local qualities: perfectly natural intrinsic properties which need nothing bigger than a point at which to be instantiated. For short: we have an arrangement of qualities. And that is all. There is no difference without difference in the arrangement of qualities. All else supervenes on that. (1986b, pp. ix–x)

Lewis does not explicitly claim that conformity to the thesis of Humean supervenience is what distinguishes a Humean theory of causation, laws, or counterfactuals from an anti-Humean theory. Nonetheless, the very name he gives to the thesis certainly suggests that this is what he has in mind. In any case, other philosophers have explicitly articulated the view that adherence to the thesis of Humean supervenience is the crucial test of loyalty for an empiricist of the Humean persuasion. (See John Carroll (1994), John Earman (1984), and Michael Tooley (1987; 1993).)

Let us note a number of features of this thesis that will be important for our future discussion. First, the thesis, as Lewis formulates it, is supposed to be a global metaphysical thesis, which is supposed to apply to everything from laws to personal identity to mental content. However, the application which will be of primary concern to us is its application to the case of causation: thus, we will be primarily concerned with the specific thesis that singular causal relations supervene on spatiotemporal relations and the local, intrinsic qualities of spacetime points and point-sized bits of matter.

Secondly, the thesis of Humean supervenience, as Lewis formulates it, is an *interworld* supervenience thesis. To say that causal relations supervene on spatiotemporal relations and local, intrinsic qualities is to say that there can be no variation in causal relations without a variation in spatiotemporal relations or local, intrinsic qualities. Lewis's thesis formulates this in terms of worlds: it states that if worlds differ in terms of their causal relations, they must differ with respect to the spatiotemporal relations or local, intrinsic qualities they instantiate. As such, it differs from an *intraworld* supervenience thesis which would state that a difference in causal relations within a world must be accompanied by a difference in spatiotemporal relations or local, intrinsic qualities in the same world.

Thirdly, Lewis's thesis is supposed to be a contingent thesis. After the passage quoted above, Lewis adds by way of qualification that the thesis is intended to be a contingent truth about the actual world. So, strictly speaking, it is not true that *any* world which differs in causal relations from the actual world must differ from it in the arrangement of spatiotemporal relations or local, intrinsic qualities. To see this suppose that the only perfectly natural relations instantiated in the actual world are spatiotemporal relations: in particular there are no perfectly natural causal relations instantiated in the actual world in addition to the spatiotemporal relations. Then a world which does include such additional causal relations among its perfectly natural relations may agree with the actual world with respect to the arrangement of spatiotemporal relations and local, intrinsic qualities, but differ from the actual world with respect to causal relations.

As a contingent truth, the thesis of Humean supervenience does not hold for the whole class of worlds, but only a restricted class of worlds. Lewis characterises the relevant class of worlds as those worlds that do not instantiate any perfectly natural properties and relations alien to the actual world: in other words, the thesis holds only for those *non-alien worlds* that do not instantiate any perfectly natural properties and relations that are not instantiated in the actual world. Given the assumption that the actual world does not include additional causal relations among its perfectly natural relations, any possible world like the one envisaged just now, which does include such relations, must be considered an alien world and so must be excluded from the relevant class of worlds. The restriction of the thesis to the non-alien worlds makes it contingent because it thereby depends on contingent assumptions, such as the one made above, about which perfectly natural properties and relations are instantiated in the actual world.

What is the relation between the two characterisations of the fundamental division between Humean and anti-Humean theories of causation – the characterisation in terms of extrinsic and intrinsic relations and the characterisation in terms of the thesis of Humean supervenience? I shall argue that these two ways of characterising the division do not coincide, but rather crosscut each other. I shall try to show this by arguing that the characterisation of causation as an intrinsic relation is compatible both with claiming that it conforms to the thesis of Humean supervenience and with claiming that it fails to conform to this thesis; and the same is true for the characterisation of singular causation as an extrinsic relation.

Let us suppose that causation is an intrinsic relation in the sense explained in the last section: in other words, it is a relation determined by the perfectly natural properties of its relata and the perfectly natural relations holding between them. From this supposition, taken by itself, nothing follows as regards conformity to the thesis of Humean supervenience. But if we make a further assumption about the precise nature of the intrinsicality of causation, we can see that it follows that causation conforms to Humean supervenience. Let us assume that there are no irreducible causal relations among the perfectly natural relations instantiated in the actual world, over and above the spatiotemporal relations; and that causal relations are intrinsic in the specific sense that they supervene on the perfectly natural properties of spacetime points and point-sized bits of matter that make up their relata and on the spatiotemporal relations holding among them.

Then it follows that these causal relations conform to Humean supervenience. Take any world in the restricted set of non-alien worlds that instantiate only the perfectly natural properties and relations instantiated in the actual world. Now suppose in conformity with the antecedent hypothesis of the thesis of Humean supervenience that this world agrees with the actual world in its distribution of local, intrinsic properties and spatiotemporal relations. Then if we take any pair of events from the actual world and a duplicate pair of events from the other world, we know that the causal relation will hold of both or neither of them. But since the distribution of local, intrinsic properties and spatiotemporal relations is the same in both worlds and neither contains irreducible, perfectly natural causal relations, we can conclude that the two worlds must also agree in their overall distribution of causal relations. But this is to say that these causal relations conform to Humean supervenience.

But notice that if we make the contrary initial assumption, we arrive at a different conclusion. Let us continue to suppose that causal relations are intrinsic in the sense explained. However, let us suppose that there are irreducible causal relations among the perfectly natural relations instantiated in the actual world; and that these causal relations are intrinsic in the specific sense that they supervene on themselves, taken in their role as perfectly natural relations. Then it is simple to see that even if two worlds agree in their distribution of local, intrinsic properties and their spatiotemporal relations, the two worlds may differ in their distribution of causal relations. The presence of the perfectly natural causal relations in addition to the spatiotemporal relations allows for variation with respect to causal relations even where there is agreement with respect to local, intrinsic qualities and spatiotemporal relations. Thus, even though these causal relations are intrinsic, they do not conform to Humean supervenience.

I have argued thus far that the claim that causation is an intrinsic relation is compatible both with conformity and non-conformity with Humean supervenience. It all depends on what further assumptions are made about the particular way in which it is intrinsic; and on what contingent assumptions are made about which perfectly natural relations are instantiated in the actual world. Now I shall show that the same is true for the claim that causation is an extrinsic relation.

Suppose that singular causation is an extrinsic relation in the sense explained in the last section: that is, it is a relation which is not determined by the perfectly natural properties of its relata, nor by these taken in conjunction with the perfectly natural relations holding between the relata. Again, nothing follows from this as regards conformity to Humean supervenience. However, let us make the further supposition that causation is extrinsic specifically in virtue of supervening on spatiotemporal relations and global regularities among occurrent events. It follows from this assumption that causation conforms to Humean supervenience. For a possible world like the actual world in its distribution of local, intrinsic qualities and its spatiotemporal relations will also be like the actual world in the global regularities that hold among its occurrent events, so that, on the current supposition, the two worlds will agree with respect to their causal relations.

This supposition about the specific way in which causation might be an extrinsic relation is certainly the natural one for a Humean to make. But it is not the only way in which causation might be extrinsic. Let us entertain a slightly far-fetched hypothesis about the way in which causation might be an extrinsic relation. Suppose that among the natural relations instantiated in the actual world there are relations of necessitation between events, much like the relations of nomic necessitation between universals which some philosophers (Armstrong (1983), Tooley (1987)) have hypothesised to constitute laws of nature; and further suppose that when a causal relation exists between a pair of events, it does so because relations of necessitation hold between that pair of events and *other similar pairs of events*.[4] In this case, the causal relation would be extrinsic, not because it depends on regularities among occurrent events, but because it depends on regularities among necessitations between similar events. Then it is possible that a world could agree with the actual world in its distribution of local, intrinsic qualities and spatiotemporal relations but differ in its distributions of event necessitations and so, on the current hypothesis, in its distribution of causal relations. This construal of the extrinsicality of causation implies failure to conform to Humean supervenience.

To be sure, the hypothesis about the extrinsicality of causation on which this argument relies is far-fetched. Since we are discussing what the Humean view of causation involves, it would be more plausible to represent this view as requiring causation to be extrinsic by virtue of supervening on spatiotemporal relations and global regularities among occurrent events. In this case, as we have seen, causation obeys Humean supervenience. So there is a reasonable and natural linkage between the view that causation is extrinsic with the view that causation does obey Humean supervenience.

Nonetheless, this is not so with the other pair of views – the view that causation is intrinsic and the view that causation does not obey Humean supervenience. We have seen that the issue whether intrinsic causal relations conform to Humean supervenience depends on the contingent issue of whether there are irreducible causal relations among the perfectly natural relations instantiated in the actual world. If there are none, then causation obeys Humean supervenience; and if there are, then it does not. How is this issue to be resolved? Are there such irreducible, perfectly natural causal relations? These are compulsory questions for a metaphysics of causation; nonetheless, questions to be taken up on another occasion. For now my intention has been simply to show that the distinction I have drawn between Humean and anti-Humean theories of causation in terms of the intrinsicality/extrinsicality distinction is orthogonal to the distinction drawn by other philosophers in terms of conformity to Humean supervenience.

6. CONCLUSION

In this paper I have been concerned with issues that arise when we ask the following ontological questions: What sort of character do the truthmakers for causal statements have? Are they intrinsic relations, as a naive conception of causation has it, or are they extrinsic relations, as Hume, conventionally interpreted, would have us believe? I have argued that the distinction between the conception of causation as an intrinsic relation and the conception of it as an extrinsic relation lies behind the now-familiar division of theories of causation into singularist theories and non-singularist theories. I have also tried to give a precise explanation of the intrinsic/extrinsic relation distinction by invoking a metaphysically élite class of perfectly natural properties and relations; and also shown how this explanation can be modified to make it more palatable to the metaphysically risk-averse. Finally, I have argued that when we frame the fundamental ontological question about causation as a question about whether it is an intrinsic or extrinsic relation, we cast a new light on theories of causation. In particular, we can see that it is less informative than hitherto thought to say that the fundamental ontological division between theories of causation is to be drawn in terms of whether they make causation conform to Humean supervenience.

Macquarie University

NOTES

* Versions of this paper have been read at Monash University, November 1995, the Research School of Social Sciences, Australian National University, December 1995, and at a special symposium on laws and causation held as part of the June 1996 meeting of the Australasian Association for the History, Philosophy, and Social Studies of Science at the University of Melbourne. I am indebted to the following people for comments and questions that have helped me clarify my ideas: David Armstrong, Simon Blackburn, Karen Green, Frank Jackson, Rae Langton, David Lewis, Daniel Nolan, Philip Pettit, Huw Price, and Denis Robinson. Thanks also to Howard Sankey for organising the special symposium on laws and causation at the AAHPSSS meeting.

[1] Does the fact that causal claims are made relative to a field of background conditions, sometimes widespread in occurrence, count against the intuition that a causal relation is an intrinsic relation that does not depend on any events occurring in its neighbourhood? I think this fact shows at best that the intuition that singularists rely on must ultimately be spelled out as part of an account of the context-dependent way in which a field of background conditions is determined for an individual causal claim. The fact does not impugn the intuition about intrinsicality; but does make it important to recognise that intrinsicality may be relative to a context.

[2] Armstrong is a special kind of anomic singularist. While he believes that there is no conceptual necessity attaching to a causal relation falling under a law, he nonetheless thinks that it is *a posteriori* necessary that causal relations are instantiations of laws. See Armstrong and Heathcote (1991).

[3] It should be noted that in this definition and subsequent ones it is allowed that objects belonging to different worlds may be duplicates of each other. Similarly, for pairs of objects and *n*-'tuples more generally.

[4] These relations of necessitation can not be exactly like those postulated by Armstrong. For him the relation of nomic necessitation between universals that constitutes a law is itself a universal, which on his account of universals is wholly present in each of its instantiations. So where he says that the causal relation is to be identified in an *a posteriori* way with an instantiation of the relation of nomic necessistation, the instantiation is to be thought of as a completely intrinsic relation. See Armstrong (this volume) for discussion of this point.

REFERENCES

Armstrong, D.M. (1978), *Universals and Scientific Realism*, Vols. I and II, Cambridge: Cambridge University Press

Armstrong, D.M. (1983), *What is a Law of Nature?*, Cambridge: Cambridge University Press

Armstrong, D.M. and Heathcote, A. (1991), 'Causes and Laws', *Nous* **25**, 63–73

Armstrong, D.M. (this volume), 'The Open Door: Counterfactual vs Singularist Theories of Causation', 175–85.

Anscombe, G. E. (1975), 'Causality and Determination', in E. Sosa (ed.) *Causation and Conditionals*, Oxford: Oxford University Press, 63–81

Bigelow, J. and Pargetter, R. (1990), *Science and Necessity*, Cambridge: Cambridge University Press

Blackburn, S. (1990), 'Hume and Thick Connexions', *Philosophy and Phenomenological Research*, suppl. **50**, 237–50

Braithwaite, R.B. (1953), *Scientific Explanation*, Cambridge: Cambridge University Press

Carroll, J.W. (1994), *Laws of Nature*, Cambridge: Cambridge University Press

Ducasse, C.J. (1926), 'On the Nature and Observability of the Causal Relation', *Journal of Philosophy* **23**, 57–68

Fales, E. (1990), *Causation and Universals*, London: Routledge

Fair, D. (1979), 'Causation and the Flow of Energy', *Erkenntnis* **14**, 219–50

Earman, J. (1984), 'Laws of Nature: The Empiricist Challenge', in R. Bogdan, (ed.), *D.M. Armstrong*, Dordrecht: D. Reidel Publishing Company

Hempel, C.G. (1965), *Aspects of Scientific Explanation*, New York: Free Press

Horwich, P. (1987), *Asymmetries in Time*, Cambridge, Mass.: MIT Press

Humberstone, I.L. (1996), 'Intrinsic/Extrinsic', *Synthese* **108**, 205–67

Hume, D. (1748), *An Enquiry Concerning Human Understanding*, ed. L.A. Selby-Bigge and P.H. Nidditch, 3rd edn., Oxford: Clarendon Press, 1975

Kim, J. (1982), 'Psychophysical Supervenience', *Philosophical Studies* **41**, 51–70

Langton, R and Lewis, D.K. (1996), "Defining 'Intrinsic'", *Philosophy and Phenomenological Research* **58**, 333–45.

Lewis, D.K. (1973), 'Causation', *Journal of Philosophy* **70**, 556–67, Reprinted in Lewis 1986b, 159–72

Lewis, D.K. (1979), 'Counterfactual Dependence and Time's Arrow', *Nous* **13**, 455–76, Reprinted in Lewis 1986b, 32–52

Lewis, D.K. (1983a), 'New Work for a Theory of Universals', *Australasian Journal of Philosophy* **61**, 343–77

Lewis, D.K. (1986a), *On the Plurality of Worlds*, Oxford: Blackwell

Lewis, D.K. (1986b), *Philosophical papers, Vol.II*, Oxford: Oxford University Press

Mackie, J.L. (1974), *The Cement of the Universe*, Oxford: Oxford University Press

Mill, J.S. (1879), *A System of Logic*, London: Longmans, Green and Company

Sider, T. (1996), 'Intrinsic Properties', *Philosophical Studies* **83**, 1–27

Strawson, G. (1989), *The Secret Connexion: Causation, Realism, and David Hume*, Oxford: Oxford University Press

Taylor, B. (1993), 'On Natural Properties in Metaphysics', *Mind* **102**, 81–100

Tooley, M. (1987), *Causation: A Realist Approach*, Oxford: Oxford University Press

Tooley, M. (1990), 'The Nature of Causation: A Singularist Account', in Copp, D. (ed.), *Canadian Philosophers: Celebrating Twenty Years of the CJP*, *Canadian Journal of Philosophy*, suppl. **16**, 271–322

Tooley, M. (1993), 'Causation: Reductionism versus Realism', in M. Tooley and E. Sosa, (eds.), *Causation*, Oxford: Oxford University Press, 172–92

Wright, J. (1983), *The Sceptical Realism of David Hume*, Manchester: Manchester University Press

HUW PRICE

THE ROLE OF HISTORY IN MICROPHYSICS[1]

I. INTRODUCTION

Historians often encounter striking correlations between events in different places at the same historical period: a new disease breaks out in several towns in the same month, for example, or a new phrase comes suddenly into widespread use. Sometimes these patterns are merely coincidental, but in general – in history as in ordinary life – we look for an explanation of the correlation in terms of a single common event, with which each of the original events may be seen to be associated. We look for the original source of infection, or the original use of the phrase. In any particular case, it may not be obvious where we should look for the linking event, and historians need a nose for these things. But one piece of advice seems infallible: Look in the past, not in the future. In other words, look for the linking event before, not after, the events it is supposed to explain.

This advice isn't simply pragmatic, of course. We're not advising historians to concentrate on the past because they don't have the techniques or training to investigate the future, for example. Rather, it seems to be a fact about the world that correlations of the relevant kind are linked in a time-asymmetric way. A common past can produce correlations in a way in which a common future cannot. It's a fact about the world that backward-looking historical explanation is the kind that works in these cases. We can imagine worlds which work differently – worlds in which forward-looking, teleological explanation has a more important role to play – but our world doesn't seem to be like that.

What sort of fact about the world is it that it exhibits this kind of temporal asymmetry – let's call it the "No Teleology Principle"? Indeed, to what extent is it a fact about the world at all, rather than something much more subjective – a feature of our notion of explanation, perhaps? These issues are of considerable interest in their own right, but my concern here lies further down the track. I want to talk about whether it is legitimate to expect this kind of time-asymmetry to prevail in microphysics, and, if not, what the consequences might be for our understanding of quantum mechanics. It turns out that the contemporary debate about the interpretation of quantum mechanics takes for granted a version of the No Teleology Principle, in a way which is so intuitive, so deeply embedded in the way in which we find it natural to think about the world, that it is easy to overlook the fact that it is there. But there are two reasons for trying to bring it into the open. First, the best account of the nature of the physical origins of the No-Teleology Principle seems to leave no place for it in microphysics

H. Sankey (ed.), Causation and Laws of Nature, 331–345.

– it seems to be associated with a different level in the physical world altogether. And second, the quantum world begins to look much more attractive, without the constraints of this intuitive asymmetry.

The rest of the paper goes like this. In the next section I sketch what seems to me to be the right account of the relationship between the No Teleology Principle and the well-known time-asymmetry of thermodynamics. This account proposes that the thermodynamic asymmetry is the objective correlate of the No Teleology Principle – the contingent objective feature of our world, in virtue of which the No Teleology Principle is true. I'll then go on to argue that some intuitively plausible applications of this principle in microphysics are not compatible with this proposal. This suggests one of two things: either the proposal is mistaken, and there is a further objective time-asymmetry in the physical world, in addition to that of thermodynamics; or these extra applications of the No Teleology Principle in microphysics are misguided. I'll argue that the first option runs counter to the apparent T-symmetry of the laws of physics – that is, to the well-established principle that the laws governing the microscopic constituents of matter are (almost entirely) insensitive to the distinction between past and future – and that the second option seems to offer considerable benefits in quantum mechanics. We thus have two reasons for taking seriously the possibility that the No Teleology Principle is an unreliable intuition in microphysics – the possibility that future history plays the same role as past history in microphysics, in what at present we find a deeply counter-intuitive way.

2. NO TELEOLOGY: THE INTUITIVE ASYMMETRY

The No Teleology Principle says that any two physical systems will be independent of one another, unless one system exerts a causal influence on the other, or both systems are influenced by some common factor in their past. More anthropomorphically, we might take it to say that when two systems encounter one another for the first time, they remain ignorant of each other's existence, until the moment at which they first interact.

Formalisations of this idea have been proposed in both physics and in philosophy. In the physical literature, Penrose and Percival (1962) formulate such a principle as what they call the *Law of Conditional Independence*. As they emphasise, it is time-asymmetric: it says that incoming systems are independent before they interact, not that outgoing systems are independent, after they interact. We don't expect outgoing processes to be independent – on the contrary, since otherwise we wouldn't expect correlations in cases in which both processes arise from some common source in their common past.

As their terminology indicates, Penrose and Percival suggest that conditional independence is a law-like principle. This view has not been popular, however. Most physicists hold that like other temporal asymmetries in the physical world, this asymmetry is "fact-like" – a nomologically contingent matter, associated with boundary conditions.

Penrose and Percival note that conditional independence is closely related to Hans Reichenbach's "Principle of the Common Cause", formulated a few years earlier in Reichenbach's posthumous book, *The Direction of Time* (Reichenbach, 1956). Reichenbach had observed that when correlations between events at different points in space are not due to a direct causal connection, they turn out to be associated with a joint correlation with a third event – what Reichenbach terms a *common cause* – in their common past. Reichenbach notes the temporal asymmetry of this principle, and explores the idea that it is connected with the direction of time. In philosophy, the temporal asymmetry these principles describe has come to be termed the *fork asymmetry*. Roughly, the fork is the ∨-shaped structure of correlations, whereby two spatially separated events are correlated in virtue of their joint correlations with a third event at the vertex (see Horwich, 1987, for example). The asymmetry consists in the fact that forks of this kind are very common with one temporal orientation (∨-shaped, or open to the future) but rare or non-existent with the opposite orientation (∧-shaped, or open to the past).

At the macroscopic level, the fork asymmetry appears to be connected with the asymmetry of thermodynamics. The easiest way to see this is to think about ordinary irreversible processes, viewed in reverse – in other words, viewed as we see them if we reverse a film. From this reversed perspective, many ordinary processes seem to involve highly correlated incoming influences from distant parts of space. Think of tiny droplets of champagne, forming themselves into a pressurised column and rushing into a bottle, narrowly escaping an incoming cork. Think of the molecules in a bath of water, organising themselves into fast and slow teams, before hurling themselves in two columns to the hot and cold taps. Or, to give a less domestic example, think of the countless fragments of the True Cross, making their meticulously coordinated journeys to meet Christ outside Jerusalem. Astounding as these feats would seem from this perspective, they are – perhaps with the exception of the last! – nothing but the mundane events of ordinary life, viewed from an unfamiliar angle. Seemingly teleological behaviour of this kind is ubiquitous in one temporal sense – when it occurs after some central event, from our usual temporal perspective – but unknown and apparently incredible in the other temporal sense.

Broadly speaking, then, the cases which look teleological when viewed in reverse seem to be the cases in which – viewed from the reverse perspective – entropy is *decreasing*. The existence of such reverse teleology seems to depend on the fact that the universe as we know it is not in thermodynamic equilibrium. If the universe were in equilibrium, reverse teleology of this kind would be as rare as forward teleology actually is – there would be no teleology in either temporal sense. This suggests that the No Teleology Principle is closely associated with the principle that entropy does not decrease towards the future. In particular, the time-asymmetry of the No Teleology Principle (the fact that it does not hold in both temporal directions) seems to be associated with the asymmetry of the Second Law – i.e., with the fact that the principle that entropy does not decrease holds towards the future but not towards the past. Indeed, a natural suggestion is that the No Teleology Principle simply *is* the principle that there are no entropy-reducing correlations towards the future.

3. AN ISSUE OF PRIORITY

There are some complexities here that I want to flag, but skirt around. The precise relationship between the No Teleology Principle and the Second Law is a controversial matter. A long tradition holds, in effect, that some version of the No Teleology

Principle *explains* (and is hence theoretically *prior to*) the Second Law. Famously, Boltzmann's so-called *H*-Theorem – a special case of the Second Law for the case of gases – relies on an asymmetric independence principle called "molecular chaos". Perhaps because the No Teleology Principle is so intuitive, it took a long time for the tradition to notice the work this asymmetric assumption does in the *H*-Theorem and its descendants. Even today, many people do not seem to appreciate the fundamental dilemma: Without some such time-asymmetric principle, the *H*-Theorem cannot yield a time-asymmetric conclusion. With such a principle, on the other hand, the *H*-Theorem cannot do more than to shift the puzzle of the time-asymmetry of the Second Law from one place to another, for the time-asymmetry of the principle in question will be equally problematic, in the light of the apparent T-symmetry of the underlying laws. (What is more, it is doubtful whether the required principle can be fact-like – a matter of boundary conditions – as the tradition also assumes. If it were, we would seem to have no reason to expect the Second Law to hold in previously unobserved regions of space and time. Only a prior observation that it does hold in a particular region of spacetime could give us reason to think that the initial boundary conditions were such that entropy does not decrease in the region in question.)

So the traditional attempt to derive the Second Law from some version of the No Teleology Principle is problematic, and produces no net reduction in the puzzles of time-asymmetry. It also has nothing to say on the issue of why entropy was low in the past. With its assumptions laid bare, the traditional approach presents us with *two* puzzling time-asymmetries: First, why is entropy low in the past, and second, what asymmetric principle causes it to increase towards the future? In my view (see Price, 1996, Ch. 2) this double counting is quite unnecessary. I think that what the statistical treatment of thermodynamics gives us is, not an asymmetric explanation of why entropy increases in one direction, but an entirely symmetric expectation that entropy will be high at any time, *ceteris paribus*. In other words, it reveals that high entropy is a normal condition, which doesn't require further explanation. It is departures from this normal condition that need to be explained, and here we do find an asymmetry: entropy decreases towards the past, but not (so far as we know) towards the future. Once we recognise that there isn't a separate puzzle as to why entropy goes up towards the future, in addition to the puzzle as to why it goes down towards the past, there's no motive to regard the No Teleology Principle as something prior to the Second Law, from which the latter may be derived. On the contrary, we should say that the default expectation is for no teleology in either direction, and that the only puzzle is that we do find teleology towards the past – in other words, that entropy does decrease in that direction.

This view of the matter is controversial, and conflicts with a long tradition in thermodynamics. But for present purposes, I don't need to insist on it. All I need is the fact that whatever one's view of the relationship between the No Teleology Principle and the thermodynamic asymmetry, there is an evident attraction in the idea that they are closely related – in particular, in the economical idea that it is essentially the same contingent feature of the world which underlies both. As long as this is so, it is plausible to hold, as most physicists seem to hold, that the No Teleology Principle is no more problematic than the thermodynamic asymmetry: and (with the above reservations) that both are of a fact-like nature. This seems to be the orthodox view in physics.

However, I want now to argue that there are some intuitively plausible applications of the No Teleology Principle in physics which cannot be accommodated within this orthodox picture, and which have nothing to do with the thermodynamics asymmetry. If so, then either there is a further objective time asymmetry in the world, in addition to that associated with thermodynamics, or the relevant intuitions are unreliable. I want to argue for the latter view.

4. THE INTUITIVE ASYMMETRY IN MICROPHYSICS

Consider a photon, passing through a polariser. According to the standard model of quantum mechanics, the state of the photon after the interaction reflects the orientation of the polariser. If you know the state of the photon after the interaction, you know the orientation of the polariser. Not so before the interaction, of course: in quantum mechanics, as elsewhere, we take it for granted that there are no *pre-interactive* correlations. The photon doesn't know the orientation of the polariser in advance. Here, as elsewhere, we assume that the No Teleology Principle holds sway.

Very few writers see this feature of the standard model as in any way problematic. Some writers are troubled by the time-asymmetry of the standard model, but their objection tends to be to the fact that the collapse of the wave function is an irreversible process, not to the asymmetry of correlations as such. And after all, why not? The fact that there is an asymmetry may be a little puzzling, but its individual components – that interactions may establish correlations, and that there are no pre-interactive correlations – surely seem plausible enough. If we were to try for symmetry, which should we give up? Besides, as we have seen, the No Teleology Principle is familiar elsewhere in physics, where it seems compatible with the T-symmetry of underlying physical laws. Thus there seems to be a precedent for the asymmetry we find in quantum mechanics, and no reason, on reflection, to doubt our initial intuitions.

I think the appearance of calm is quite illusory, however. The time asymmetry embodied in the standard model turns out to have nothing to do with the thermodynamic asymmetry, and hence to be quite distinct from its supposed analogue elsewhere in physics. This means that in taking for granted the No Teleology Principle in quantum mechanics, we are not simply applying the same well-grounded principle to new cases. It is not the same principle as before, and, I shall argue, cannot be reconciled with the T-symmetry of the laws of physics in the same way. Given T-symmetry, I want to argue, pre- and post-interactive correlations should be on the same footing in microphysics. Any reason for objecting to pre-interactive correlations is a reason for objecting to post-interactive correlations, and any reason for postulating post-interactive correlations is a reason for postulating pre-interactive correlations. Hence, I want to argue, the No-Teleology Principle is unreliable in quantum mechanics.

I emphasise that for the present, my interest is in the intuitions underlying the No Teleology Principle, not in the quantum mechanical examples. For the moment, the standard model simply provides vivid examples of the intuitions I want to challenge. For the moment, what's important is just that we find it "natural" that photons should be correlated with polarisers before but not after they interact. Later, I'll discuss the significance of a challenge to this intuition for the puzzles of quantum mechanics, but until then it is the intuition and not the quantum mechanics that matters.

We'll need a name for the kind of application of the No Teleology Principle which is involved in the photon example. As I noted earlier, we find it natural to express these intuitions in a rather anthropomorphic way, in terms of what one system may be expected to know about another. We take it to be intuitively obvious that interacting systems will be ignorant of one another until the interaction actually occurs, at which point each system may be expected to "learn" something about the other. In Price (1996) I called the microscopic case of this intuitive principle the "microscopic innocence principle", or "μInnocence", for short. Here, less anthropomorphically, I'll call it "μIndependence". The first task is to show that unlike the instances of the No Teleology Principle we discussed initially, μIndependence does not depend on the thermodynamic asymmetry – in other words, on the principle that there are no correlations of the kind there would be if entropy decreased towards the future. (I'll call the principle that there are no entropy-reducing correlations "*H*-Independence".)

5. μINDEPENDENCE AND *H*-INDEPENDENCE

I don't know of any other writer who distinguishes μIndependence explicitly from *H*-Independence. However, it is not hard to see that they are distinct. For one thing, the correlations associated with low-entropy systems are essentially "communal", in the sense that they involve the behaviour of very large numbers of individual systems. But the kind of post-interactive correlation we take to be exemplified by the photon example is individualistic, in the sense that it involves the simplest sort of interaction between one entity and another.

Secondly, the photon case is not dependent on the thermodynamic history of the system comprising the photon and the polariser, or any larger system of which it might form a part. Imagine a sealed black box containing a rotating polariser, and suppose that the thermal radiation inside the box has always been in equilibrium with the walls. Intuitively, we still expect the photons comprising this radiation to establish the usual post-interactive correlations with the orientation of the polariser, whenever they happen to pass through it. The presence of these post-interactive correlations does not require that entropy was lower in the past. By symmetry, then, the absence of matching pre-interactive correlations cannot be deduced from the fact that entropy does not decrease toward the future: a world in which photons were correlated with polarisers before they interacted (so as to violate μIndependence) would not necessarily be a world in which the Second Law did not hold. Our intuitions about μIndependence seem to be independent of the existence of a thermodynamic asymmetry.

Even if μIndependence and *H*-Independence are distinct, however, a natural suggestion is that they have the same status, in the sense that both are fact-like products of time-asymmetric boundary conditions, rather than asymmetric laws. But in the thermodynamic case, the observed thermodynamic asymmetry in our region provides evidence for the kind of asymmetric boundary conditions required to explain it. We have very good reason to accept the existence of the asymmetry, at least in our region, independently of any claim about boundary conditions. In the case of μIndependence, however, there is no *observed* asymmetry to be explained. We don't *observe* that the incoming photon is not correlated with the polariser through which it is about to pass. Rather, we rely on the asymmetric principle that interaction *produces* correlations only in one temporal direction – "towards the future", not "towards the past". In other words, we rely on a principle of No Teleology, but in a context in which it has nothing to do with the principle that entropy does not decrease.

As it guides our intuitions in quantum mechanics, then, μIndependence seems to be not an *a posteriori* principle derived from observation, but a law-like principle in its own right. It seems to be a kind of meta-law, which allows laws imposing post-interactive correlations, while excluding their pre-interactive counterparts. But why should we accept an asymmetric principle of this kind? I know of two main lines of argument, but I want to show that both are fallacious.

6. INITIAL RANDOMNESS?

The first argument goes like this. We have seen that observational evidence for *H*-Independence need not be observational evidence for μIndependence – at any rate, not directly. There might be indirect evidence in the offing, however. Perhaps the Second Law supports some hypothesis about the initial conditions of the universe, an independent consequence of which is that photons are not correlated with polarisers before they interact. For example, it is often suggested that the Second Law derives from the fact that the initial microstate of the universe is as random as it can be, given its low-entropy macrostate. Would this not also explain why photons are not correlated with future polarisers?

More directly, it has been suggested that μIndependence simply depends on the plausible principle that all initial conditions be treated as equally like, other things being equal. After all, wouldn't pre-interactive correlations require that the initial condition of the system concerned be chosen from a very special subset of its phase space. In other words, doesn't μIndependence simply embody a contingent but highly plausible hypothesis about the initial states of physical systems, namely that they be as random as possible? (This argument is offered in defence of μIndependence by Lebowitz, 1997, for example.)

In my view, however, this argument simply fails to recognise the law-like character of what is prohibited by μIndependence. Suppose, contrary to μIndependence, that there were laws imposing pre-interactive correlations. The phase space of a physical system is defined by the operative physical laws: in effect, the phase space just *is* the set of states allowed by the laws. Hence if a law imposed pre-interactive correlations then μIndependence would fail for any nomologically possible choice of initial conditions, and wouldn't require any special choice. After all, if the standard model of quantum mechanics is correct, then no special choice of phase space trajectories is required to ensure that a photon is correlated with a polariser after they interact, for *all* trajectories exhibit this correlation. This objection gets things the wrong way around, then. It begs the question in favour of μIndependence, by assuming that the phase space is such that only a special subset of trajectories would display pre-interactive correlations.

7. COLLIDING BEAMS?

The second argument turns on the claim that by postulating μIndependence, we are able to explain certain otherwise puzzling observable phenomena (and hence, again, on the idea that these phenomena provide indirect observational evidence for μIndependence). For example, Penrose and Percival themselves argue that their Law of Conditional Independence explains a variety of otherwise inexplicable irreversible processes. Although few physicists agree with Penrose and Percival that Conditional Independence is a law-like principle, it does seem a common view that their examples provide indirect observational evidence for pre-interactive independence.

A typical example involves the scattering produced when two tightly organised beams of particles are allowed to intersect. It is argued that this scattering is explicable if we assume that there are no prior correlations between colliding pairs of particles (one from each beam), and hence that the scattering reveals the underlying independence of the motions of the incoming particles.

But μIndependence is neither necessary nor sufficient here. The explanation rests entirely on the absence of entropy-reducing correlations between the incoming beams – i.e., on *H*-Independence – not on μIndependence at the level of individual particle pairs. In other words, these cases involve nothing more than the familiar thermodynamic asymmetry, from which μIndependence is supposed to be distinct.

I will offer short and long arguments for this conclusion. The short argument simply appeals to cases in which it seems clear that there is no microscopic asymmetry – Newtonian particles, for example. Here there is nothing to sustain any asymmetry at the level of individual interactions, and yet we still expect colliding beams to scatter. This suggests that the scattering is associated with the lack of some global correlation, not with anything true of individual particle pairs.

The longer argument goes like this. We suppose that there is a microscopic asymmetry of μIndependence, distinct from the correlations associated with the thermodynamic asymmetry, and yet somehow compatible with the T-symmetry of the relevant dynamical laws. We then construct a temporal inverse of the scattering beam experiment, and show that it displays (reverse) scattering, despite the assumed absence of the post-interactive analog of μIndependence. By symmetry, this shows that μIndependence is not necessary to explain the scattering observed in the usual case. Finally, a variant of this argument shows that μIndependence is also insufficient for the scattering observed in the usual case.

If μIndependence were necessary for scattering, in other words, then scattering would not occur if the experiment were run in reverse. It is difficult to replicate the experiment in reverse, for we do not have direct control of final conditions. But we can do it by selecting the small number of cases which satisfy the desired final conditions from a larger sample. We consider a large system of interacting particles of the kind concerned, and consider only those pairs of particles which emerge on two tightly constrained trajectories (one particle on each), having perhaps interacted in a specified region at the intersection of these two trajectories (though not with any particle which does not itself emerge on one of these trajectories). We then consider the distribution of initial trajectories, before interaction, for these particles. If the dynamical laws are T-symmetric, the predicted distribution must mirror that in the usual case.

This can be made more explicit by describing a symmetric arrangement, subsets of which duplicate both versions of the experiment. Consider a spherical shell, divided by a vertical plane. On the inner face of the left hemisphere are particle emitters, which produce particles of random speed and timing, aimed at the centre of the sphere. In the right hemisphere is a matching array of particle detectors. Dynamical T-symmetry implies that if the choice of initial conditions is random, the global history of the device is also T-symmetric: any particular pair of particle trajectories is equally likely to occur in its mirror-image form, with the position of emission and absorption reversed.

We can replicate the original experiment by choosing the subset of the global history of the device containing particles emitted from two chosen small regions on the left side. Similarly, we can replicate the reverse experiment by choosing the subset of the history of the entire device containing particles absorbed at two chosen small regions on the right side. In the latter case, the particles concerned will in general have been emitted from many different places on the left side. This follows from the fact that the initial conditions are as random as possible, compatible with the chosen final conditions. Thus we have scattering in the initial conditions, despite the assumed lack of post-interactive μIndependence between interacting particles.

Thus if there were post-interactive correlations of the kind denied to the pre-interactive case by μIndependence, they would not prevent scattering in the reverse experiment, which is guaranteed by the assumption that the initial conditions are as random as possible, given the final constraints. By symmetry, this implies that μIndependence is not necessary to produce scattering in the normal case. We get scattering without any extra assumption in the role that μIndependence is supposed to play, provided that the choice of trajectories is as random as possible, given the initial constraints. (This cannot be the same thing as μIndependence, for otherwise μIndependence would not fail in the post-interactive case, and there would not be the assumed microscopic asymmetry.)

A third version of the experiment shows that μIndependence is also not sufficient to explain what happens in the normal case: Assume μIndependence again, and consider the subset of the first experiment in which we have collimation on the right, as well as the left – i.e., in which we impose both final and initial conditions. Here we have no scattering, despite μIndependence. (Again, the imposition of the final condition cannot amount to a denial of μIndependence: if so, the asymmetry of μIndependence in the normal case would amount to nothing more than the presence of a low-entropy initial condition, in conflict with the supposition that μIndependence differs from *H*-Independence.)

Thus μIndependence is both insufficient and unnecessary to explain the scattering phenomena. The differences between the various versions of the experiment are fully explained by the different choices of initial and final boundary conditions. The asymmetry of the original case stems from the fact that we have a low-entropy initial condition (the fact that the beams are initially collimated) but no corresponding final condition. The issue as to why this is the case that occurs in nature is a sub-issue of that of the origins of the thermodynamic asymmetry in general. It has nothing to do with any further asymmetry of the kind described by μIndependence. In other words, contrary to popular opinion, these cases provide no indirect observational support for μIndependence.

8. WHAT TO DO ABOUT μINDEPENDENCE

Thus in its role as the intuition that underpins the asymmetry of the photon's behaviour in the standard model of quantum mechanics, μIndependence is not an *a posteriori* principle derived from observation, but a kind of tacit meta-law in its own right. We do not *observe* that the incoming photon is not correlated with the polariser through which it is about to pass. Rather, we rely on the prior principle that laws enforcing pre-interactive correlations would be unacceptably teleological. We allow dynamical principles producing post-interactive correlations, but not their pre-interactive twins.

Conceding that μIndependence is law-like does not improve its prospects, of course. It simply owns up to the principle's current role in guiding our intuitions in microphysics. Indeed, it makes its prospects very much worse, for as a law-like principle, μIndependence conflicts with T-symmetry. We might be justified in countenancing such a conflict if there were strong empirical evidence for a time-asymmetric law, but the supposed evidence for μIndependence turns out to rely on a different asymmetry altogether.

What are the options? First, we might look for other ways of defending μIndependence. Unless this support takes the form of *a posteriori* evidence, however, its effect will be simply to deepen the puzzle about the T-asymmetry of microphysics. Moreover, although there is undoubtedly more to be said about the intuitive plausibility of μIndependence, I suspect that the effect of further investigation is to *explain* but not to *justify* our intuitions. For example, the intuitive appeal of μIndependence may rest in part on a feature of human experience, the fact that in practice our knowledge of things in the physical world is always post-interactive, not pre-interactive. The explanation of this asymmetry is tricky. It seems to depend in part on our own time-asymmetry as structures in spacetime, and in part on broader environmental aspects of the general thermodynamic asymmetry. Whatever its exact provenance, however, it seems to provide no valid grounds for extending the intuitions concerned to microphysics.

Similarly, as I have argued elsewhere (see Price 1996, pp. 181–4), some apparent post-interactive dependencies turn out to be associated with a temporal asymmetry in counterfactual reasoning – roughly, the fact that we "hold fixed" the past, when considering the consequences of counterfactual conditions. Given a conventional account of this aspect of counterfactual reasoning, the asymmetries concerned are demystified, in the sense that they are shown to require no independent asymmetry in the physical systems concerned. Again, some of the intuitive appeal of μIndependence is thereby accounted for, but in a way which does nothing to solve the puzzle of the photon case.

Another response would be to try to ensure T-symmetry in microphysics by disallowing post-interactive correlations, rather than by admitting pre-interactive correlations. The standard model of quantum mechanics would thus be ruled unacceptable, for example. But the move seems misguided. It does nothing to justify μIndependence, and restores symmetry by creating two puzzles where previously we had one.

In my view, the only option which really faces up to the problem is that of admitting that our intuitions might be wrong, and that μIndependence might indeed fail in microphysics – in other words, admitting that we have no good reason to exclude teleology in microphysics, so long as it is not of the entropy-reducing kind. I want to finish with a few remarks on the possible relevance of this option in quantum mechanics. In order to clarify the force of these remarks, I emphasise again that thus far, my references to quantum mechanics have been inessential. The standard model provides vivid examples of an asymmetry we take for granted, but the case against this asymmetry is essentially classical. The main point is that despite common opinion to the contrary, it is not associated with the classical asymmetry of thermodynamics. In effect, then, the case against μIndependence is a prior constraint on the interpretation of quantum mechanics. We have reason to doubt the intuition concerned on symmetry grounds, independently of any consequences this might turn out to have in quantum mechanics.

9. μINDEPENDENCE AND QUANTUM MECHANICS

When we do turn to quantum theory, we find that μIndependence is a fundamental assumption of most, if not all, of the range of arguments which have been taken to show that the quantum world is non-classical in puzzling ways. The most striking such argument is Bell's Theorem (1964). Bell's result has been taken to establish that quantum reality involves non-locality. Although the nature of this non-locality and the extent of its conflict with Special Relativity have been a matter for much debate, the consensus has been that quantum mechanics is committed to it in some form. As Bell and others have pointed out, however, his argument depends on the assumption that quantum systems are not correlated with the settings of measurement devices, prior to their interaction. Thanks to μIndependence, this assumption has normally seemed uncontentious. Bell himself considered relaxing it, but even he tended to think about this possibility in a way which doesn't conflict with μIndependence. (His suggestion, which he called "superdeterminism", was that the required correlation might be established by an additional common cause in the past, not simply in virtue of the existing interaction in the future; see Bell, 1987; Bell *et al.*, 1985.)

The upshot is that without μIndependence, there seems to be no firm reason to think that quantum mechanics commits us to non-locality. This conclusion applies not simply to Bell's Theorem, but to other more recent arguments for non-locality, such as the Greenberger-Horne-Zeilinger (GHZ) cases (see, e.g., Mermin, 1990; Clifton, Pagonis and Pitowsky, 1992); these too depend on μIndependence, in the form of the assumption that the states of incoming particles are not correlated with the settings of instruments they are yet to encounter.

Non-locality is not the only non-classical consequence thought to flow from quantum mechanics, of course. Since the early years of quantum theory, many physicists have been convinced that quantum measurements do not simply reveal a pre-existing classical reality. In Bohr's Copenhagen Interpretation, which remains very influential in physics, the view was that reality is somehow "indeterminate" until a measurement is made – measurement was said to force reality to take on a definite condition, where none existed before. Later, and rather more precisely, a range of mathematical results (the so-called No Hidden Variable theorems) seemed to establish that no system of pre-existing properties could reproduce the predictions of quantum mechanics, at least in certain cases; see Kochen and Specker (1967), for example.

These interpretations and results also take for granted μIndependence, however. Otherwise, they would not have been entitled to assume that the pre-existing reality could not depend on the nature of a later measurement. In place of Bohr's indeterminate reality, one might have postulated a reality which, while fully determinate before a measurement is made, is partly constrained by the nature of that measurement. In the case of the No Hidden Variable theorems, similarly, μIndependence serves to justify the assumption that a single hidden state should be required to reproduce the quantum predictions for any possible next measurement. If the hidden state is allowed to vary with the nature of the measurement, the problem is relatively trivial, at least in principle. (In David Bohm's 1952 hidden variable theory, the trick is to allow measurement to have an instantaneous effect on the hidden variables; again, however, it is μIndependence which underpins the assumption that the effect must be instantaneous, rather than advanced.)

μIndependence thus plays a crucial role in the arguments which are taken to show that quantum mechanics has puzzling non-classical consequences. We might symbolise the logical relationships like this:

QM + μIndependence $\rightarrow$ Non-locality + Indeterminacy + . . .

On the right hand side we have a list of the problematic non-classical consequences of quantum mechanics. The contents of the list vary a little with one's favoured interpretation of quantum theory, but virtually everyone agrees that there are some such consequences.

To understand the significance of these connections, try to imagine how things would have looked if we had considered abandoning μIndependence on symmetry grounds, before the development of quantum mechanics. Quantum mechanics would then have seemed to provide a dramatic confirmation of the hypothesis that μIndependence fails. Given quantum mechanics, the assumption that μIndependence does not fail turns out to imply absurdities such as non-locality and indeterminacy. Against this imagined background, then, experimental confirmation of the Bell correlations would have seemed to provide empirical data for which the only reasonable explanation is that μIndependence does fail, as already predicted on symmetry grounds.

From a contemporary standpoint it is difficult to see the issue in these terms, of course. We are so used to talk of non-locality and indeterminacy in quantum mechanics that they no longer seem entirely absurd. And of course we are still so strongly committed to μIndependence that it is hard to see that rejecting it could provide a more plausible way to understand the lessons of quantum mechanics. But I think it is worth making the effort to challenge our preconceptions. I have argued that we have reason to doubt μIndependence on purely classical grounds – simply for symmetry reasons, in effect, once we appreciate that we have no empirical reason to question T-symmetry. With this new conception of the proper form of a classical microphysics, it seems unwise to continue to insist that quantum mechanics is a radically non-classical theory, in what have become the orthodox ways.

10. μINDEPENDENCE AND BACKWARD CAUSATION

Thus the hypothesis that μIndependence fails seems to throw open the conventional debate about quantum mechanics in a rather appealing way – it suggests that quantum mechanics might be a very much less non-classical theory than almost everybody has assumed. But is the hypothesis really one to be taken seriously? The appeal to symmetry notwithstanding, many will feel that there is something fundamentally absurd about the suggestion that μIndependence might fail.

This is not the place to attempt a comprehensive response to all such doubts – I discuss these issues at greater length in Price (1996) – but I want to finish by mentioning one surprising consequence of abandoning μIndependence, which turns out to be very much less objectionable than it seems at first sight. In the process, I think, it provides further ammunition for the claim that quantum theory provides precisely the kind of picture of the microworld we should have expected, if we had accepted in advance that T-symmetry requires us to abandon μIndependence.

The consequence concerned would manifest itself in a case in which we ourselves had influence over one member of a pair of interacting systems – over the setting of a polariser before its encounter with an incoming photon, for example. If the state of the incoming photon were correlated with that of the polariser, then if we could control the polariser without preventing the interaction, we would be able to control the photon. This would not be action at a distance – the correlation would be conducted continuously, via the interacting worldlines of the two systems involved – but it would seem to amount to a kind of backward causation. This consequence might well seem absurd, and potentially paradoxical.

The usual objection to advanced causation involves the bilking argument. The essential idea is that in order to disprove a claim of advanced causation, we need only arrange things so that the claimed later cause occurs when the claimed earlier effect has been observed not to occur, and vice versa. It might be thought that an argument of this kind will be sufficient to defend μIndependence. Any claimed pre-interactive correlation looks liable to be defeasible in this fashion – we need only ensure that the properties of one system affect those of the other, in such a way as to conflict with the claimed correlation.

But is the bilking argument effective in the kind of case we are considering? Consider the photon example. In order to set up the kind of experiment just outlined, we would need to observe the relevant state of the photon before it reaches the polariser, so as to orient the polariser in such a way as to defeat the claim that the incoming photon is correlated with polariser setting. But how would we set about making such an observation? Presumably we would have to place a second polariser, or some other measuring device, in the path of the photon, before it reaches the original polariser. But if we are entertaining the hypothesis that μIndependence fails, we have two reasons to dispute the relevance of the information yielded by this measurement procedure. Firstly, if μIndependence is in doubt then we are not entitled to assume that the state revealed by this measurement is the state the photon *would have had*, had the measurement not been made. After all, if measurements can affect the earlier states of the systems measured, then what measurement reveals is not in general what would have been there, in the absence of the measurement in question. Even if we found that the correlation required for backward causation failed in the presence of the measuring device, in other words, we would not be entitled to conclude that it would have failed in its absence.

Secondly, and more importantly, what the failure of μIndependence requires is that there be a correlation between the polariser setting and the state of the incoming photon *from the time of that photon's last interaction with something else*. For think of the usual case: How long do we expect a correlation established by interaction to survive? Not beyond the time at which the system in question interacts with something else, a process which may destroy the initial correlation. In the case we are considering, then, the effect of interposing a second measuring device in the photon's track will be to ensure that the correlation is confined to the interval between this measurement and the main interaction. The presence of the measurement ensures that the advanced effect is more limited in extent than it would be otherwise, and again the required contradiction slips out of reach.

These objections could be evaded if it were possible to observe the state of the incoming photon without disturbing it in any way – if the presence of the measuring device made no difference to the object system, in effect. Classical physics is often said to have taken for granted that perfectly non-intrusive measurements of this kind are possible, in principle, or at least approachable without limit. If this view was ever assumed by classical physics, however, it was decisively overturned by quantum mechanics. One of the few things that all sides agree on about quantum mechanics is that it shows that we cannot separate the processes of measurement from the behaviour of the systems observed. This provides a further respect in which quantum mechanics is the kind of microphysical theory we might have expected, if we had questioned μIndependence on symmetry grounds, within the classical framework. The bilking argument suggests that classical non-intrusive measurement is incompatible with the kind of symmetry required if we abandon μIndependence. Transposing, then, it seems that symmetry considerations alone might have led us to predict the demise of classical measurement.

11. CONCLUSION

Summing up, we began with the No Teleology Principle. We saw that on the small scale, as on the large, we take for granted that interacting systems do not behave in a teleological way – they do not become acquainted before they actually meet, so to speak. I argued that on the large scale, this time-asymmetric principle seems closely associated with the time-asymmetry of thermodynamics. Not so on the small scale, however. As it guides our intuitions in microphysics, the No Teleology Principle is not only independent of the thermodynamic asymmetry, but conflicts with the widely-accepted doctrine that the laws of microphysics are insensitive to the distinction between past and future. As such, I suggested, any applications of the No Teleology Principle in microphysics should be suspect. We have good reason to think that the intuitions concerned are simply unreliable. Microscopic teleology should be no more unacceptable in one temporal direction than in the other. In looking for explanations of present correlations, microphysicists should be encouraged to consider the future of the systems concerned, as much as their past. History should play a time-symmetric role in microphysics.

Finally, I have suggested that there is already some empirical evidence in favour of this symmetric alternative, albeit of a very indirect and incomplete kind. From a classical standpoint quantum mechanics itself is naturally taken to provide such evidence, on the grounds that if combined with the principle of μIndependence, it leads to such conceptual horrors as non-locality and indeterminacy. From a contemporary standpoint, however, these ideas have lost their capacity to shock. Familiarity has bred a measure of contentment in physics, and the *reductio* has lost its *absurdum*. Regaining a classical perspective would not be an easy step, nor one to be attempted lightly, but it does seem worth entertaining. In abandoning a habit of thought which already conflicts with well-established principles of symmetry, we might free quantum mechanics of metaphysical commitments which once seemed intolerable in physics, and might yet seem so again.

University of Sydney

NOTES

[1] This paper owes much to comments on related talks by audiences at ANU, Western, Columbia, Pittsburgh and Cambridge, at the AAHPSSS Conference at Melbourne in June, 1996, and at the PSA meeting in Cleveland, Ohio in November, 1996. I am especially grateful to John Norton, who sat through three versions of the talk, and helped me to understand what I was trying to say.

REFERENCES

Bell, J.S. (1964), 'On the Einstein-Podolsky-Rosen Paradox', *Physics* **1**, 195–200

Bell, J.S. (1987), *Speakable and Unspeakable in Quantum Mechanics*, Cambridge: Cambridge University Press

Bell, J., Clauser, J., Horne, M. and Shimony, A., (1985), 'An Exchange on Local Beables', *Dialectica* **39**, 86–110

Bohm, D. (1952), 'A Suggested Interpretation of Quantum Theory in Terms of Hidden Variables', *Physical Review* **85**, 166–93

Clifton, R., Pagonis, C. and Pitowsky, I. (1992), 'Relativity, Quantum Mechanics and EPR', in Hull, D., Forbes, M. and Okruhlik, K., eds., *PSA 1992, Volume 1*, 114–28

Horwich, P., (1987), *Asymmetries in Time*, Cambridge, Mass.: MIT Press

Kochen, S. and Specker, E.P. (1967), *Journal of Mathematics and Mechanics* **17**, 59–87

Lebowitz, J. (1997), Review of Price, H., *Time's Arrow and Archimedes' Point, Physics Today* **50**, 1, 68–9

Mermin, N. David (1990), 'What's Wrong With These Elements of Reality?', *Physics Today* **43**, 6, 9–11

Penrose, O. and Percival, I.C. (1962), 'The direction of time', *Proceedings of the Physical Society* **79**, 605–16

Price, H. (1996), *Time's Arrow and Archimedes' Point: New Directions for the Physics of Time*, New York: Oxford University Press

Reichenbach, H. (1956), *The Direction of Time*, Berkeley: University of California Press

KEITH HUTCHISON

NO INTERACTION WITHOUT PRIOR CORRELATION: COMMENT ON HUW PRICE

The essence of Huw Price's piece is a recommendation – that his so-called 'no teleology' principle be abandoned within microphysics. For if that is done (he argues) many of the problems in interpreting quantum mechanics will evaporate. There are (he adds) no major obstacles to following his recommendation, for the standard reasons for imposing the principle are restricted to macrophysics. Price does not examine these reasons in any detail, and relies mainly on cited literature, plus a couple of illustrative examples – for he clearly regards the macroscopic principle as beyond all reasonable doubt, and quite familiar to his audience. It is this fact that gives weight to his recommendation, for if the principle did not apply to the macroworld, there would be no temptation to apply it to the microworld. It thus becomes vital to ask if the macroscopic principle is trustworthy, and such an enquiry is to be the main burden of this commentary. I aim to show that despite its wide endorsement, the principle is not plausible. Where this leaves Price's claims about the microworld is unclear to me.

A brief commentary like this is not the place to pursue this question exhaustively, so I shall restrict myself to a few illustrations of my objections. In essence, I view the macroscopic principle as reliable in exactly the same illusory sense as it is 'true' that rare hands in cards are rare. I.e. it is a psycho-social fact, not a physical one – a reflection of what gets noticed by people from a particular background, not what actually happens. For as is well-known, *every* hand in cards is equally rare – but we are far more inclined to notice the rarity in special cases, cases that have warranted human attention, and (perhaps) been graced with an identifying name (like 'all aces').

Price's principle (as I understand it) asserts that systems which share a common past have more in common than those sharing a common future. As a result, if we are confronted with a group of entities having much in common, we often regard the fact that the systems have so much in common as good evidence of a prior interaction – and we explain the similarities by appeal to that interaction. But we feel no temptation to explain it by appeal to some future interaction – hence the name 'no teleology'.

There is no doubt that this procedure is sometimes valid. If two students in my class hand in identical essays, I am justified in believing that, at some time in the past, the students have not behaved honestly: they have either collaborated, or both borrowed from a third source, etc. 'No correlation, without prior interaction', we might say to summarise this intuition.

But before placing any general confidence in such a procedure of inference, we need to weigh it against situations that are less friendly to it – and here it fares poorly.

H. Sankey (ed.), Causation and Laws of Nature, 347–348.

Consider indeed the following example, drawn from a particularly trite game of luck, in which each member of a family of four is given a card at random from a pack. If it turns out that all four members of the family get an ace, the family wins a prize. Since this game is unequivocally dull, it is deemed suitable for early evening television, and is seized upon by a soap-powder company as sponsor. But it is soon discovered that the audience loses interest, and sales of the soap-powder declines. An hypothesis is made that the reason for the loss of interest is the rarity of wins, so it is decided to 'fix' the game, by deliberately giving *some* of the families four aces.

In a rigged game, the correlation between the cards held by a fortunate family can be explained by the common prior cause; in an unrigged game, the correlation between the cards held by a genuinely lucky family cannot be explained: it is just a coincidence. But among families that win the prize, the degree of correlation between the cards held by the two families is absolutely identical. All that is required for this degree of correlation to exist is a common future (the prize). Having a common past does not increase the correlation, it merely serves to add an explanation to it. (And furthermore, there is exactly as much correlation between the cards of unlucky families, those that win nothing. For what makes them unlucky is not a lack of correlation between the cards they individually hold, but the lack of correlation between the hand they collectively hold and the hand given preference by the rules of the game.)

Possession of a common future here does not (it is true) 'generate' the prior correlation between the cards, but it does provide a test for its existence. The test is of high reliability too, and certainly better than that provided by the common generating cause in the past. For the correlation required by a common future includes both coincidental and non-coincidental sources of the correlation; but possession of a common past omits all the coincidences.

This example is not in any way an isolated one, and clearly models many other causal processes. Systems that have a common future are *routinely* correlated before that future eventuates. If one of the asteroids in the gap between Mars and Jupiter is going to collide with the Earth some time in the 21st century, there will be some subtle relationship in existence now between its orbit and that of the Earth. Were such a relationship identified, we would probably not have any idea why it existed, and may well accept that it arises from pure coincidence – but our failure to explain it would not mean that it is less real.

Indeed, it would seem that systems cannot possibly share a common future, unless there is suitable coordination between their present states. Yet present states can sometimes be coordinated by pure coincidence, without a common cause. In consequence, there will often be that slight (?) tendency (noted above) for systems sharing a common future to be more correlated than those sharing a common past. I agree with Price that this fact cannot be used to *explain* the correlations, so I am happy to accept a 'no explanatory teleology' principle. But Price interprets his principle far more broadly, as denying the existence of the correlations that are a pre-condition for explanation. This broader interpretation does not convince me at all. To it, I oppose my own alternative principle: 'No Interaction Without Prior Correlation'.

University of Melbourne

NOTES ON CONTRIBUTORS

D.M. ARMSTRONG was born in 1926. His undergraduate years in Sydney from 1947 to 1950 under the remarkable Scotsman, Professor John Anderson, gave him his intellectual formation. After taking the BPhil degree at Oxford, he taught briefly at Birkbeck College, London University, and then at Melbourne University, before returning to Sydney in 1964 as Challis Professor. He retired in 1991. His most recent book is *A World of States of Affairs*.

ALAN BAKER read philosophy at Emmanuel College, Cambridge and then entered the doctoral program at Princeton University in 1992. In between he spent a year teaching philosophy as an Assistant Lecturer at the University of Ghana. His PhD thesis focuses on indispensability arguments for the existence of mathematical objects; his research interests include philosophy of mathematics, philosophy of science, metaphysics and logic.

JOHN BIGELOW came from Canada and New Zealand. He has a BA from the University of Canterbury and MA from Simon Fraser University. He got a doctorate from Cambridge, under the supervision of Ian Hacking, and contracted from that a grudging but lasting sympathy for French philosophers such as Foucault. His first position was at Victoria University of Wellington, where he fell in love with the possible worlds semantics of Hughes and Cresswell. He thrived for twelve years at La Trobe University, and is now Professor of Philosophy at Monash University. Bigelow is author of *The Reality of Numbers*, and joint author with Robert Pargetter of *Science and Necessity*.

ALAN CHALMERS is Associate Professor in History and Philosophy of Science at the University of Sydney. He graduated with a BSc in Physics from the University of Bristol, before taking a MSc in physics from the University of Manchester and a PhD in History and Philosophy of Science from the University of London. Chalmers is author of *What Is This Thing Called Science?*, *Science and Its Fabrication*, and some forty articles on history and philosophy of physics.

F. JOHN CLENDINNEN taught in the Department of History and Philosophy of Science at the University of Melbourne, and held visiting appointments at the Universities of Indiana and Pittsburgh. The main focus of his research has been on inference in science, especially the rationality of induction, the relation of inductive thinking to theorizing, and the question of formal versus non-formal standards. The place of causation in scientific explanation is a recent interest.

H. Sankey (ed.), Causation and Laws of Nature, 349–352.

JOHN D. COLLIER studied planetary science and philosophy at MIT, and did an MA at UCLA and PhD at the University of Western Ontario. He has taught at the Universities of British Columbia, Calgary and Melbourne, and is currently a researcher at the University of Newcastle. His interests lie in the foundations and applications of information theory and statistical mechanics, evolutionary theory, evolutionary ethics, meaning change, cognitive science and dynamical systems theory. Collier's major publications include the papers: 'Entropy in Evolution', 'Supervenience and Reduction in Biological Hierarchies', 'Intrinsic Information', 'Evolutionary Ethics and the Objectivity of Morality', 'The Necessity of Natural Kinds' and 'Complexly Organised Dynamical Systems'.

PHIL DOWE did his PhD at the University of Sydney, and is currently Senior Lecturer in Philosophy at the University of Tasmania. He has published papers on causation in various journals including *Philosophy of Science*, *The British Journal for the Philosophy of Science*, *Synthese*, and *Mind*, and has a forthcoming book entitled *Physical Causation*. He is currently working on backwards in time causation, and chance.

BRIAN ELLIS is Emeritus Professor of Philosophy at La Trobe University and Professorial Associate of the Department of History and Philosophy of Science at the University of Melbourne. He is the author of books on measurement theory (*Basic Concepts of Measurement*), epistemic foundations of logic (*Rational Belief Systems*), and systematic metaphysics (*Truth and Objectivity*), and of a large number of philosophical papers, mainly in these three areas, published in journals and collections. Ellis has a long-standing interest in, and has done a great deal of work on, the philosophy of space and time. He is currently dividing his time between work on scientific essentialism, which can wait, and much more urgent work on public policy and the philosophy of economics, which cannot.

JOHN FORGE works at Griffith University in Queensland, where he teaches Science, Technology and Society. His research interests are mostly in the philosophy of physical science, especially explanation, measurement, quantity, law and theory. He also works on philosophy of quantum mechanics and ethics of science. Forge has been a visiting fellow at the Center for Philosophy of Science at the University of Pittsburgh, and is Editor of the journal *Metascience*.

KEITH HUTCHISON graduated in mathematics at the Australian National University, before taking a DPhil in modern history at Oxford. He now teaches history of science in the Department of History and Philosophy of Science at the University of Melbourne. He has published on early modern philosophies of nature, the development of thermodynamics in the nineteenth century and temporal asymmetries in classical mechanics. His current research interests include renaissance cosmology and the foundations of statistics and probability.

KEVIN B. KORB studied philosophy at Berkeley and philosophy of science at Indiana University. In between times he obtained an MS in computer science and worked as a systems programmer. He now combines his philosophical and computational interests as a Senior Lecturer in computer science at Monash University. His research is primarily in the foundations of, and automation of, inductive inference and in the application of Bayesian methods to machine learning, including learning causal models. In his spare time he edits *PSYCHE: An Interdisciplinary Journal of Research on Consciousness* and worries about the mental lives of some machines.

MARTIN LECKEY completed a BSc Honours in physics and MA in History and Philosophy of Science at the University of Melbourne, before completing a PhD in the philosophy of physics at Monash University. Apart from his interest in laws of nature and metaphysics, he works on philosophy of quantum mechanics, space and time, and the arrow of time. He co-authored a paper with John Bigelow entitled 'The Necessitarian Perspective: Laws as Natural Entailments'.

CATHY LEGG did her undergraduate degree at the University of Melbourne, her Masters at Monash University, and is currently finishing a PhD in philosophy at the Australian National University while working at the University of Sydney. Her acquaintance with Australian realism is, therefore, extensive, yet her relationship to it has been somewhat "stormy". She is now, however, reconciled to realism, of a Peircean persuasion.

DAVID LEWIS studied philosophy at Swarthmore (BA, 1962) and Harvard (MA, 1964; PhD, 1967). He taught at UCLA before moving to Princeton, where he has been since 1970. Lewis has held numerous visiting appointments at universities around the world, including Oxford, Uppsala, Witwatersrand, Berkeley, Harvard and Stanford, as well as several universities in Australia and New Zealand. He is a regular visitor to Australia, where he frequently passes the rainy winter months in the environs of the University of Melbourne. He works on philosophical logic, metaphysics, epistemology, philosophy of science, and ethics and social philosophy. His many publications include the books: *Convention: A Philosophical Study*, *Counterfactuals*, *Philosophical Papers* (Volumes I & II), *On the Plurality of Worlds*, *Parts of Classes*, *Papers in Philosophical Logic*.

CAROLINE LIERSE is currently completing a PhD on the philosophy of time at La Trobe University. She has taught philosophy at La Trobe, Melbourne and Monash Universities, and has published papers on metaphysics in various journals and volumes of collected papers.

PETER MENZIES is Senior Lecturer in Philosophy in the School of History, Philosophy, and Politics at Macquarie University. He has an undergraduate degree from the Australian National University and graduate degrees from St Andrews University and Stanford University. He has held positions at the University of Sydney and the Research School of Social Sciences, Australian National University. He has published many articles on causation in journals such as *Mind*, *Philosophy of Science*, and *The British Journal for the Philosophy of Science*, and is working on a book on the subject entitled *The Metaphysics of Causation*.

HUW PRICE is Professor of Natural Metaphysics at the University of Sydney, and Research Affiliate of the Philosophy Program in the Research School of Social Sciences, Australian National University. He studied for a BA in Pure Mathematics and Philosophy at the Australian National University, an MSc in Mathematics at Merton College, Oxford, and a PhD in Philosophy at Darwin College, Cambridge. His publications include *Facts and the Function of Truth*, *Time's Arrow and Archimedes' Point*, and a wide range of articles in journals such as *The Journal of Philosophy*, *Mind*, *The British Journal for the Philosophy of Science* and *Nature*.

J.J.C. SMART was educated at the University of Glasgow (MA, 1946) and the University of Oxford (BPhil, 1948). He was a Junior Research Fellow at Corpus Christi College, Oxford (1948–50), Hughes Professor of Philosophy, University of Adelaide (1950–72), Reader, La Trobe University (1972–76), and Professor of Philosophy, Research School of Social Sciences, Australian National University (1976–85). His research interests are in metaphysics, philosophy of science and ethics. His most recent books are *Our Place in the Universe* and (with J.J. Haldane) *Atheism and Theism*.

BARRY TAYLOR studied at the Universities of Melbourne (MA, 1970) and Oxford (DPhil, 1974), and is currently a Reader in Philosophy at the University of Melbourne, where he has taught since 1974. He is the author of *Modes of Occurrence*, and of a number of papers on metaphysics and the philosophy of language.

The Editor

HOWARD SANKEY is Senior Lecturer in philosophy of science in the Department of History and Philosophy of Science at the University of Melbourne. A native of southern California, he completed a PhD at Melbourne after undergraduate studies in philosophy in New Zealand at the University of Otago. He has held research positions at the Center for Philosophy of Science at the University of Pittsburgh and Saint David's University College, Lampeter. His main interests lie in the general philosophy of science, with special emphasis on semantic aspects of conceptual change, scientific realism, rationality of scientific theory choice and naturalized accounts of the epistemic warrant of methodological norms. His major publications include two books, *The Incommensurability Thesis* and *Rationality, Relativism and Incommensurability*.

INDEX

Abduction 138–9, 141
Achinstein, P. 31
Almeder, R. 129–30, 140
Ampere's law 108
Anderson, D. 141
Andromeda galaxy 79
Anscombe, G.E.M. 314–5
Antirealism 125, 128, 130, 132
A posteriori 28, 30, 66, 85, 131, 171, 268, 328, 339–40
A priori 28, 30–2, 38, 66, 103, 187, 267, 315
Aquinas, T. 15
Archimedes 166
Archimedes' lever 241
Archimedes' principle 6
Aristotle 10, 135, 224, 237
Aristotelian 11
Armstrong, D.M. xii–xiii, xv, 15, 39, 40, 62, 70, 74, 81, 83, 88, 108–9, 111–2, 115–8, 120–3, 149–50, 162–4, 171–2, 194, 199, 256–7, 262, 314–5, 317–8, 326, 328
Aronson, J. 258, 262
Artificial intelligence (AI) xiv, 265, 268, 278, 297, 305
Aspect, A. 281
Assertibility 56, 73
Assertibility conditions 73, 163, 193, 210
at–at theory 190, 195, 230, 247, 252, 254
Atkins, P.W. 225
Averill, E.W. 32
Ayer, A.J. 132

Backtracker 145, 175
Backtracking conditional 175, 199
Bain, C. 131
Baker, A. xiii
Banaschewski, B. 239
Bannerjee, S. 216
Bayes' theorem 303
Bayesian 269, 278, 297, 302, 304–5,
Bealer, G. 31–2
Behavioural dispositions 70–1, 77–8, 84–5
Bell, J.L. 239
Bell, J.S. 341
Bell's theorem 279, 281
Bennett, C. 218, 229, 233–4, 241
Bennett, J. 56
Bentley, R. 20
Berkeley, G. 20, 24, 54, 80
Bhaskar, R. 11, 45, 62, 74
Bigelow, J. xii, xiv, 12, 14, 53, 57–8, 61–2, 64–5, 68–9, 72, 74, 77, 81, 83, 86, 88, 111, 118–9, 167, 262, 314–5
Bigelow and Pargetter (BP) 111–2, 118–9, 122
Blackburn, S. 313
Blanshard, B. 32
Bohm, D. 223, 342
Bohr, N. 341
Boler, J. 140
Boltzmann, L. 225
Bondi, H. 164
Boyle, R. 3, 5, 11, 13
Boyle's question 3, 5, 11, 14
Boyle's law 92
Braithwaite, R.B. 315
Brentano, F. 35
Brillouin, L. 225–6, 240–1
Bromberger, S. 267
Brooks, D. 217, 225, 241
Brown, G.S. 239

Campbell, N.R. 123
Capacities 10, 19–34 *passim*, 61, 70, 293–4
Carnap, R. 132, 168, 188, 266, 269
Carroll, J. 107–8, 324
Carter, B. 165
Cartesian 47, 216
Cartwright, N. 8–10, 12, 184, 282, 284–8, 290–3, 297, 306, 308
Categorical properties 36–7
Categorical base 24–5, 46, 51–2, 150
Categorical realism 24, 39, 69–70
Catholic 53, 65
Cauchy–Riemann technique 219
Causal base 63, 67, 70–1, 147–9, 153, 157–8
Causal connection 46, 64, 187, 190–1, 194, 196, 200, 202, 204, 209–10, 212, 222, 238, 265, 272, 302
Causal dependence 175, 178, 187, 194, 198–9
Causal factor 293, 295–6
Causal law xv, 13–4, 19, 21–2, 25, 27, 30, 42–3, 57, 62–5, 68–9, 73, 95, 233, 248, 293
Causal model 297–305
Causal platitudes xv, 182–3
Causal powers xii, xv, 14, 35, 39–41, 63–4, 69–71

Causal process xiv, 13, 19, 187, 191–3, 215, 221–2, 224, 232, 236–7, 240, 247–8, 250, 252–5, 257, 260–1, 306
Causal relation 63, 172, 189, 191, 194, 200, 213, 224, 265, 305, 313–5, 323–4, 326–7
Causal theory of identity 86, 250, 253, 255–7
Causation xi, xiii, 5, 23, 28, 36, 62, 145, 147, 163, 175–6, 178, 181–3, 187–8, 191–3, 195, 204–5, 208, 215–8, 222, 224, 229, 236–7, 247, 249, 255–8, 265–329 *passim*, 342
Causation, regularity theory of 215, 235, 238, 248, 315
Cavendish, H. 7
Chalmers, A. xii, xvi, 15
Chandrasekhar, S. 216
Chickering, D. 299
Christ 26, 333
Christensen, W.D. 217, 233, 237, 241
Christian 53
Christie, M. 15
Clarke, S. 32
Clendinnen, F.J. xiv–xv, 168
Clifton, R. 341
Collier, J. xiv–xv, 215, 217–8, 223, 226–7, 233–4, 236–9, 241, 296
Common cause 195, 197–8, 224, 248, 272, 276–7, 281, 298, 306, 332–3
Complexity 217–220, 233; informational 218, 220, 223–4
Compton scattering effect 279–80
Concrete particulars 215, 218, 221, 234, 238
Conditional analysis 142–7, 152, 157
Conditionals 29, 49–51, 57, 74, 93, 143, 158, 179, 195, 209
Confirmation 95–8, 100, 102, 104–6
Conjunctive fork 272, 277, 279–80, 307
Conserved quantity 191, 215, 236–7, 248, 250, 252–5
Constitution thesis 51
Contingency 4, 20, 23, 28–9, 33, 47, 54, 63, 67, 71, 77, 83, 86, 121, 127–9, 133, 158, 172, 176, 184, 188, 324, 326, 332
Contingency theory of laws 83–6, 88
Copenhagen interpretation 341
Cosmic coincidence 161–2, 164, 171–2
Costa, M.J. 32
Coulomb force 12
Coulomb's law 14
Council of Trent 26, 65
Counterfactual xiii–xiv, 30, 46, 56–8, 112, 116–7, 121, 145, 147–9, 163, 175, 177–9, 182–3, 191–2, 195–6, 198, 200, 203, 212, 215, 223, 229, 231, 234–5, 242, 260–2, 285, 291, 316, 323, 340
Counterfactual conditional 29, 33, 72, 143; vacuous 55, 57, 71–2
Counterfactual dependence 195–6, 199, 201, 221
Counterfactual theory of causation 175–6, 178, 180

Dai, H. 302–3
Dauben, J. 141
Davies, P. 108
Davis, W. 280
De dicto 30, 54, 68
De re 30, 48–9, 54, 68
Demopoulos, W. 237
Denbigh, K. and J. 241
Descartes, R. 63, 131, 166, 224, 237
Dewey, J. 126
Disjunctive generalisations 96–8, 103–7, 109
Disjunctive laws 104–6
Disparateness 96, 99, 100
Disposition xii–xiii, xv, 6, 10, 35, 40, 52–3, 68–70, 85, 87, 143–50, 153–4, 157–8; behavioural 70–1, 77–8, 84–5; Finkish xiii, 143–5, 148, 153–4, 157
Divine command theory 19, 21
Dombel, F. 278
Dowe, P. xiv–xv, 178, 185, 191, 215, 221, 232, 236, 247, 250–2, 254, 256–7, 307
Dretske, F. 81, 83, 88, 108
Ducasse, C.J. 314–5
Dummett, M. 158
Dupré, J. 294–5, 307
Dyck, L. 216

Earman, J. 324
Eddington, A. 166
Eells, E. 286, 288, 292–3, 295–7, 307–8
Ehring, D. 258
Einstein, A. 164, 166, 249
Einstein, Podolsky and Rosen paradox (EPR) 279, 281
Elder, C. 32
Ellis, B. xii, xiv, 12, 14, 32, 35–8, 45–8, 50–8, 70, 74, 77–9, 122–3, 165, 167, 257–8
Ellis-eliminativism 258, 260
Empiricism xi, 5, 21, 234, 265, 324
Entailment question 91, 101, 106–7
Entailment thesis 51
Essence 35, 80, 165; individual 67–8; kind 66–9; real 23, 67
Essential passivity of matter 20, 24, 46–7, 63–6
Essentialism (scientific) xii–xv, 15, 19, 20–1, 23, 26–8, 30–2, 45, 47, 51, 61–3, 66, 68–9, 71–2, 77, 83–4
Euclid 108, 166
Euthyphro 51, 57, 74
Experiment 7–9, 131, 135–7, 265, 278, 285, 287, 291, 300, 306–7, 339; Peirce's Harvard stone experiment 137–9
Explanation 92, 95, 106–7, 112, 118, 138–9, 331; causal 217, 237; statistical relevance 189, 204–6
Extrinsic(-ality) 52, 66, 68, 313–4, 316–7, 321, 326

Fair, D. 258, 262, 314–5
Fales, E. 31–2, 183, 314–5
Feynman, R. 216

Finkish disposition xiii, 143–5, 148, 153–4, 157
Fodor, J. 109, 217
Force 7, 9, 64, 66–7, 70, 74, 79, 86, 183, 192, 258, 313; Coulomb 12
Forge, J. xiii, 111, 118, 123
Fragility 70, 143, 146–50, 157, 179, 181, 200–1
Fredkin, E. 241
Freeman, P. 232
Freudenthal, H. 168

Gale, G. 217
Gatlin, L. 241
Geiger, D. 309
General relativity 92, 101, 165
Georgeff, M. 303
Giere, R.N. 271, 285, 296–7, 307
Glymour, C. 297, 299, 301, 305
God 4, 5, 20–1, 32, 64, 80–1, 132, 222, 232, 242
Gods 257
Goldman, A.I. 258
Goldstein, H. 239
Good, I.J. 267, 306
Goodman, N. 56, 108
Graves, J. 216
Gravitation 79, 164; law of 7, 8, 73, 138
Gravitational attraction 8, 11, 138
Gravitational mass 25, 37–8, 69, 79, 86
Gravity 10, 20, 24
Greenberger–Horne–Zeilinger (GHZ) 341
Greeno, J.G. 189

H-independence 336
Hacking, I. 9
Hall, N. 175, 180
Halley's comet 10
Hamiltonian 13
Harré, R. xii, xvi, 11, 31, 45, 62
Hawking, S. 167
Heathcote, A. 262, 328
Heckerman, D. 278, 309
Hegel, G. 127
Heisenberg's uncertainty principle 92
Hempel, C.G. 95, 107, 122, 315
Henry, J. 15
Heracleitean 166–7
Hertz, H. 7
Hesslow, G. 282–5, 287–8, 294, 300
Hirsch, E. 262
Hitchcock, C.R. 251, 306
Hobbes, T. 11, 217, 237
Holzmüller, W. 217, 232
Hooker, C. 217, 234, 239, 241
Horwich, P. 223, 315, 333
Howson, C. 304
Hoyle, F. 236
Humberstone, L. 52, 317
Hume, D. xi–xii, 5, 6, 8, 21, 23–6, 28, 32, 42, 46–7, 50–5, 63–6, 72, 74, 163, 171, 176, 179, 183, 190, 192, 194, 215, 223, 235, 248, 252–3, 265–6, 274, 292, 313–6, 320, 323–6
Hume world 46, 55, 63–5, 74
Humean supervenience 25, 32, 215, 237, 323–7
Humphreys, P. 200, 206–7
Hunt, I. 143
Hutchison, K. xiv, 4

Identity xiv, 19, 22, 27, 30, 38, 41, 61, 65, 67, 71, 77–8, 138, 184, 222, 232, 234, 237, 248–9, 252, 254–7, 293–4; causal theory of identity 86, 250, 253, 255–7; kind-identity 67
Iltis, C. 32, 74
Imaginability 29, 72
Impossibility 55; logical 47, 77
Individual 49, 67, 73, 79, 113, 118, 288–9
Inertia 24
Inertial mass 25, 32, 38
Inference to the best explanation 45–6, 65, 138, 168
Information xiv, 215–45 *passim*
Information transfer 215, 222, 224, 237, 240
Informational complexity 218, 220, 223–4
Instantiation problem 119, 122
Interactive forks 222–3, 278–9, 281
Intrinsic (-ality) 52–3, 57–8, 63, 66, 81, 147, 154, 158–9, 324–5
Irzik, G. 309

Jackson, F. 32, 46, 70, 75, 147, 154, 157–8
James, W. 125–6
Jeffrey, R.C. 189
Johnston, M. 150, 154

Kepler, J. 166
Kepler's laws 108
Kestin, J. 225
Kim, J. 74, 317, 321
Kind 38, 77, 202
Kistler, M. 253
Kitcher, P. 221
Kochen, S. 341
Kolak, D. 262
Kolmogorov, A.N. 229, 239
Kolmogorov complexity 220
Korb, K. xiv–xv, 269, 302, 304, 309
Kripke, S. xii, 33, 67, 144, 154, 184
Küppers, B. 217, 232

Lagrangian 13
Landauer, R. 229
Landsberg, P.T. 229
Langton, R. 52, 159, 317, 321–2
Laplace 203, 267
Laudan, L. 32, 74
Law 42, 77, 82, 86, 100–3, 107, 115–6, 135, 138, 166–7, 183–4, 187, 190, 192–3, 197, 209–10, 212, 215, 218, 234–5, 265–6, 323, 333–4, 336; causal law xv, 13–4, 19, 21–2, 25, 27, 30, 42–3, 57, 62–5, 68–9, 73, 95, 233, 248, 293; closure of laws 92; Coulomb's law 14; disjunctive law 104–6;

Law (*cont.*)
Galileo's law 5, 266; Lorentz force law 7, 9; maximal view of law 91, 100, 103, 106; minimal view of law 91–2, 100–3, 106; naturalness theory of law xii–xiii, 77, 83; Newton's laws 6, 9–10, 13, 139, 249; phenomenological law 8–10; physical law 114; probabilistic law 285; regularity theory of law xiii–xiv, 5–7, 32, 94, 112, 161, 316; statistical law 190; stipulational view of law 91–2, 101
Law of conservation of energy 14, 279
Law of gravitation 7, 8, 73, 138
Law of Dulong and Petit 15
Law of nature xi, xiii, xv, 3–31 *passim*, 62, 72, 78, 80, 83, 91, 93, 111–2, 121, 125, 164, 176, 180, 209
Lawhood 83, 101, 147
Layzer, D. 224, 229, 241
Lebowitz, J. 337
Leckey, M. xii, xiii, 53, 57–8, 81, 83
Leibniz, G.W. 32, 63, 74, 217–8, 224, 232
Levine, M. 45
Lewis, D. xiii–xv, 30, 50–2, 56, 72, 77–8, 109, 154, 157, 159, 168, 175–6, 178, 180–3, 187, 194–6, 199–201, 203–4, 208, 215, 235, 262, 269, 316–9, 321–3
Li, M. 220–1, 229, 233, 239
Lierse, C. xiii–xiv, 12, 14, 32, 36, 61, 69–70, 74, 77, 79, 85, 88, 165, 167
Locke, J. 24, 33, 54, 63

μ-independence 336–45
Mach, E. 225
Mackie, J. 33, 163, 266–7, 315
Madden, E.H. xii, xvi, 11, 31, 45, 62
Malcolm, N. 176
Malebranche, N. 80
Mark 190, 215, 221, 229
Mark transmission 253
Markov property 273, 278, 281, 287, 289, 299
Martin, C.B. xiii, 35–7, 143–5, 147, 151, 153–4, 157–8, 185, 223
Martin, R. 262
Mass 40, 42, 79, 86, 116, 118, 171, 318; gravitational 25, 37–8, 69, 79, 86; inertial 25, 32, 38
Matter 3–5, 47, 63, 250
Maxwell, J.C. 3, 13
Maxwell's equations 162
Maxwell's demon 226, 241–2
McDermott, M. 185
McGinn, C. 33
Mechanism 19, 20, 28, 42, 63
Meinong, A. 35
Meinongian 39–40
Menzies, P. xiv, xvi, 182–3, 185
Merck 268
Mermin, N.D. 341
Meyer, E. 309
Michotte, A. 185
Mill, J.S. 50, 107, 153, 176, 315
Minkowski space 164, 167
Miracle 73, 77–8, 81
Molnar, G. 35–6, 108
Mumford, S. 32, 155
Musgrave, A.E. 252

N-dependence 196–202, 205, 207, 210, 212
Natural kind xi, xiv–xv, 21, 28, 40, 45–8, 52–5, 57, 66–7, 69, 99, 113, 165, 223, 234, 236; categorical 55; dispositional 55; property 53–4
Natural kind hierarchy 66
Naturalness 78, 80–1, 84, 323
Naturalness property 80, 88
Necessary connection 23, 63, 313, 323
Necessity xi, 22, 27–8, 35, 69, 77, 83–4, 92–3, 163–4, 167, 184, 192, 223, 237; epistemic 31; logical 30, 56, 80, 82, 84; metaphysical 30, 82; natural 21, 25, 64, 69, 77, 81–2, 121, 161; physical 30, 62, 94, 108
Negentropy Principle of Information (NPI) 216, 218, 224–6, 230–2, 237–8
Neil, J. 304, 308
Neopolitan, R. 274, 297
Newton, I. 20, 32, 37, 64–5, 74, 79, 92, 128, 138–9,164, 223, 249
Newton's laws 6, 9–10, 13, 139, 249
Newtonian mechanics 13
Neyman, J. 306
No teleology principle xiv, 331–7, 344, 347
Nolan, D. 150
Nomic connection 201
Nomic dependence 193, 195, 200, 203, 209
Nomic relationship 200
Nomic necessity 83, 163
Nominalism 113, 135–8, 141; class 111–4, 118, 123
Non-vacuousness of laws 94–5, 102, 106

Objectivity 92, 101–2
Ohm's law 108, 111, 164–5
Olbers, H. 164–5
Omission xiv, 177, 248, 257–61
Owens, D. 96–8, 108

Pagonis, C. 341
Parfitt, D. 262
Pargetter, R. 32, 64, 70, 111, 118–9, 147, 154, 157–8, 262, 314–5
Parmenidean 166
Particulars xiv–xv, 11–2, 35, 38, 40, 235, 242; concrete 215, 218, 221, 234, 238
Patrick, J. 303
Pearl, J. 273–4, 278, 297–9, 301–2, 308
Peirce, C.S. xiii, 125–42 *passim*
Peirce's limit theory of truth 129–30
Peirce's Harvard stone experiment 137–9
Peirce's Scholastic realism xiii, 125, 135, 137
Penrose, R. 167, 332, 338
Percival, I.C. 332, 338
Petty, W. 11

Pitowsky, I. 341
Plato 51, 127, 215–6, 237
Plato's forms 188
Platonic heaven 167
Platonic TOE 167–8
Platonism 127
Popper, K.R. 6, 269, 271
Possible world 26, 42, 56, 73, 77, 79, 82, 84, 86, 127, 130, 175–6, 235, 259–60, 316
Possibility 49, 54, 62, 67, 72, 77, 84; epistemic 29, 31, 33, 72; logical 55, 72, 78; metaphysical 29; physical 72
Power xv, 6, 10–1, 19–34 *passim*, 35, 43, 52, 61, 63, 192, 313
Pragmatism 125–6, 128–9, 131–2
Preemption xiv, 179, 182, 194–5
Prevention xiv, 178, 248, 257–9, 261, 275
Price, H. xiii–xiv, 166, 334, 336, 340, 343, 347–8
Prior, E.W. 31–2, 70, 147, 149–50, 154, 157–8
Prior, Pargetter and Jackson (PPJ) 70–1
Probabilistic causation 179, 185, 190, 203, 207, 265–311 *passim*
Probability xv, 97, 205–8, 210, 212, 222–3, 231, 236, 265–311 *passim*; objective 207; physical 269, 281, 306; raising of 176, 182, 273
Propensity 19–34 *passim*, 41, 52, 61, 70, 78, 306–8
Property 4, 35–7, 43, 47, 67, 86–7, 99, 111, 116, 118–9, 123, 147, 149, 150–1, 195, 200, 205–6, 221, 251, 313–29 *passim*; accidental 84; active 4, 24; categorical 4, 27, 36–7, 41, 43, 54–5, 64, 83; dispositional xiii, 4, 11, 22–5, 27–8, 45–7, 49, 52, 54, 57–8, 63, 69, 70, 73, 83; essential 14, 21, 45–8, 53, 55, 66, 78–9; extrinsic 66, 68, 152; internal 74; intrinsic, 19, 21–3, 52, 65, 67, 153; modal 25; natural 318, 320–1, 324–5; naturalness xiii, 80; occurrent 70; passive 4; relational 80
Protestant 53, 65
Pseudoprocess 190, 230, 247, 249–50, 255–6
Pythagoras 109

Quantity 111–3, 118–20, 122, 191, 193
Quantum mechanics 9, 12, 217, 221, 223, 230, 267–8, 279, 281, 331–2, 335, 339–45
Quine, W.V.O. 109, 130, 161, 163, 168
Quinean 47, 57, 162, 164
Quinlan, J.R. 303

Ramsey, F. 107, 176
Ramseyfication 167
Rayleigh's criterion 92
Realism xi, xiv, 8, 48, 53–4, 113, 128, 136–9, 161, 175, 235; categorical, pragmatic 125; Scholastic xiii, 125, 135, 137; scientific 58, 65, 161
Reeder, N. 32
Regularity theory of causation 215, 235, 238, 248, 315
Regularity theory of laws xiii–xiv, 5–7, 32, 94, 112, 161, 316
Reichenbach, H. 163, 190, 209, 211, 215, 222, 249, 262, 267, 272–3, 275–8, 281, 306, 332–3
Reid, T. 266
Relation 36–7, 42, 111–2, 114–5, 117–20, 122, 162, 247, 313–29 *passim*; external 320; extrinsic 323; intrinsic 313, 322; natural 318, 321–2, 324–6
Rissanen, J. 232
Rivest, R.L. 303
Robinson, H. 166
Rorty, R. 125–8, 131, 133–4
Rosen, D. 273
Rosen, G. 109
Ruby, J. 3, 15, 166
Russell, B. 190, 215, 230, 247–50, 253, 254, 256, 265, 305
Ryle, G. 50

Salmon, W. 187, 189–93, 203–5, 207–8, 211, 215, 221–2, 230, 236, 247–55, 262, 267, 271, 274–80, 283, 306–7
Schaffer, S. 11
Schlesinger, G. 99–100, 108
Schneider, T.S. 217, 232
Scotus, D. 125, 135–6, 140
Schlick, M. 132
Schrödinger, I. 225
Schrödinger's equation 92, 122, 161
Schrödinger's Negentropy Principle of Information 216, 218, 224–6, 230–2, 237–8
Scriven, M. 268
Seagar, W. 96–8, 108–9
Shannon, C.E. 225, 303
Shapin, S. 11
Shoemaker, S. 31, 35–6, 262
Shope, R. 154
Sider, T. 322
Simpson, E.H. 289, 291
Simpson's paradox 284–5, 300
Singular causation 176, 178–9, 181–2, 184, 315
Singularism xiii, xv–xvi, 175, 177, 179, 181, 194, 314–6, 320, 327
Skagestad, P. 137–8
Sklar, L. 231, 236
Skyrms, B. 307
Smart, J.J.C. xiii–xv, 35, 62, 74, 122, 161–2, 171, 211, 238
Smith, A.D. 155
Snell's law 92
Snow, C.P. 226
Sober, E. 282, 286, 288, 293, 296, 306, 308
Socrates 136
Sosa, E. 283
Special relativity 101, 249
Specker, E.P. 341
Spinoza, B. 237
Spirtes, P. 280, 297, 301–2, 306

Stalnaker, R. 56, 154
Star algorithm 287–8, 292
Statistical relevance 204–5, 208, 269–70, 273–5, 277, 307
Strawson, G. 313
Suppes, P. 267, 274–5, 277–8, 306
Swinburne, R. 35–7, 41
Swoyer, C. 31, 35–6

T-symmetry 332, 334–5, 338–40, 342
Taylor, B. xiii, 158, 322–3
Temporal asymmetry xiii, 213, 221, 271, 321–45 *passim*
TETRAD II 301–2, 304, 308
Thayer, H.S. 20
Theory of Everything (TOE) 165–7
Thom, R. 216
Thomson, J.J. 7, 12, 167
Thompson, D. 217
Thompson, I.J. 31
Thompson, M. 137
Toffoli, T. 241
Tooley, M. 81–3, 88, 252, 283, 307, 314, 324, 326
Transubstantiation 26–7, 53–5, 57, 63, 65
Truth 126–9, 132, 134, 135, 148, 151, 165, 175, 192, 211–2; Peirce's limit theory of 129–30
Truth conditions 73, 163, 199, 211
Truthmakers 176, 178, 258, 261–2, 313, 316

Ulanowicz, R. 217
Universality 93–5, 102, 106
Universals xi, xiii, 32, 38, 47, 48, 62, 81, 99, 106–7, 115–20, 161–2, 171–2, 215, 236, 242, 318, (non-)instantiation of 115–6, 119–21
Urbach, P. 304

Vacuous counterfactuals 55, 57, 71–2
Van Fraassen, B. 74, 107–8, 162, 165, 168
Venn, J. 247
Verification 193, 209
Verificationism 65–6, 131–2, 235
Verma, T. 299, 301
Vis viva 63, 74
Vitànyi, P. 220–1, 229, 233, 239
Voyager spacecraft 10

Wallace, C.K. 303, 308
Wallace, C.S. 232
Walters, R. 107
Wang, D. 309
Weaver, W. 225, 233, 303
Wegener, A. 286
Whittaker, E.T. 166
Wicken, J. 225
Wiley, E.O. 217
Wisdom, J. 162
Wittgenstein, L. 176
Wright, J. 313
Wright, S. 297–8

Yagil, G. 232, 240
Yule, G. 284

Zeno's paradoxes 190

AUSTRALIAN STUDIES IN HISTORY AND PHILOSOPHY OF SCIENCE

1. R. McLaughlin (ed.): *What? Where? When? Why?* Essays on Induction, Space and Time, Explanation. Inspired by the Work of Wesley C. Salmon. 1982 ISBN 90-277-1337-5
2. D. Oldroyd and I. Langham (eds.): *The Wider Domain of Evolutionary Thought.* 1983 ISBN 90-277-1477-0
3. R.W. Home (ed.): *Science under Scrutinity.* The Place of History and Philosophy of Science. 1983 ISBN 90-277-1602-1
4. J.A. Schuster and R.R. Yeo (eds.): *The Politics and Rhetoric of Scientific Method.* Historical Studies. 1986 ISBN 90-277-2152-1
5. J. Forge (ed.): *Measurement, Realism and Objectivity.* Essays on Measurement in the Social and Physical Science. 1987 ISBN 90-277-2542-X
6. R. Nola (ed.): *Relativism and Realism in Science.* 1988 ISBN 90-277-2647-7
7. P. Slezak and W.R. Albury (eds.): *Computers, Brains and Minds.* Essays in Cognitive Science. 1989 ISBN 90-277-2759-7
8. H.E. Le Grand (ed.): *Experimental Inquiries.* Historical, Philosophical and Social Studies of Experimentation in Science. 1990 ISBN 0-7923-0790-9
9. R.W. Home and S.G. Kohlstedt (eds.): *International Science and National Scientific Identity.* Australia between Britain and America. 1991 ISBN 0-7923-0938-3
10. S. Gaukroger (ed.): *The Uses of Antiquity.* The Scientific Revolution and the Classical Tradition. 1991 ISBN 0-7923-1130-2
11. P. Griffiths (ed.): *Trees of Life.* Essays in Philosophy of Biology. 1992 ISBN 0-7923-1709-2
12. P.J. Riggs (ed.): *Natural Kinds, Laws of Nature and Scientific Methodology.* 1996 ISBN 0-7923-4225-9
13. G. Freeland and A. Corones (eds.): *1543 and All That.* Image and Word, Change and Continuity in the Proto-Scientific Revolution. 1999 ISBN 0-7923-5913-5
14. H. Sankey (ed.): *Causation and Laws of Nature.* 1999 ISBN 0-7923-5914-3

KLUWER ACADEMIC PUBLISHERS – DORDRECHT / BOSTON / LONDON

